AF761822

Coastal Shellfish – A Sustainable Resource

Developments in Hydrobiology 160

Series editor
H. J. Dumont

Coastal Shellfish
A Sustainable Resource

Proceedings of the Third International Conference on Shellfish Restoration, held in Cork, Ireland, 28 September–2 October 1999

Edited by

Gavin Burnell
University College Cork, Ireland

Reprinted from Hydrobiologia, volume 465 (2001)

Springer-Science+Business Media, B.V.

Library of Congress Cataloging-in-Publication Data

A C.I.P. Catalogue record for this book is available from the Library of Congress.

ISBN 978-94-010-3900-0 ISBN 978-94-010-0434-3 (eBook)
DOI 10.1007/978-94-010-0434-3

Printed on acid-free paper

TABLE OF CONTENTS

THE REAL AND POTENTIAL IMPACTS OF SEA URCHIN EXPLOITATION

ENVIRONMENTAL IMPACT OF SHELLFISH EXPLOITATION AND AQUACULTURE

DISEASES AND HARMFUL ALGAL EVENTS

SHELLFISH AS ENVIRONMENTAL WATCHDOGS AND MONITORS

Hydrobiologia **465:** ix–x, 2001.
G. Burnell (ed.), Coastal Shellfish – A Sustainable Resource.

Preface

As we begin the 21st century there is a growing, worldwide commitment to restore our planet's degraded coastal ecosystems. Governments at all levels have renewed their interest in preserving and enhancing coastal resources. Volunteer organizations concerned with declines in environmental quality and fisheries have developed successful programs to identify problem areas, recommend improvements, and monitor progress. In response to this commitment, a small group of government, industry, and community representatives from the world community, anxious to turn the tide on declining environmental quality, came together in 1995 to plan the first International Conference on Shellfish Restoration (held on Hilton Head Island, South Carolina in 1996). The conference was designed to recognize the importance of shellfish as indicators of environmental quality and to examine how the health of coastal ecosystems could be improved through shellfish restoration. Consumed raw in many cultures, shellfish serve as excellent indicators of environmental health. When concentrations of bacterial or viral pathogens, heavy metals or toxic chemicals in waters or shellfish meats exceed safe levels, harvest waters are closed. Heavy loading of nutrients from runoff decrease oxygen levels causes fish kills and a die-off of benthic organisms and submerged aquatic vegetation. Such effects indicate that the quality of these environments is compromised and the waters may he unsuitable for other marine life or human activities.

The second conference returned to Hilton Head in 1998 where over 200 scientists, resource managers, and government officials met to consider how community, regional, national and international partnerships can help the shellfish restoration process.

In October 1999, the ICSR conference was held in Cork, Ireland, where it took on a European dimension, and its scope was broadened to include non-molluscan shellfish such as sea urchin and lobsters. The emphasis throughout the conference was on the marriage of science and management. While highlighting shellfish as a valuable coastal resource, ICSR Europe also addressed important new restoration topics such as artificial reefs, interactions with birds, the carrying capacity of bays, red tides, and the genetic implications of sea ranching.

Most shellfish species have a broadcast spawning strategy which results in a variable and often unpredictable recruitment. As a result they are very vulnerable to over-fishing or natural catastrophic events which reduce the critical mass of their spawning stock. One remedy to this problem is to use aquaculture technology to produce large number of juveniles which, with suitable nursery protection, could then be released back into areas which have been depleted of their native populations. Countries such as Japan have been practicing this type of integrated aquaculture-fisheries management for several decades and the time is ripe for its implementation on a global scale. There is now evidence from several research groups that if a critical spawning stock is established in a hydrographically suitable area the populations should, in time, become self-recruiting.

Recent concerns about the over estimation of world fish production, particularly in China, have highlighted the need to develop aquatic industries which are sustainable and do not rely on a diminishing supply of fishmeal for their production. Once again shellfish scores in that many commercially exploited species of molluscs and echinoderms are primary producers, feeding low down the food chain on phytoplankton or macroalgae. This makes them ecologically and philosophically more desirable candidates for coastal farming and fishing operations.

So, there is a greater need than ever for the enhancement of our coastal ecosystems though shellfish restoration. Not only does the public have a keen interest in restoring water quality to shellfish-growing water standards, but so does the aquaculture industry, which has focused on advancing the quality assurance and safety of aquaculture products. The demand for seafood is expected to increase significantly in the next ten years, and the industry will need new sites with good water quality. Our goal is to provide forums throughout the world to draw attention to shellfish, shellfish water quality, and related issues, and to foster partnerships and the exchange of information to further the science of restoration and aquaculture.

G. BURNELL
D. LEONARD
R. DEVOE

List of referees

P.S. Kristensen
A. Brand
P. Britz
P. Cook
A. Taylor
K. Ramsay
E. Fahy
M. Kaiser
F. O'Beirn
J. Lewis
G. Flimlin
G. Thayer
O. Tully
J. Moore
J. O'Halloran
M. Gaspar
A. Jensen
R. Furness
M. Luckenbach
M. Kelly
P. Dolmer
S. Jenkins
W. Arnold
R. Carnegie
E. McKnight
C. Bradshaw
A. Smaal
C. Wilson
P. Goulletquer
D. Barnes
B. Ball
G. Burnell
S. Shumway
S. De Grave
S. Robinson
N. Dankers
L. Coen
G. Savidge

Acknowledgements

We would like to thank NOAA for their generous support for both the ISCR'99 Conference in Cork and also the publication of this volume.

Hydrobiologia **465:** 1–5, 2001.
G. Burnell (ed.), *Coastal Shellfish – A Sustainable Resource.*

Shellfish ranching in the UK

C. A. Burton, J. T. MacMillan & M. M. Learmouth
Sea Fish Industry Authority, SeaFish Aquaculture, Marine Farming Unit, Ardtoe, Acharacle, Argyll PH36 4LD, U.K.

Key words: mussel, *Mytilus edulis*, flat oyster, *Ostrea edulis*, Manila clam, *Tapes philippinarum*, Pacific oyster, *Crassostrea gigas*, king scallop, *Pecten maximus*, Palourde or native clam, *Ruditapes decussatus*, lobster, *Homarus gammarus*, cockle, *Cerastoderma edule*, shellfish, ranching, UK

Abstract

There are six main species of shellfish of commercial importance within the UK (mussels, flat oysters, cockles, king scallops, Manila clams and lobsters) which can be thought of as ranched to varying degrees. A seventh (Pacific oysters) can be included, since a very small proportion of the production may undergo a period of growth on natural habitats in some areas. An eighth species (Palourdes or native clams) appeared in the statistics for the first time in a number of years in 1998. Seafish, as part of its role in promoting the sustainable management of resources, has examined the current state of the shellfish aquaculture industry in the UK.

Introduction

During 1998 and 1999, Seafish undertook a major review of the UK shellfish industry, its markets, its potential and its future (Lake, 1999). As part of this work, the shellfish aquaculture industry was examined in detail. Culture methods vary from species to species, but most are reared either semi-intensively or extensively. It is within the realm of extensive culture that aquaculture and fisheries interact and some of the distinctions begin to blur. The terms 'extensive' and 'ranched' are often used interchangeably, which can lead to confusion. For the purposes of this review 'extensive' is taken to mean grown at low or moderate densities with some limited husbandry or containment. Whereas to be described as 'ranched' some extensively grown species are placed for all or part of their life onto natural habitats, usually the seabed, and grown to market size with little or no further human input or husbandry. Harvesting can be carried out in a number of ways, using either dedicated systems or conventional fishing methods.

As the review progressed, it became apparent that the size and economic importance of the 'ranched' sector in particular was poorly reported and underestimated within the UK and elsewhere. The original poster, presented at the third International Conference on Shellfish Restoration, sought to redress the balance by highlighting the contribution of ranched shellfish to the UK's output of fish and fish products.

For some species, the culture industry is well established (e.g., mussel, *Mytilus edulis*; flat oyster, *Ostrea edulis*; Manila clam, *Tapes philippinarum*; and Pacific oyster, *Crassostrea gigas*) or in its infancy (e.g., king scallop, *Pecten maximus*; and Palourde or native clam, *Ruditapes decussatus*), whereas, for others, exploitation in this way is only just commencing (e.g., lobster, *Homarus gammarus*; and cockle, *Cerastoderma edule*).

Data sources

Published fishery and aquaculture production statistics were drawn from a number of official sources; predominantly the Ministry of Agriculture, Fisheries and Food (MAFF) (for England and Wales), the Scottish Executive Rural Affairs Department (SERAD; formerly, the Scottish Office Agriculture, Environment and Fisheries Department, SOAEFD) and the Department of Agriculture for Northern Ireland (DANI). International statistical data (e.g., FAO) compiled for the UK were used for comparison.

Unfortunately, the official statistics are always published in arrears with a delay of up to 2 years. For the purposes of the Seafish review, the base year was 1996, the last year for which complete statistics were available (Lake, 1999). To update the data as far as possible, a number of other sources were approached; all the Sea Fisheries Committees (only established in England and Wales) were interviewed about activity in their area; local council statistics were examined where they are collected; Internet searches were performed; and lastly, practitioners within the industry were canvassed. This allowed the picture for 1997 to be completed.

In addition, since the original data were prepared, some preliminary production figures for 1998 have become available. They are by no means complete, but they have been added in order to present a more up to date data set.

Methods

Information from the above sources was combined and, in addition, Seafish staff used their experience and knowledge of the industry in order to produce the overall picture presented in the review.

For some species either the quantity produced or its monetary value was not stated. Where this occurred, the value of the harvest (UK£) or the quantity produced (T) have been back-calculated from either the average industry prices or the production figures

Results

The estimated figures for the production of ranched shellfish in the UK for 1996–1997 are given in Table 1.

Similar results for 1998 are presented in Table 2.

Ranching techniques

Whilst some mussels and flat (or native) oysters are farmed semi-intensively, the majority of the UK non-wild fishery output is from relayed stock. Usually, natural mussel spatfall or part-grown oysters (known as 'halfware') are harvested from wild beds, transported and relayed on to extensive on-growing sites. However, recently it has proved possible to settle hatchery-produced flat oyster spat on to cockle shells before broadcasting them at low density onto the seabed for later harvest (Howard, 1999a). Both mussels and oysters remain on the bottom with no further input until they are harvested. For mussels, this can be 18 months to 2 years later, whereas for halfware oysters it is around 12 months. No time to harvest is given for the seed oysters, but it is usually 3–4 years in semi-intensive conditions.

The harvest techniques employed are often similar to those used by the wild fishery. A majority of the beds in England and Wales are subject to Several Fishery Orders, which confer property rights to the relayed species to the leasee, or other fishery controls. A recent development has been the relaying of cockles in a similar manner. At present in the UK, it is confined to two collaborating partners, who have been reluctant to provide too much information about the scope or economics of the undertaking for commercial reasons. On-growing is reported to be short; a matter of months, with harvest quality high and prices greater than those obtained for wild-fishery stocks (Howard, 1999b). There is interest in pursuing this technique elsewhere in the country and abroad (e.g., The Netherlands, Smaal, pers. commun.).

Despite advances in the hatchery production of spat (Millican, 1997), the culture of king scallops in the UK is largely dependant upon the active collection of natural spatfall. The principle settlement area is Raasay Sound adjacent to the Isle of Skye (Scotland). Settled spat are transferred from the mesh-filled collector bags to pearl or lantern nets, or more recently, stackable shallow plastics trays and on-grown for 2 years in either suspended culture or, for some tray systems, on-bottom stacks. The nets may be changed twice per year and the trays are cleaned and serviced regularly. Once they are large enough, the juvenile scallops are seeded on to the seabed in suitable areas. It is the transfer of the juveniles to seabed sites which moves scallop culture from semi-intensive farming to extensive ranched systems. Further input during the on-growing period is limited; some operators have no further involvement until harvest, whereas others will visit the site regularly to remove predators. Harvest is usually by diver collection, although dredging may also be considered under certain circumstances.

Ranching techniques used in the UK for the European Lobster are well documented in the literature (Beard & Wickins, 1992; Burton, 1992; Cook, 1995). The initial hatchery rearing cycle for larvae and young juveniles is highly intensive, but once the animals attain a size suitable for release they are liberated

Table 1. Estimated annual average UK production of ranched shellfish (based on production figures from 1996 to 1997)

Species	Amount of seed relayed	Value of seed relayed (£)	Amount harvested	Value of harvest (£)
Mussel (*Mytilus edulis*)	11 000 T	1 100 000	18 200 T	7 371 000
Flat oyster (*Ostrea edulis*)	100 T	180 000	482 T	1 106 000
King scallop (*Pecten maximus*)	500 000	100 000	36 T	90 000
Manila clam (*Tapes philippinarum*)	13 150 000	79 000	12 T	22 000
Pacific oyster (*Crassostrea gigas*)	66 020 000*	660 000*	74 T	107 000
Cockle (*Cerastoderma edule*)	>10 T	Not divulged	Not divulged	Approx £600 T^{-1}
Lobster (*Homarus gammarus*)	35 000	17 500	Not harvested	Not harvested

Key: Not harvested, industry has yet to make its first harvest from animals seeded out; Not divulged, official statistics not available and industry practitioners unwilling to provide data.
*Total seed production values.

Table 2. Estimate of UK ranched shellfish production for 1998

Species	Amount of seed relayed	Value of seed relayed (£)	Amount harvested	Value of harvest (£)
Mussel	Not available	Not available	9295 T	3 587 000
Flat oyster	Not available	Not available	726 T	1 074 000
King scallop	500 000	100 000	41 T	102 500
Manila clam	20 000 000	160 000	19 T	35 000
Pacific oyster	110 035 000*	1 486 000*	32 T	16 000
Palourde (*Ruditapes decussatus*)	4 010 000	40 000	12 T	37 000
Cockle	Not divulged	Not divulged	43 T	26 000
Lobster	40 000	20 000	Not harvested	Not harvested

Key: not available = Figures not yet published. Otherwise as Table 1.

on to natural rocky habitats and the programme becomes very extensive in character, perhaps the closest to what most people think of when they hear the term 'ranching'. There is no further human input from the time of release until the animals are recaptured by the commercial fishery at market size. This may take from 5 to 7 years, depending upon the size of juvenile released and their growth rate.

The Manila clam and Pacific oyster are non-native species, but both are well-established aquaculture subjects within the UK (Spencer, 1990; Spencer et al., 1991). The industry is almost entirely dependent upon hatcheries for the production of spat and the supply of seed. Juvenile Manila clams are seeded onto natural intertidal banks around the extreme low water mark (ELWST) to minimise exposure to the air. The plots are netted to offer protection from predation, predominantly by starfish, crabs and wading birds, and to provide some containment of the stock, which may be required under the Wildlife and Countryside Act 1981. On-growing takes place under natural conditions until the stock is harvested approximately 3 years later. Most Pacific oyster cultivation is semi-intensive, using the conventional bag and trestle system, but in some areas, part-grown stock is laid unrestrained onto the seabed to complete its growth to market size. It is this switch to bottom culture and dredge harvest which constitutes a period of extensive culture, although landings from this sector form only a very small proportion of the total production.

Palourde, also known as the native clam, carpet shell or butterfish (*Ruditapes* = *Tapes* = *Venerupis, decassatus* = *decussata*), has been subject to culture attempts in the past (Walne, 1976). It is cultivated in a similar manner to the Manila clam, but its slower growth to market size, approximately 4 years, was thought to be a disadvantage despite its higher market value. After a steady decline in production during the early 1990s, efforts have recommenced, culminating in an output of almost 12 T in 1998; the first recorded production since 1994, when it was only 43 kg (Howard, 1999a). Seed for culture is produced in a hatchery, although Seafish noted, during the course of Manila clam cultivation trials in Scotland (Paul, 1988), that natural settlement of Palourde seed occurred on the plots and that the density was greater than that observed in the surrounding beach (MacMillan, unpublished observation). This may be attributable to the protection from predation offered by the mesh enclosures.

Concluding remarks

It rapidly became apparent that definitive statistics were not available for this sector and that there were many gaps in the knowledge. Some statistics include ranched shellfish with the Fishery Order landings whilst others classify them as farmed. There appears to be little consistency between species or agencies. In addition, some official figures were at variance with the perception and experience of the industry, so all statistics quoted should be treated with caution.

It has been noted and is apparent from the tables that the value of the harvest in any one year can be less than the value of the seed layed in that year. This arises because the time taken for the relayed stock to reach marketable size varies from species to species and between areas around the country. Consequently, it is the amount and value of seed relayed up to 5 years previously which relates to the production figures for harvested marketable stock presented in the statistical tables. However, in many cases, alterations or variations in the way the statistics were compiled or reported prevents direct comparisons between the data relating to the seed layed and the resultant harvest.

For mussels and scallops, the availability of seed is a constraint upon the expansion of the sectors, whereas for both oyster species it is competition from producers elsewhere, particularly France, which controls price and hence profitability and production. With the clams species the decision whether to enter or expand production is determined by the price of the harvested stock. In marked contrast, lobsters are non-commercial at present and the decisions as to whether to proceed with stocking initiatives is largely political.

From our experience of compiling the production figures and values from 1996 to 1998, it is evident that further research is required to provide a clearer picture of this under-reported sector of the UK shellfish industry.

References

Beard, T. W. & J. F. Wickins, 1992. Techniques for the production of juvenile lobsters (*Homarus gammarus* (L.)). Fisheries Research Technical Report 92. MAFF Directorate of Fisheries Research, Lowestoft. 22 pp.

Burton, C. A., 1992. The techniques of lobster stock enhancement. Seafish Report 396. Sea Fish Industry Authority, Edinburgh. 55 pp.

Cook, W., 1995. A lobster stock enhancement experiment in Cardigan Bay. Final report. North Western & North Wales Sea Fisheries Committee, Lancaster. 33 pp.

Howard, A. E., 1999a. Farmed shellfish production in England and Wales, 1998. Shellfish News 8: 31–32.

Howard, A. E., 1999b. Cockle farming takes off. Shellfish News 8: 5–6.

Lake, N. C. H., 1999. The UK Shellfish Cultivation Industry – A strategy for development. Sea Fish Industry Authority, Edinburgh. 414 pp. (Also available on CD-ROM)

Millican, P. F., 1997. The hatchery rearing of king scallop (*Pecten maximus*). CEFAS, Lowestoft. 40 pp.

Paul, J. D., 1988. Experimental Cultivation of Manila Clams in Northwest Scotland. Technical Report 325. Sea Fish Industry Authority, Edinburgh. 28 pp.

Spencer, B. E., 1990. Cultivation of Pacific Oysters. Laboratory Leaflet 63. MAFF Directorate of Fisheries Research, Lowestoft. 47 pp.

Spencer, B. E., D. B. Edwards & P. F. Millican, 1991. Cultivation of Manila Clams. Laboratory Leaflet 65. MAFF Directorate of Fisheries Research, Lowestoft. 29 pp.

Walne, P. R., 1976. Experiments on the culture in the sea of the butterfish *Venerupis decussata* L. Aquaculture 8: 371–381.

Statistical sources

FAO: via Internet http://apps.fao.org/

UK: UK Sea Fisheries Statistics
SeaFish Key Indicators

England & Wales: Farmed shellfish production in England and Wales. Annual returns of production/landings of shellfish from Fishery Orders in England and Wales.
Scotland: Scottish Sea Fisheries Statistics
Scottish Shellfish Farms, Annual Production Survey

Northern Ireland: Occasional publications

Hydrobiologia **465:** 7–19, 2001.
G. Burnell (ed.), Coastal Shellfish – A Sustainable Resource.

Bivalve enhancement and restoration strategies in Florida, U.S.A.

William S. Arnold
Florida Marine Research Institute 100 Eighth Avenue SE St. Petersburg, FL 33701-5020, U.S.A.
Tel: 727-896-8626. Fax: 727-893-1374. E-mail: bill.arnold@fwc.state.fl.us

Key words: restoration, enhancement, bivalve, population, culture, *Mercenaria*, *Argopecten*, Florida, larvae, seeding, spawner transplant, cage, genetics, dispersal, fishery, metapopulation, seagrass, Indian River lagoon, habitat, harmful algal bloom, recruitment, fertilization

Abstract

In Florida, U.S.A. populations of recreationally and commercially important bivalve molluscs are stressed by a variety of factors, including habitat degradation, overfishing and development. Bivalves such as the bay scallop, *Argopecten irradians*, and the hard clam, *Mercenaria* spp., are uniquely positioned to benefit from restoration and enhancement activities because of the physical setting in which they live or because of recent efforts to restore habitat. For bay scallops, a major restoration effort has been implemented on the west coast of Florida. Adult scallops are collected from the target site and spawned in the laboratory. The offspring are cultured in ponds until they reach an average shell height of approximately 30 mm and then are planted in cages at the target site from which the parents were collected. Parents and offspring are genetically monitored, as are new recruits sampled at a variety of sites along the coast. The ultimate goal of this restoration program is to create concentrated patches of spawners that will supply recruits to west Florida seagrass beds. For hard clams, three different strategies are being tested to determine the most ecologically and economically feasible approach to use to enhance harvestable clam populations. Hard clams are occasionally abundant in the Indian River lagoon on the east central coast of the state, but transportation causeways fragment the lagoon into a series of basins that are largely isolated from one another. At any time, only a subset of the basins that compose the lagoon may be environmentally suitable for supporting the sensitive larval and juvenile stages of hard clams, and the members of that subset change constantly as environmental conditions change in the lagoon. Transplanting spawner stock, seeding juveniles under a variety of protective treatments, and injecting fertilized eggs directly into the lagoon are each being tested for their effectiveness in exploiting environmentally appropriate conditions and producing successful settlement events. Each strategy is being tested for its ecological and economic suitability, and the most appropriate strategy will be implemented in an effort to maintain clam abundance at a level where licensed clammers in the Indian River region can realize a minimum income from the fishery, even when naturally recruiting clam stocks are depleted.

Introduction

In developed and developing countries throughout the world, the health and ecology of coastal marine communities are being affected by such anthropogenic activities as commercial fishing (Safina, 1995), recreational and commercial watercraft activities (Sargent et al., 1995), upland runoff (White, 1996), and increased occurrence of harmful algal blooms (Cosper et al., 1987; Hallegraeff, 1993). The success of efforts to mitigate the impacts of these activities often depends upon the nature of the activity and the nature of its resultant impact. For example, the scarring of seagrass beds by boat propellers is one problem that may be minimized, and the impacts essentially reversed, by implementing the four-point approach advocated by Sargent et al. (1995). Other problems, such as the hypothesized increase in the frequency or severity of harmful algal blooms, are much more complex and the solutions more controversial. Between these two extremes are problems such as commercial fishing efforts and upland runoff, the effects of which may be reduced by changing activity patterns or by advancements in technology (e.g. bycatch reduction devices, retention ponds). Unfortunately, reducing the effect will not necessarily eliminate the symptoms, especially in the case of biological communities.

In Florida, U.S.A., population growth has been substantial and has had an undeniable impact on coastal marine communities. Between 1950 and 1990, the population of Florida grew from fewer than 2.8 million inhabitants to almost 13 million inhabitants, and much of that growth has occurred along the approximately 1200 miles of Florida's marine coastline (Marth & Marth, 1995). As a result, seagrass beds have declined by more than 80% in some areas of the state (Haddad & Harris, 1985; Lewis et al., 1985). Similar impacts have been noted in other habitats, such as mangrove forests, and in other areas of the world (e.g. see Eleuterius, 1987). This loss of habitat has had a substantial impact on coastal marine species, such as redfish (*Sciaenops ocellatus*), seatrout (*Cynoscion nebulosus*), and bay scallops (*Argopecten irradians*), that are dependent upon seagrass during some phase of their life cycle (Thayer & Stuart, 1974; Gilmore, 1987). In the case of bay scallops, the impact has been substantial and has been exacerbated by coastal development, degradation in water quality, and overfishing. As a result, bay scallop populations in some areas of Florida have collapsed and others may soon follow (Arnold et al., 1998).

The bay scallop occupies nearshore marine environments and is dependent upon seagrass throughout its lifetime (Thayer & Stuart, 1974; Eckman, 1987, 1989). Thus, loss of seagrass habitat has adversely affected the distribution and abundance of bay scallops in Florida. Historically, bay scallops in Florida occur from West Palm Beach on the southeast coast to Pensacola in the western Florida panhandle (Waller, 1969), but recent quantitative and anecdotal information suggests that the range in Florida has contracted considerably. Substantial efforts have been made during the 1990s to enhance water quality and increase seagrass abundance in many areas of Florida, including Tampa Bay, and one benefit of these habitat restoration efforts is the reestablishment of suitable habitat for bay scallops (Blake, 1998). The State of Florida has initiated a program to reestablish bay scallop populations in Tampa Bay and in several other sites along the Gulf of Mexico coast of the state. I herein describe the strategy that we are implementing in an effort to restore bay scallop populations in Florida.

Hard clams of the genus *Mercenaria* support an important commercial fishery in Florida and along the entire eastern seaboard of the United States. That fishery is characterized by drastic fluctuations in landings: several good years of fishing are often followed by several to many years of very poor fishing. Fishery participants suffer extreme hardship during years in which fishing is poor, and hard clam fishermen have requested that researchers work toward devising a strategy for enhancing clam populations so that at least a minimum level of income can be earned even during poor years. In most areas where hard clams are commercially harvested, this request would be almost impossible to fulfill because the open nature of the hard clam fishery allows for the over-exploitation of clams at any level of abundance. In Florida, however, the fishery for naturally occurring hard clams is concentrated in Brevard County, on the east central coast of the state. There, participation in the fishery is limited to a maximum of 500 licensed clammers. This allows for an estimation of the number of clams required to provide an acceptable annual income. As a rough estimate, if each clammer requires an income of $500 per week and clams sell for an average of $0.15 each, then approximately 87 million clams will be required over the course of a year to meet that income goal. Of course, the difficulty is in providing those 87 million clams, but at least in this limited-entry fishery we can establish such a goal. I will describe the three alternative strategies that we are testing to determine the best approach to hard clam enhancement in Florida.

Bay scallops

Background

The bay scallop population in Florida appears to function as a metapopulation, with local populations occupying discrete embayments and seagrass beds from West Palm Beach (Fig. 1) to the Chandeleur Islands in Louisiana (Waller, 1969; Arnold et al., 1998). Alternatively, it is possible that what appears to be metapopulation structure is simply the structure of a population that was once continuous but is now in the throes of extinction (Hanski & Simberloff, 1996). However, available evidence argues against this latter alternative. First, along much of the Florida coastline, available bay scallop habitat is confined within bay systems such as Biscayne Bay, Florida Bay, Pine Island Sound, and Tampa Bay in the south and St. Joseph Bay, St. Andrew Bay, and Pensacola Bay in the north. Similarly, the 'Big Bend' region between

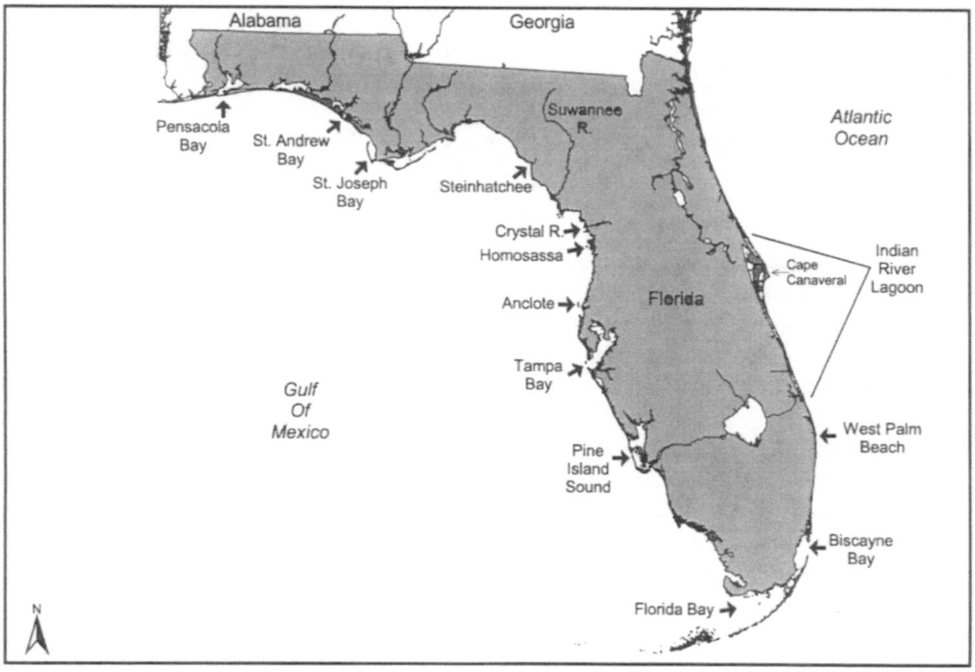

Figure 1. Map of Florida, U.S.A., illustrating locations of sites for bay scallop (*Argopecten irradians*) adult sampling and restoration programs and the location of the Indian River lagoon on the Atlantic coast of Florida.

Tampa Bay and St. Joseph Bay (Fig. 1) may also function as a single large bay defined along its seaward edge by a coastal 'jet' that flows between those two points (Yang et al., 1999; Li & Weisberg, 1999). Even within the Big Bend, scallop populations are discretely distributed at sites such as Anclote, Homosassa and Steinhatchee (Arnold et al., 1998). Second, commercial fisheries landings and anecdotal information suggest that local populations have experienced large fluctuations in size, as would be expected in a short-lived animal such as the bay scallop. Thus, it appears that the bay scallop population in the eastern Gulf of Mexico has always been composed of discrete local populations and that extinction and recolonization are common features of the local populations that compose the bay scallop metapopulation. We have based our restoration strategy upon this fundamental precept.

The local populations that compose the purported bay scallop metapopulation in Florida have collapsed south of the Suwanee River, and the historic pattern of this collapse appears to have been from south to north (Arnold, 1994). For example, a major scallop fishery occurred in Pine Island Sound throughout the 1940s and 1950s, but that fishery essentially disappeared after 1963 (Murdock, 1955; Florida State Board of Conservation, 1964). In Tampa Bay, available information suggests that scallops were extremely abundant until the early 1960s (Haddad, 1988), but scallops are now practically nonexistent in Tampa Bay. Recreational bay scallop fishing activities were targeted and successful in the Anclote and Homosassa areas during the 1970s and 1980s, respectively, but by the mid-1980s (Anclote) and the early 1990s (Homosassa) those populations had essentially ceased to exist (Arnold et al., 1998). Only areas north of the Suwanee River, particularly St. Joseph Bay and the nearshore waters around Steinhatchee, still support substantial bay scallop populations (Arnold et al., 1998). The Florida bay scallop restoration program is designed to reverse that trend, and the initial focus is on the area between Tampa Bay and Crystal River, where the collapse has been most recent.

If habitat degradation has occurred, as we suspect it has in areas such as Pine Island Sound and Tampa Bay, then it is probable that directly enhancing scallop populations in those areas will be fruitless. However, if one or more 'source' (Pulliam, 1988) populations have been adversely affected by habitat degradation or other factors, with the result that other (neutral or sink) populations in otherwise suitable areas have been adversely affected, then it may be possible to rebuild multiple local populations by rebuilding the key source populations. Tampa Bay appears to function as a key source population on the Florida west coast. Historically, Tampa Bay was a very large, pristine environment composed of numerous seagrass-dominated areas. Each of those areas may have originally harbored a distinct local population. As the land area around Tampa Bay was developed, dredge-and-fill and bridge-building activities radically altered the seascape, causing degradation of water quality and loss of scallop habitat. If those trends could be reversed, it might be possible to rebuild this key scallop population. In recent decades, efforts to restore habitat in Tampa Bay have resulted in a considerable increase in water quality (Blake, 1996, 1998) and a concomitant increase in the habitat available for scallops. This provides an ideal opportunity to test the feasibility of bay scallop restoration in Florida and, under the proper experimental conditions, also may provide a test of the hypotheses that Florida bay scallops do function as a metapopulation and that the Tampa Bay scallop population is a source population.

In areas north of Tampa Bay, including Anclote, Homosassa and Crystal River, the loss of scallop populations may be the result of several factors acting individually or in concert. Harmful algal blooms (HABs) are common in this area of Florida, and by all accounts, HABs are inimical to the survival of scallops (Cosper et al., 1987; Summerson & Peterson, 1990). Harmful algal blooms have been occurring along the Florida west coast for centuries, so the bay scallop metapopulation should be adapted to survive HABs. But if the frequency or intensity of HABs is increasing (Hallegraeff, 1993), then the stability of the metapopulation may be compromised. However, firm evidence to support an increased frequency or intensity of HABs in the eastern Gulf of Mexico is lacking. Alternatively, seagrass habitat at sites such as Anclote and Homosassa may be degraded to the point that scallop populations are no longer supported. Development continues along the west coast of Florida, but the area north of Anclote still appears to be relatively pristine and the seagrass beds vast and abundant. Finally, it may be that each local population between Anclote and Crystal River simply experienced collapse in the normal course of its life history. As with any annual species, population fluctuations tend to be drastic even in apparently healthy populations. During years of very low population abundance, reproduction in the local population may no longer be successful (Levitan, 1995; Marelli et al., 1999). In that circumstance, the self-seeding necessary for year-to-year maintenance of the population (Arnold et al., 1998) may fail, and the population will not recover without allochthonous larval inputs. Such allochthonous inputs and subsequent population recovery may require many years (Peterson & Summerson, 1992), and in the case of Florida bay scallops, recovery may never occur if neighboring local populations also have collapsed.

Our strategy for bay scallop restoration on the Florida west coast exploits the above considerations to test the bay scallop metapopulation hypothesis and to rebuild local populations within that framework. We hypothesize that restoration of Tampa Bay, Anclote, Homosassa and Crystal River scallop populations will require the input of locally generated recruits. The plan we have implemented involves capturing wild animals from the target site and spawning them in the laboratory. We culture the offspring in saltwater ponds until they are approximately 20–30 mm shell height (SH=maximum distance from umbo to ventral margin) and then plant them in cages at the target site from which the parents were originally collected. Our goal is to create dense concentrations of spawners that will produce large numbers of successfully fertilized gametes. A substantial portion of the resultant offspring will then settle in the surrounding seagrass beds, thereby providing the spawner stock for future generations.

Spawning

We carefully monitor the genetic composition of each bay scallop cohort throughout the selection, culture and growout process. As mentioned, adult scallops to be used as spawner stock are collected from the site at which the offspring eventually will be planted. We attempt to collect up to 100 adults per target site, but that is not always feasible because scallops usually are difficult to find within the restoration area. Adults are returned to the laboratory, and an immediate spawning is attempted. Each spawner batch consists of approximately 6–8 adult scallops, and each adult is labeled so that we can track the genetic linkage between adult and offspring. Adults are then sacrificed and genetically analyzed by using allozymes and nuclear markers. The offspring are similarly isolated and tracked throughout the culture process, and samples for genetic analysis are collected from the offspring when they are transplanted for nursery growout, at the time of field planting, and during fall when spawning is imminent.

Laboratory culture

Offspring are held in 500-l tanks in a laboratory throughout the larval and post-settlement phases. As the larvae reach competency, a suitable settlement substrate is provided (Lu & Blake, 1997), and the juvenile scallops are held in the laboratory and fed *Isochrysis galbana* and *Pavlova lutheri* until they reach a shell height of approximately 1 mm (Lu & Blake, 1997). The scallops are then transferred to 800-μm-mesh culture bags and grown in the natural environment at an initial density of 6000 scallops per bag, which is eventually culled to a density of 2000 scallops per bag. At a shell height of approximately 5 mm, the juvenile scallops are transferred to 0.1-hectare ponds at the Florida Marine Research Institute Stock Enhancement Research Facility (SERF) for nursery growout.

Nursery culture

At the SERF facility, juvenile scallops are held in the growout ponds and exposed to 400-μm-filtered seawater pumped from nearby Tampa Bay at a rate of approximately 10 000 l per hour. The scallops are maintained in the 800-μm-mesh culture bags, but the bags are placed in pairs in 6-mm-mesh plastic-coated wire cages. Scallop densities are adjusted to approximately 2000 scallops per cage, and the cages are mounted on polyvinyl chloride (PVC) racks that hold each cage at a depth of 30–50 cm below the water surface. As soon as the scallops reach a shell height that would prevent them from slipping through the cage mesh, they are removed from the 800-μm-mesh bags and planted directly into the 6-mm-mesh cages. Growth rate tends to vary widely among scallops within each brood, but scallops transplanted to the holding ponds in January generally grow large enough by April or May to be transplanted to the field.

Field growout

Transplanting the scallops to the field growout site requires that the animals be transported up to 200 km from the SERF facility. Thus, transport conditions are a critical determinant of the health of the animal when it is introduced into the field growout site. Bay scallops tend to be especially sensitive to salinity and dissolved oxygen (Tettelbach & Rhodes, 1981; Leverone, 1993), whereas temperature is less important as long as transplant operations are not conducted during midsummer, when water temperature becomes excessively hot in Florida (Mercaldo & Rhodes, 1982). Scallops are placed in a trailer equipped with a 3500-l-capacity water tank. The tank is filled with 5-μm-filtered seawater pumped from Tampa Bay. Compressed oxygen gas is used to obtain a minimum level of 5.0 mg per l dissolved oxygen in the tank, and battery-operated submersible pumps are employed to recirculate the water during transport. At the launch facility, the scallops are transferred to a 650-l tank mounted on the research vessel and then transported with constant aeration to the planting site. Salinity and dissolved oxygen are closely monitored, and adjusted if necessary, throughout the transport process.

Each planting site consists of ten stations, with five cages per station. Cages are constructed of 12.7-mm-mesh plastic-coated wire and are anchored to the bottom with cinder blocks. During both 1998 and 1999, scallops were planted at a density of 300 per cage. An additional cage at one station is planted at the same density and used to provide samples for reproductive monitoring. Every 6 weeks, shell growth is estimated by measuring 30 individuals from one cage at each of nine stations and by measuring 30 individuals from each of the five cages at the tenth station. Mortality is estimated by counting all empty valves (at times of high scallop density, early in the season) or by counting all live scallops (at times of low scallop density, late in the season) in each cage where scallops were measured. Fifteen scallops are collected from the additional cage and returned to the laboratory for histological preparation and subsequent determination of reproductive status. For reproductive analyses, the entire gonad is excised from each animal, fixed in a 5% formalin-seawater mix, and dehydrated through increasing concentrations of ethanol (70–95%). A 4-mm subsection of the whole gonad is then embedded in glycol methacrylate, and 3.5-μm-thick cross-sections are cut from the sample block at two points separated by a minimum of 35–70 μm. Sections are mounted on acid-cleaned slides and stained with hematoxylin and eosin. Resultant samples are quantitatively analyzed for egg density and egg diameter (Heffernan & Walker, 1989) and qualitatively analyzed for reproductive development stage (Table 1) following the procedures of Arnold et al. (1997).

Table 1. Reproductive staging values, and a description of the characteristics of each stage, for bay scallops (*Argopecten irradians*) sampled from restoration cages on the Florida west coast. Staging scheme from Arnold et al. (1997)

Reproductive	Description of tissue	Stage
No data	Tissue unreadable	0
Inactive, immature	Gonad tissue undifferentiated	1
Developing	Tissue differentiated, eggs and sperm present	2
Ripe	Tissue full of gametes	3
Early spawning	Gametes being shed, but still full in appearance	3.5
Spawning	Many gametes shed, follicles partially empty	4
Spent	Follicles nearly empty	5

Additional research

Smaller experiments are being conducted to test specific hypotheses or as pilot projects for alternative restoration strategies. One experiment involves testing various planting densities within our 0.6-m × 0.6-m cages. Within a small seagrass bed, we deployed three groups of cages at densities of 50, 150 and 300 scallops per cage, with three replicate cages per group. In a separate experiment, we planted three replicate cages for each of four treatments to test the effect of habitat on scallop growth and survival. Treatments included placing the cages directly on a sand substrate, directly on a seagrass substrate, on cinder blocks above a sand substrate, and on cinder blocks above a seagrass substrate. We have also deployed mesh panels of various construction (coconut mesh, plastic mesh) to act as recruitment-enhancement substrates. These panels are designed to capture bay scallop larvae directly from the water column and to provide a protected environment where the juvenile scallops have an opportunity to grow to a size at which they are relatively immune from predation (Tettelbach, 1986). The latter experiment may provide an alternative strategy for bay scallop restoration in Florida and throughout the range of the animal.

Hard clams

Hard clams of the genus *Mercenaria* support important commercial fisheries throughout the animal's natural range along the Atlantic and Gulf of Mexico coasts of North America. Fishery landings at any site tend to fluctuate considerably from year to year (MacKenzie et al., 2001), and even within sites, clam abundance is extremely patchy and unpredictable (e.g. Godcharles & Jaap, 1973; Haven & Loesch, 1973). The relationship between adult abundance and juvenile recruitment is poorly understood in hard clams, but many instances have been recorded of large sets of hard clams occurring in areas that previously supported few if any clams (e.g. see MacKenzie, 1997). On the Florida east coast, the clam fishery expanded and collapsed twice between 1980 and 1995, and the collapse in both cases was attributed to decreased salinity over the beds (MacKenzie et al., 2001). However, the fishery that developed during the early 1980s was separated by approximately 60 km from the fishery that developed in the early 1990s. The Indian River is a hydrographically complex lagoon, composed of numerous basins defined by the causeways that traverse the lagoon at various points along its length (Fig. 2). As a result, the physical environment may be poorly suited for hard clams within some basins and well suited for hard clams in other basins. This mismatch between clam biology and the physical environment may explain why hard clam recruitment events are so rare in the lagoon despite the apparent abundance of clams (Dierberg, 1986) and their frequency of spawning (Hesselman et al., 1989). Our strategy for enhancing hard clam populations in the lagoon is based upon overcoming this mismatch – we focus our enhancement efforts in areas of suitable water quality and shift focus as water quality changes. We are taking a three-pronged approach to clam enhancement: (1) transplanting adult clams to create spawner sanctuaries, (2) seeding juvenile clams under a variety of treatment conditions to create spawner sanctuaries, and (3) directly injecting fertilized eggs into the water column to circumvent the spawning process.

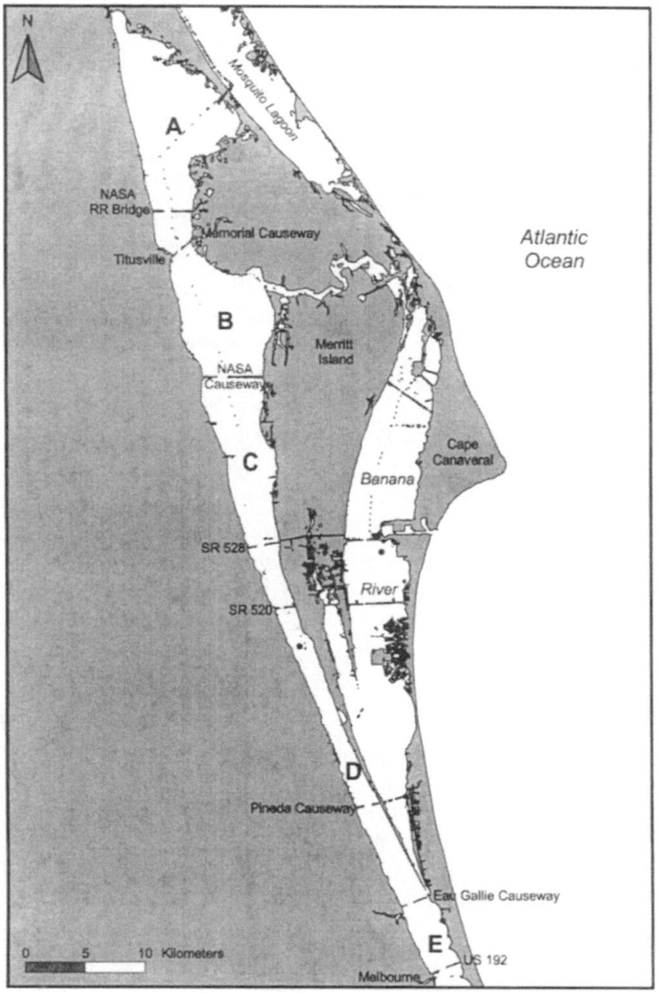

Figure 2. Indian River lagoon, Florida, with locations of Shellfish Harvesting Areas A through E. Filled circles represent locations of the adult clam relay and seeding studies. Note the locations of causeways traversing the lagoon throughout its length.

Adult clam transplants

Shellfish harvesting areas in the Indian River lagoon are variously classified depending upon their suitability for harvest at various times of the year and in accordance with extant environmental conditions. We employed professional clammers to assist in the harvest of clams from legally accessible harvest areas for transplant into areas closed to shellfish harvesting. On each of four dates (November 8, 1998, February 2, 1999, April 27, 1999, and August 9, 1999), the clammers used long rakes or hand rakes to harvest 5000 clams of all sizes. The clams were partitioned into two groups of 2500 each, and the shell of each clam was painted in a color that indicated the season of harvest. The clams were then transplanted into closed shellfishing waters at each of two sites, one in the Indian River lagoon and one in the nearby Banana River lagoon. Each seasonal group was poured over the side of the research vessel as it traversed the extent of a preestablished 10-m × 10-m plot marked at the corners with small styrofoam floats. This planting technique was designed to emulate the methods that will be used if a large-scale transplantation program is implemented.

Transplant plots were sampled after two weeks and again once every season for a year. Sampling consisted of hand-raking all clams from each of 20 randomly selected 0.25 m^2 quadrats. To ensure that no location was sampled twice, the position of each quadrat sample was marked with a short length of PVC tubing. All harvested clams were returned to the research vessel, the clams were counted and measured, and a random subsample was collected for analysis of reproductive condition. An additional sample of 200 clams, collected during each original seasonal harvest, was similarly analyzed for size and reproductive condition to provide a baseline for comparison.

The goals of the transplant study included (1) determining the survival rate of transplanted hard clams and assessing differences in survival among clam cohorts planted during each season of the year, (2) monitoring the reproductive condition of the clams and comparing that with the reproductive condition of naturally occurring hard clams in the Indian River lagoon, and (3) comparing these parameters between the two study sites. Ultimately, we intend to develop an estimate of seasonal egg production from each transplant cohort relative to undisturbed clams and to develop an estimate of the cost to the clammers of creating that concentrated egg production. A key assumption of this study is that concentrating clam spawners at relatively high densities will increase fertilization efficiency (Levitan, 1995; Southworth & Mann, 1998), thus increasing the residual reproductive value of each clam relative to its pre-transplant value. Clams were originally harvested from an area where clam density is approximately 1 per m^2, whereas post-transplant density within each plot averaged 25 clams per m^2.

Seeding

Previous research in the Indian River lagoon (Marelli & Arnold, 1996) and elsewhere (e.g. Menzel & Sims, 1962; Arnold, 1984) indicates that survival of small seed clams planted in unprotected plots is very low, but historic (Castagna & Kraeuter, 1977) and recent (Peterson et al., 1995) research suggests that planting seed clams at low densities under some sort of protective cover considerably increases survival. Those protective measures form the basis of the hard clam aquaculture industry in the United States, but those measures also may provide an effective approach for restoring public shellfish beds (Peterson et al., 1995). At a site in the Indian River lagoon, we are planting seed clams of various sizes under various protective conditions to test the economics and feasibility of using this strategy in creating spawner sanctuaries. For a variety of reasons (e.g. cost, logistics), we are not considering this strategy as a means of increasing clam abundance for direct harvest.

Three size-classes of seed clams (2, 8 and 16 mm mean SH) were purchased from Harbor Branch Oceanographic Institution in Ft. Pierce, Florida. The two smaller size-classes were labeled with tetracycline (Marelli & Arnold, 1996) and the shells of the larger clams were painted to allow for identification after harvest. Equal dollar values of clams in each size class (equivalent to 199 2-mm-SH, 65 8-mm-SH, and 23 16-mm-SH clams per 0.0278 m^2) were planted in replicate under each of four treatment conditions. Treatments included no protective cover, an oyster shell aggregate cover, a 12.7-mm-plastic-mesh cover, and a cover of oyster shell laid over the plastic mesh. We sampled each plot after 2 weeks and again after 1 year by partitioning each plot into 0.0278-m^2 subplots and suction-dredging the contents of five randomly selected subplots within each plot to a depth of 5 cm. Each replicate sample was returned to the research vessel, preserved with formalin, and returned to the laboratory for sorting and analysis. Only those clams for which a tetracycline or paint mark could be verified were included in the final data set.

Larval injection

As an alternative to clam transplants or seeding, we are testing the concept of circumventing the natural spawning process by spawning clams in the laboratory, fertilizing the resultant eggs, and releasing the eggs at a predetermined location in the Indian River lagoon within several hours of spawning. Using the considerable expertise of the hard clam culture group at Harbor Branch Oceanographic Institution, we were able to produce over 250 million fertilized eggs during July 1998 for transplantation into Shellfish Harvesting Area A in the northern Indian River lagoon (Fig. 2). One day prior to releasing the eggs, we sampled the study area to estimate the density of naturally occurring clam larvae. One-hundred-liter water samples were collected from each of the 25 nodes of a 5 km × 5 km grid (node spacing 1 km; Fig. 3), filtered through a 63-μm-mesh plankton net, distilled into small jars, and frozen for later analysis. Similar 'background' samples were collected the following morning, and then the larvae were introduced at station 9. We tracked these larvae by surrounding the larval mass with three small subsurface drifters and placing a fourth subsurface drifter at the center of the mass. The drifters were patterned after the design of Davis (1985) but were scaled to 0.7× the size of Davis'

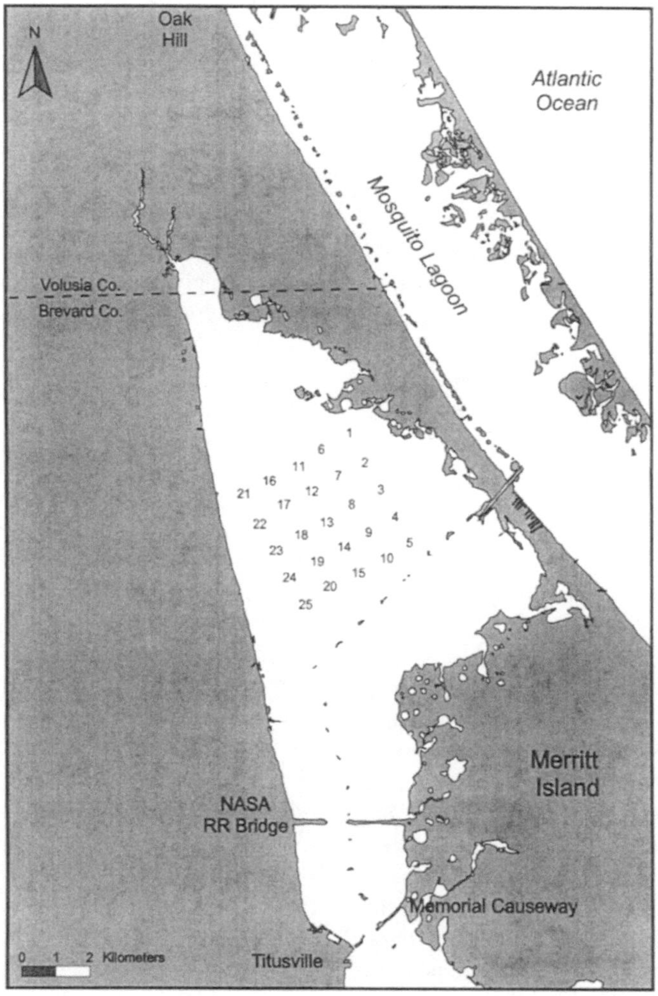

Figure 3. Indian River Shellfishing Harvesting Area A, where larval *Mercenaria mercenaria* were released at station 9 and tracked over four days. Grid node stations are numbered.

drifters to account for the shallow Indian River waters (Hitchcock & Arnold, submitted).

Immediately following larval release, and continuing for the next 3 days, additional plankton samples were collected at various times and at various points surrounding the tracked water mass. Plankton samples collected throughout the 5-day duration of the study were analyzed for the presence of *Mercenaria* larvae by using a genus-specific larval probe that allows for the identification and quantification of bivalves such as hard clams (e.g. Frischer et al., 2000). Additional samples collected from larval traps (Yund et al., 1991), deployed in triplicate at each of the grid nodes, are being analyzed with the genetic probe to determine the abundance of settling larvae throughout the study area.

Contemporaneous with larval-mass tracking, we released and tracked the diffusion of Rhodamine WT dye to estimate particle diffusion within the larval mass. Rhodamine was released approximately 1 km away from the actual larval mass to emulate hydrodynamic conditions within the mass without subjecting the larvae to this potentially toxic chemical. In combination, the drifter-tracking and Rhodamine-dye studies allow us to estimate advection and diffusion of the larval mass within this coastal lagoon. Preliminary results indicate that diffusion within such an enclosed water body may determine the ultimate distribution and fate of the larval mass and that the advective term becomes less significant as the larvae diffuse throughout the basin.

Discussion

The process of restoring shellfish populations requires the consent and cooperation of numerous interest groups, including fishermen, business and community representatives, environmental organizations, funding agencies, and other user groups. The collision of these various and potentially antagonistic perspectives will influence, and to some degree dictate, the options available for restoration. Financial considerations are also important, and it is likely that in any shellfish restoration effort, funding limitations will restrict effort to only the most essential tasks. Our work on shellfish restoration in Florida is no exception. For example, our funding for bay scallop restoration was provided by a federal grant with a 4-year life-span, and all aspects of our restoration study were required to fit within that time frame. Thus, we were not able to conduct a thorough assessment of the genetic structure of bay scallop populations in Florida prior to initiating restoration. Instead, the restoration study was designed using the best available information, including the results of studies which suggested that local populations of bay scallops were largely self-seeding in Florida (Arnold et al., 1998) and elsewhere (Peterson & Summerson, 1992) and that a dearth of recruits was preventing recovery of collapsed populations (Marelli et al., 1999). In the hard clam restoration project, the fishery participants' financial support and active involvement were paramount to success, and those participants have been and remain involved in the development and execution of each component of the research plan. To maintain that communication, we meet with every licensed Brevard County hard clam fishery participant at least once per year to discuss issues associated with the fishery.

Bay scallops should benefit from recent efforts to restore seagrass habitat on the west coast of Florida (e.g. Hoffman et al., 1982), but this benefit will not be realized if the abundance of naturally occurring scallops is so low that successful reproduction and recruitment are impeded. For example, although seagrass coverage in Tampa Bay has expanded in the last decade (Blake, 1998), by-catch information from the commercial shrimp fishery and the results of various SCUBA surveys suggest that scallop abundance remains extremely low. In other areas of Florida where bay scallop populations have collapsed, average scallop density is less than 1 per 100 m^2, whereas average scallop density in Steinhatchee and St. Joseph Bay may exceed 25 per 100 m^2 (Arnold et al., 1998). At very low population density, fertilization success decreases substantially (Levitan & Petersen, 1995). Our restoration strategy is designed to concentrate spawning scallops, thereby increasing fertilization success, larval supply, and the resultant availability of competent recruits. Those recruits will then occupy the surrounding seagrass beds, increasing the density of naturally occurring scallops and facilitating the long-term maintenance of the population. For our approach to be successful, multiple local populations must be restored and maintained at a reasonable level of stability. Otherwise, in a short-lived animal such as the bay scallop, extreme fluctuations in population density will eventually lead to the collapse of local populations. Those populations will not recover without allochthonous larval inputs from neighboring populations.

The fragmented nature of the Indian River lagoon provides an opportunity to exploit environmental variation among basins within the lagoon to maximize the success of hard clam population enhancement efforts. We are comparing the ecological and economic suitability of three approaches for enhancing the Indian River clam population, but these approaches share a common goal of increasing the number of available recruits. First, adult clams are being transplanted from diverse areas of the lagoon into spawner sanctuaries where densities of mature adults may be increased 25-fold, thereby considerably increasing fertilization success and larval production. Second, seeding of juvenile scallops is being tested as an alternative means of creating dense concentrations of spawning clams. Because seeding maximizes the residual reproductive value of the clams, this approach has the potential to provide many more clams for less money than does the adult-transplant approach. Third, we will circumvent the natural spawning process entirely by spawning clams in the laboratory and releasing the resultant eggs directly into the water column. We expect that losses of larvae will be substantial, but the combination of the vast numbers of fertilized eggs, the low cost per egg, and the flexibility that this approach provides to exploit suitable habitats over the short-term, creates the conditions under which this larval-injection strategy may prove successful. Our first test of larval injection resulted in a 3- – 4-fold increase in larval clam abundance. Even after three days of dispersal, peak densities were within a range of 3500-4000 larvae per 100 l, densities that are similar to the peak densities of oyster and clam larvae recorded by Carriker (1951; 1961) in New Jersey inshore waters. We conclude that our small-scale release substantially increased the abundance of larval hard clams in a sub-basin of the Indian River lagoon. We anticipate that future releases will involve more larvae and eventually will be conducted over multiple days, with the goal of releasing billions to tens-of-billions of larval clams. An inherent component of our ongoing research is to determine the success of these larvae in settling and surviving to harvest size.

The strategies that we are employing to enhance or restore shellfish populations are not unique. Numerous shellfish restoration efforts are underway in the United States and throughout the world, and those efforts involve a variety of strategies that are designed to exploit the unique life-history attributes of the selected species. In North Carolina, U.S.A., adult bay scallops were transplanted into areas where scallops had been virtually eliminated by red tide (Peterson et al., 1996). Results suggested that recruitment was limiting population recovery and that spawner transplants did enhance recruitment and subsequent population size. Similar programs have been undertaken for bay scallops in New York (Tettelbach & Wenczel, 1993) and Connecticut (Tammi & Turner, 1998), in each case concentrating juvenile or adult scallops to increase spawning success and larval production. Again, preliminary results are promising, but the ultimate success of these efforts remains to be determined. Seeding of juveniles has been conducted for the king scallop (*Pecten maximus*) in the Bay of Brest, France (Dao et al., 1994). More than 20% of the seeded scallops were recovered by the commercial fishery, and an additional 6% of the seeded scallops survived the fishing season and potentially contributed to future generations. Larval releases have been conducted to enhance abalone stocks in South Australia, and preliminary results suggest that the density of juvenile abalone resulting from these small-scale (<3 million larvae total) larval releases was about 0.6 per m^2 (Preece et al., 1997). Obviously, restoration of many shellfish species is being attempted, but most of those restoration efforts are in the early stages of development and results are not yet available on which to assess their success or even to determine which strategies are the most suitable.

Perhaps the most extensive and longest-running bivalve-enhancement programs have involved various species of oysters. In the United States, restoration and enhancement of *Crassostrea virginica* reefs were underway in the late 1800s (Kennedy & Sanford, 1999) and continue today both in the Gulf of Mexico (e.g. Perret et al., 1999) and along the Atlantic Ocean coast of North America (e.g. Wesson et al., 1999). However, oyster restoration efforts primarily target habitat rather than the animal itself because unlike the techniques used in the inshore clam and scallop fisheries, which apparently do little harm to the ambient habitat, the techniques used in the oyster fishery tend to alter the very habitat upon which the oyster is dependent. The lack of suitable habitat, rather than the availability of recruits, appears to limit the abundance of oysters although efforts are underway to seed oyster reefs to speed the rate of recovery and to increase overall production (Supan et al., 1999).

In summary, our bay scallop and hard clam restoration efforts in Florida reflect the spectrum of methodologies being applied to shellfish restoration projects throughout the world. We are targeting a variety of life stages, including larvae, juveniles and adults, in an attempt to identify the life stage during which restoration efforts should be conducted to realize the best return on the logistical and financial investment. Preliminary results appear promising, particularly for bay scallop restoration, but all of this work has been initiated since 1997 and final results remain to be analyzed. All members of the research team remain optimistic that our efforts will assist in the development of effective strategies for molluscan shellfish restoration, but only if suitable environmental conditions exist for the survival and proliferation of the species. If the conditions that led to the demise of the population are not corrected, all attempts at shellfish restoration will fail.

Acknowledgements

Restoration and enhancement of hard clam and bay scallop populations involves the efforts of numerous people and institutions. Gary Hitchcock and Chris Kelble (University of Miami) assisted with tracking the hard clam larval mass, and Marc Frischer (Skidaway Institute of Oceanography) developed the clam genetic probe and analyzed resultant samples. David Vaughan and associates (Harbor Branch Oceanographic Institution) generously provided the hundreds of millions of clam larvae necessary for the larval tracking study. Kate Hagner, Melissa Harrison, Philip Hoffman, Micah Humphrey, Tracy Idocks, Dan Marelli, Melanie Parker, Sarah Peters and Winnie

White assist with all aspects of field and laboratory research for both the hard clam and bay scallop programs. Ami Wilbur and Terrie Bert guide and conduct the bay scallop genetic program, Norm Blake and Candice Way (University of South Florida) provide the expertise for bay scallop culture, and Bill Halstead and associates assist with scallop growout and field planting.

References

Arnold, W. S., 1984. The effects of prey size, predator size, and sediment composition on the rate of predation of the blue crab, *Callinectes sapidus* Rathbun, on the hard clam, *Mercenaria mercenaria* (Linne). J. exp. mar. Biol. Ecol. 80: 207–219.

Arnold, W. S., 1994. Bay scallops: summary report to the Florida Marine Fisheries Commission. Internal Report, Florida Marine Research Institute: 1–13.

Arnold, W. S., D. C. Marelli, C. P. Bray & M. M. Harrison, 1998. Recruitment of bay scallops *Argopecten irradians* in Floridan Gulf of Mexico waters: scales of coherence. Mar. Ecol. Prog. Ser. 170: 143–157.

Arnold, W. S., D. C. Marelli, K. Hagner, M. Harrison & P. Hoffman, 1997. Annual report of the bay scallop project, 1996. Unpublished report to the Florida Department of Environmental Protection Marine Research Institute: 1–50.

Blake, N. J., 1996. Demonstration of large-scale reintroduction of the southern bay scallop to Tampa Bay, Florida. Final Report, Tampa Bay National Estuary Program: 1–28.

Blake, N. J., 1998. The potential for reestablishing bay scallops to the estuaries of the west coast of Florida. Trans. 63rd No. Am. Wildl. and Natur. Resour. Conf.: 184–189.

Carriker, M. R., 1951. Ecological observations on the distribution of oyster larvae in New Jersey estuaries. Ecol. Monogr. 21: 19–38.

Carriker, M. R., 1961. Interrelation of functional morphology, behavior, and autecology in early stages of the bivalve *Mercenaria mercenaria*. J. Elisha Mitchell Sci. Soc. 77: 168–241.

Castagna, M. & J. N. Kraeuter, 1977. *Mercenaria* culture using stone aggregate for predator protection. Proc. Natl. Shellfish. Assoc. 67: 1–6.

Cosper, E. M., W. C. Dennison, E. J. Carpenter, V. M. Bricelj, J. G. Mitchell, S. H. Kuenstner, D. Colflesh & M. Dewey, 1987. Recurrent and persistent brown tide blooms perturb coastal marine ecosystems. Estuaries 10: 284–290.

Dao, J-C., J. Barret & M. L. Muzellec, 1994. Seed-recapture of the scallop, *Pecten maximus*, by commercial dredging in the Bay of Brest during 1991–92. In Bourne, N. F., B. L. Bunting & L. D. Townsend (eds). Proceedings of the 9th International Pectinid Workshop, Nanaimo, B.C., Canada, April 22–27, 1993. Vol. 2. Can. Tech. Rep. Fish. aquat. Sci. 1994: 16–22.

Davis, R., 1985. Drifter observations of coastal surface currents during CODE: the method and descriptive view. J. Geophys. Res. 90: 4741–4772.

Dierberg, F., 1986. Knowing how much is there. In Busby, D. (ed.), An Overview of the Indian River Clamming Industry and the Indian River Lagoon. Florida Sea Grant Extension Program, Technical Paper No. 44: 23–25.

Eckman, J. E., 1987. The role of hydrodynamics in recruitment, growth and survival of *Argopecten irradians* (L.) and *Anomia simplex* (D'Orbigny) within eelgrass meadows. J. exp. mar. Biol. Ecol. 106: 165–191.

Eckman, J. E., 1989. Effects of flow speed, turbulence, and orientation on growth of juvenile bay scallops *Argopecten irradians concentricus* (Say). J. exp. mar. Biol. Ecol. 132: 123–140.

Eleuterius, L., 1987. Seagrass ecology along the coasts of Alabama, Louisiana, and Mississippi. In Durako, M. J., R. C. Phillips & R. R. Lewis III (eds), Proceedings of the Symposium on Subtropical-Tropical Seagrasses of the Southeastern United States. Fla. Dept. Nat. Res., Bur. Mar. Res., Fla. Mar. Res. Pub. 42: 11–24.

Florida State Board of Conservation, 1964. Summary of Florida commercial marine landings, 1963. Florida Board of Conservation, Salt Water Fisheries Division, Marine Fisheries Research.

Frischer, M. E., J. M. Danforth, L. C. Tyner, J. R. Leverone, D. C. Marelli, W. S. Arnold & N. J. Blake, 2000. Development of an *Argopecten*-specific 18S rRNA targeted genetic probe. Mar. Biotechnol. 2: 11–20.

Gilmore, R. G., 1987. Subtropical-tropical seagrass communities of the southeastern United States: fishes and fish communities. In Durako, M. J., R. C. Phillips & R. R. Lewis, Jr. (eds), Proceedings of the Symposium on Subtropical–Tropical Seagrasses of the Southeastern United States. Fla. Dept. Nat. Res., Bur. Mar. Res., Fla. Mar. Res. Pub. 42: 117–137.

Godcharles, M. F. & W. C. Jaap, 1973. Exploratory clam survey of Florida nearshore and estuarine waters with commercial hydraulic dredging gear. Fla. Dept. Nat. Res., Mar. Res. Lab., Prof. Pap. Ser. No. 21: 1–77.

Haddad, K. D., 1988. Habitat trends and fisheries in Tampa and Sarasota Bays. NOAA Estuary of the Month Series, # 11, Tampa and Sarasota Bays: 113–128.

Haddad, K. D. & B. A. Harris, 1985. Assessment and trends in Florida's marine fisheries habitat: an integration of aerial photography and thematic mapper imagery. In Mengel, S. K. & D. B. Morrison (eds), Proceedings of Machine Processing of Remotely Sensed Data with Special Emphasis on Quantifying Global Process: Models, Sensor Systems and Analytical Methods. Lab. Appl. Remote Sensing, Purdue Univ.: 130–138.

Hallegraeff, G. M., 1993. A review of harmful algal blooms and their apparent global increase. Phycologia 32: 79–99.

Hanski, I. & D. Simberloff, 1996. The metapopulation approach, its history, conceptual domain, and application to conservation. In Hanski, I. A. & M. E. Gilpin (eds), Metapopulation Biology: Ecology, Genetics, and Evolution. Academic Press, San Diego: 5–26.

Haven, D. S. & J. G. Loesch, 1973. An investigation into commercial aspects of the hard clam fishery and development of commercial gear for the harvest of molluscs. Annual contract report for the period 1 July, 1972 through 30 June, 1973, Contract No. 3-124 R, Virginia Institute of Marine Sciences, Gloucester Point, Virginia: 1–91.

Heffernan, P. B. & R. L. Walker, 1989. Quantitative image analysis methods for use in histological studies of bivalve reproduction. J. Moll. Stud. 55: 135–137.

Hesselman, D. M., B. J. Barber & N. J. Blake, 1989. The reproductive cycle of adult hard clams, *Mercenaria* spp. in the Indian River lagoon, Florida. J. Shellfish Res. 8: 43–49.

Hitchcock, G. & W. S. Arnold, submitted. Role of diffusion in the dispersion of larval *Mercenaria mercenaria* (Linne) in the Indian River lagoon, Florida, U.S.A. Mar. Ecol. Prog. Ser.

Hoffman, W. E., M. J. Durako & R. R. Lewis III, 1982. Habitat restoration in Tampa Bay. In Treat, S.-A. F., J. L. Simon, R. R. Lewis III & R. L. Whitman Jr. (eds), Proceedings Tampa Bay Area Scientific Information Symposium, Florida Sea Grant College Rep. 65: 636–657.

Kennedy, V. S. & L. P. Sanford, 1999. Characteristics of relatively unexploited beds of the eastern oyster, *Crassostrea virginica*, and early restoration programs. In Luckenbach, M. W., R. Mann & J. A. Wesson (eds), Oyster Reef Habitat Restoration: A Synopsis and Synthesis of Approaches. Proceeding from the Symposium, Williamsburg, Virginia, April 1995: 25–46.

Leverone, J. R., 1993. Environmental requirements assessment of the bay scallop *Argopecten irradians concentricus*. Technical Publication #01-93, Tampa Bay National Estuary Program: 1–61.

Levitan, D. R., 1995. The ecology of fertilization in free-spawning invertebrates. In McEdward, L. (ed.), Ecology of Marine Invertebrate Larvae. CRC Press, Boca Raton: 123–156.

Levitan, D. R. & C. Petersen, 1995. Sperm limitation in the sea. Trends Ecol. Evol. 10: 228–231.

Lewis, R. R. III, M. J. Durako & R. C. Phillips, 1985. Seagrass meadows of Tampa Bay-a review. In Treat, S.-A. F., J. L. Simon, R. R. Lewis III & R. L. Whitman Jr. (eds), Proceedings Tampa Bay Area Scientific Information Symposium, Florida Sea Grant College Rep. 65: 210–246.

Li, Z. & R. H. Weisberg, 1999. West Florida shelf response to upwelling favorable wind forcing: kinematics. J. Geophys. Res. 104: 13507–13527.

Lu, Y. & N. J. Blake, 1997. The culture of the southern bay scallop in Tampa Bay, an urban Florida estuary. Aquaculture International 5: 439–450.

MacKenzie, C. L., Jr., 1997. The U.S. molluscan fisheries from Massachusetts Bay through Raritan Bay, N.Y. and N.J. In MacKenzie, C. L., Jr., V. G. Burrell, Jr., A. Rosenfield & W. L. Hobart (eds), The History, Present Condition, and Future of the Molluscan Fisheries of North and Central America and Europe, Volume 1, Atlantic and Gulf Coasts. United States Department of Commerce, National Oceanic and Atmospheric Administration NOAA Technical Report NMFS 127: 87–117.

MacKenzie, C. L., Jr., D. L. Taylor & W. S. Arnold, 2001. A history of quahoging. In Kraeuter, J. & M. Castagna (eds), The Biology of the Hard Clam, *Mercenaria*. Elsevier Scientific Press, New York: 651–673.

Marelli, D. C. & W. S. Arnold, 1996. Growth and mortality of transplanted juvenile hard clams, *Mercenaria mercenaria*, in the northern Indian River lagoon, Florida. J. Shellfish Res. 15: 709–713.

Marelli, D. C., W. S. Arnold & C. Bray, 1999. Levels of recruitment and adult abundance in a collapsed population of bay scallops (*Argopecten irradians*) in Florida. J. Shellfish Res. 18: 393–399.

Marth, D. & M. Marth, 1995. Florida Almanac. 11th edn. Suwannee River Press, Branford, Florida: 1–441.

Menzel, R. W. & H. W. Sims, 1962. Experimental farming of hard clams, *Mercenaria mercenaria*, in Florida. Proc. Natl. Shellfish. Assoc. 53: 103–109.

Mercaldo, R. S. & E. W. Rhodes, 1982. Influence of reduced salinity on the Atlantic bay scallop *Argopecten irradians* (Lamarck) at various temperatures. J. Shellfish Res. 2: 177–181.

Murdock, J. F., 1955. Investigation of the Lee County bay scallop fishery. The Marine Laboratory, University of Miami, Report #55-13: 1–10.

Perret, W. S., R. Dugas, J Roussel, C. A. Wilson & J. Supan, 1999. Oyster habitat restoration: A response to Hurricane Andrew. In Luckenbach, M. W., R. Mann & J. A. Wesson (eds), Oyster Reef Habitat Restoration: A Synopsis and Synthesis of Approaches. Proceeding from the Symposium, Williamsburg, Virginia, April 1995: 93–99.

Peterson, C. H. & H. C. Summerson, 1992. Basin-scale coherence of population dynamics of an exploited marine invertebrate, the bay scallop: implications of recruitment limitation. Mar. Ecol. Prog. Ser. 90: 257–272.

Peterson, C. H., H. C. Summerson & J. Huber, 1995. Replenishment of hard clam stocks using hatchery seed: combined importance of bottom type, seed size, planting season, and density. J. Shellfish Res. 14: 293–300.

Peterson, C. H., H. C. Summerson & R. A. Luettich, Jr., 1996. Response of bay scallops to spawner transplants: a test of recruitment limitation. Mar. Ecol. Prog. Ser. 132: 93–107.

Preece, P. A., S. A. Shepherd, S. M. Clarke & J. K. Keesing, 1997. Abalone stock enhancement by larval seeding: effect of larval density on settlement and survival. Moll. Res. 18: 265–273.

Pulliam, H. R., 1988. Sources, sinks and population regulation. Am. Nat. 132: 652–661.

Safina, C., 1995. The World's imperiled fish. Sci. Am. 273: 46–53.

Sargent, F. J., T. J. Leary, D. W. Crewz & C. R. Kruer, 1995. Scarring of Florida's seagrasses: assessment and management options. Fla. Dept. Env. Prot., FMRI Technical Report TR-1: 1–43.

Southworth, M. & R. Mann, 1998. Oyster reef broodstock enhancement in the Great Wicomico River, Virginia. J. Shellfish Res. 17: 1101–1114.

Summerson, H. C. & C. H. Peterson, 1990. Recruitment failure of the bay scallop, *Argopecten irradians concentricus*, during the first red tide, *Ptychodiscus brevis*, outbreak recorded in North Carolina. Estuaries 13: 322–331.

Supan, J. E., C. A. Wilson & K. J. Roberts, 1999. Economics of augmentation of natural production using remote setting techniques. In Luckenbach, M. W., R. Mann & J. A. Wesson (eds), Oyster Reef Habitat Restoration: A Synopsis and Synthesis of Approaches. Proceeding from the Symposium, Williamsburg, Virginia, April 1995: 359–366.

Tammi, K. A. & W. H. Turner, 1998. Mind your E's and C's! An effective approach to shellfish restoration. J. Shellfish Res. 17: 1314.

Tettelbach, S. T., 1986. Dynamics of crustacean predation on northern bay scallops, *Argopecten irradians irradians*. Ph.D. Dissertation, Univ. of Connecticut, Storrs, Connecticut: 1–229.

Tettelbach, S. T. & E. W. Rhodes, 1981. Combined effects of temperature and salinity on embryos and larvae of the northern bay scallop *Argopecten irradians irradians*. Mar. Biol. 63: 249–256.

Tettelbach, S. T. & P. Wenczel, 1993. Reseeding efforts and the status of bay scallops *Argopecten irradians* (Lamarck, 1819) populations in New York following the occurrence of 'brown tide' algal blooms. J. Shellfish Res. 12: 423–431.

Thayer, G. W. & H. H. Stuart, 1974. The bay scallop makes its bed of seagrass. Mar. Fish. Rev. 36: 27–30.

Waller, T. R., 1969. The evolution of the *Argopecten gibbus* stock (Mollusca: Bivalvia), with special emphasis on the tertiary and quaternary species of eastern North America. J. Paleont. 43: 1–125.

Wesson, J., R. Mann & M. Luckenbach, 1999. Oyster restoration efforts in Virginia. In Luckenbach, M. W., R. Mann & J. A. Wesson (eds), Oyster Reef Habitat Restoration: A Synopsis and Synthesis of Approaches. Proceeding from the Symposium, Williamsburg, Virginia, April 1995: 117–129.

White, N. M., 1996. Spatial analysis of fecal coliform bacteria fate and transport. Ph.D. Dissertation, North Carolina State University, Raleigh, North Carolina: 1–145.

Yang, H., R. H. Weisberg, P. P. Niiler, W. Sturges & W. Johnson, 1999. Lagrangian circulation and forbidden zone on the west Florida shelf. Continental Shelf Res. 19: 1221–1245.

Yund, P. O., S. D. Gaines & M. D. Bertness, 1991. Cylindrical tube traps for larval sampling. Limnol. Oceanogr. 36: 1167–1177.

Hydrobiologia **465:** 21–30, 2001.
G. Burnell (ed.), Coastal Shellfish – A Sustainable Resource.

Recovery of intertidal mussel beds in the Waddensea: use of habitat maps in the management of the fishery

N. Dankers, A.G. Brinkman, A. Meijboom & E. Dijkman
ALTERRA, P.O. Box 167, 1790 AD Texel, The Netherlands
Tel: 00-31-222-369702. Fax: 00-31-222-319235. E-mail: ndankers@alterra.wag-ur.nl

Key words: mussel bed, fishery-impact, *Mytilus edulis*, mussel bed restoration, Waddensea, habitat

Abstract

In the Waddensea, musselbed distribution and abundance has decreased. Management is aiming at increase of area of beds. Fisheries have been regulated. Mussel beds are slowly recovering, and localities where recovery occurs can be predicted reasonably well. Active restoration of beds has not been necessary, as long as newly formed beds are protected from fisheries. Potential suitable sites should be protected from bottom-disturbing activities which might influence the substrate for spatfall such as cockle beds and tube-building polycheates.

Introduction

During the first International Conference on Shellfish Restoration in South Carolina, in 1996, attention was drawn to the declining quality of coastal environments, the key-role of filterfeeding bivalves, and the possibilities of amelioration through shellfish restoration. Also, in Europe the loss of biogenic structures such as mature mussel beds was recognised. Bed area has seriously declined in The Wash, Morecambe Bay, the southern estuaries in the U.K. (McGrorty, pers. comm.), and in the Waddensea (Dankers et al., 1999). Possibilities were considered to restore the area of these beds, either for sustainable fisheries or nature conservation purposes.

In the Waddensea, the main management goal is Nature Conservation. After the extent of the decrease of distribution and extent of the beds had been realised, mussel beds became an issue. In the Trilateral governmental conference in Esbjerg (Esbjerg, 1991), it was decided to close parts of the Waddensea for shellfish fisheries in order to allow mussel beds to re-establish. Goals were set at the trilateral ministerial conference (Stade, 1997), and for the Waddensea, '*an increased area and a more natural distribution and development of natural musselbeds....*' has been declared an '*ECOTARGET*' (Stade, 1997).

In this paper, we will first discuss the role of mussel beds and the factors important for their development and survival. This understanding is important for policies or management programs aiming at recovery or restoration. Management strategies in the different Waddensea countries will be dealt with briefly, and the Dutch approach will be considered in detail.

The role of mussel beds

From the point of view of nature conservation, mussel beds have an intrinsic value of their own. As early as 1877, Möbius (1877) described an oyster bed in the Waddensea as a community, and introduced the term biocoenosis. Much of what he described is also valid for a mature mussel bed. Like in other congregates of living organisms, such as oyster reefs, coral reefs and subtidal seagrass beds, individual organisms influence each other and create new habitats for other species or increase survival chances for themselves, These biocoenoses are formed and maintained by biotic processes. A mature musselbed can be regarded a 'superorganism' (Dankers, 1993) that only rarely develops. Juvenile beds develop regularly, but their survival rate is low. As beds grow old, they stabilise and a mature bed has a high survival rate as mortality of individual mussels is balanced by survival of spat in the shelter of the bed. A mussel bed provides a habitat for a number of species that seek shelter, organic enriched sediment, permanent water in pools at low tide or hard substrate. For example, Asmus (1987) and Dittman (1990) found 41 and 96 allied faunal species respectively in German mussel beds.

Mussels increase the sedimentation of suspended matter by filtration and production of faeces and pseudofaeces in and around mussel beds. From the habitat map of Dijkema et al. (1989), it is clear that the surroundings of mussel beds are characterised by a relatively small grain size of the sediment. Kröncke (1996) showed higher organic content in the sediment surrounding the beds. During storms, part of these biodeposits are resuspended and may be deposited in salt marshes (Linke, 1954; Kamps, 1962). On the basis of exchange measurements (Dame & Dankers, 1988; Prins & Smaal, 1990; Dame et al., 1991) one can conclude that mussel beds speed up the cycle of production and breakdown of organic matter on ecosystem level.

A juvenile musselbed collects considerable amounts of fine silt between August and November, and beds can rise 30–40 cm above the surrounding sandflat (Hilgerloh et al., 1997). These beds are unstable, and often disappear due to winterstorms. If the bed survives the winter, sand and shells are washed in and settle between the mussels. The silt consolidates and forms a clay layer. In old mussel beds, the majority of the underlying sediment consists of sand and clay. Old beds are very resistant to storms and may enhance the stability of gullies. The importance of beds in the sediment budget of the Dutch Waddensea has been quantified by Oost (1995). In Schleswig Holstein, the most stable beds occur on gravel in areas with enough current or wave action which removes fine silt (Ruth, pers. comm.).

The vertical structure of the Waddensea is enhanced because of the mussel beds, and mussels fall dry well before the surrounding sandflat emerges. Also if the mussel population is (temporally) absent, remnants are visible as elevations of clay banks or shell layers. These remnants often are a good basis for new spatfall and should be considered important for restoration of the bed. Very old beds may stabilise creek patterns, because clay and shell layers are erosion resistant.

As has been said, mussel beds serve an important function as habitat for a range of organisms which may be consumed by non-resident species, such as a variety of bird species. Zwarts (1991) found 25% of the wading birds in the Waddensea feeding on mussel beds that (in the eighties) covered 3% of the intertidal area. In addition, Ens et al. (1993) also showed a good correlation between bird-density and musselbed-area.

Factors influencing development of mussel beds

Mussel larvae are planktonic for several weeks and can, therefore, be transported over considerable distances. The distribution of larval settlement is stochastic and within one tidal inlet spatfall may be successful on one bed and not on another (Pulfrich, 1995). As larvae almost always seem to be in sufficient supply (Bryant, 1999; Vooys, 1999), it is assumed that predation after settlement is an important factor for the success of spatfall. Habitat complexity within mussel beds may reduce this predation. Suitable substrate for spatfall are hydroid polyps, algae (*Ceramium rubrum, Gracilaria verrucosa)*, tubes of *Lanice conchilega* and shell debris in niches in mussel beds where the settlers are protected against predation (Verwey,1952; Dare, 1976; McGrorty et al., 1990) only in years with extraordinary heavy spatfall will new beds develop outside old mussel beds on less suitable sustrates and even on bare sand (Dare, 1976; Beukema, 1982; Dankers & Koelemaij, 1989; McGrorty et al., 1990). Predation on newly settled mussels is mainly by small crabs and flatfish, and can be up to 30–53% of the settlement within a mature bed and up to 100% outside those beds (Dare, 1976; McGrorty et al., 1990). Larger mussels form an important food item for Herring Gull (9–25 mm), Knot (<15 mm), Eider and Oystercatcher (>20 mm) (Hilgerloh et al., 1997; Kam et al., 1999). Predation by birds may be responsible for a yearly mortality of 25–40% of the mature mussels on a bed (McGrorty et al., 1990; Zwarts, 1991)

Newly formed beds are vulnerable to environmental disturbances like storms and ice shear. The beds are unstable because they consist of a monolayer of mussels on a bed of pseudofaeces and silt (Tsuchiya & Nishihira, 1986). The mussels can easily be washed away by waves or currents (Verwey, 1952; Nehls & Thiel, 1993; Flemming & Delafontaine, 1994). As beds mature, they develop more stable structure (Kuenen, 1942; Maas Geesteranus, 1942; Tsuchiya & Nishihira, 1985, 1986) because sandy sediments are washed in by waves and currents and a decrease in water content of the silt, which might change into consolidated clay. In addition, shell debris, entangled by byssus threads forms a solid foundation (Seed & Suchanek, 1992). Older beds exhibit a structured patchiness in which the edges shelter the rest of the bed from currents and waves (Tsuchiya & Nishihira, 1985; Obert & Michaelis, 1991; Svane & Ompi, 1993).

Mussel dredgers can damage this structure by either removing the entire bed or by making the structure more open and exposed to wave action (Nehls & Thiel, 1993). Fishermen claim that they keep mussel beds young by giving space to new recruitment and provide better drainage that may enhance the removal of unstable silt from underneath the mussels, however this claim is not backed by research or other evidence. A German study (Herlyn et al., 1999; Herlyn & Millat, 2000) on the impact of fisheries on a few mussel beds in Lower-Saxony, indicated that even removal of a small percentage of mussels caused almost complete destruction of the beds within one year after the fisheries took place.

Present status of mussel beds in the Waddensea

The area of intertidal mussel beds has decreased strongly in estuaries in Europe. The problem seems to be caused by continued fisheries during a long period with little or no spat surviving to a few cm length (Dijkema, 1992, 1993; Vine, 1999; Herlyn et al., 1999), even when mussel larvae seem to be abundant and settle on suitable substrates like collector ropes (Bryant, 1999). But also in the area between Weser and Elbe, all beds disappeared, and in this area no fishing took place (Ruth, pers. comm.).

In the Waddensea, the decrease has been well documented. For different sections of the Waddensea, information on locality and size of mussel beds is available in notebooks of fisheries inspectors, scientists, bird watchers, Waddensea guides and fishermen, since the 1950's or even earlier (Dankers, unpublished). Maps were sometimes produced based on surveys on foot and occasionally on aerial photographs. The first comprehensive inventory of mussel beds, based on aerial photographs, was made by Dijkema et al. (1989) and covered tidal flats in the entire Waddensea. For the different sub-regions, the photographs were from various years between 1968 and 1976. The presence of the mussel beds was verified by groundtruthing in 1978. As only beds larger than 10 ha were included in the map, the inventory is only a minimum estimate. In most investigations which were carried out after 1978, the map of Dijkema et al. (1989) was used as a reference. In recent years, a combination of image analysis and ground truth assessment has been developed to a successful and generally applicable method in Germany (Millat, 1996).

For the Dutch Waddensea, the map of Dijkema indicates that in the late seventies 4120 ha of mussel beds were present in the Dutch Waddensea. In the 1980s, the area decreased because of intensive fisheries and severe winters (Higler et al., 1998). In spring 1987, only 650 ha covered by mussels was left, but on the remains of the old beds new spatfall resulted in the formation of new beds. These beds were removed by fisheries in 1988, 1989 and 1990 (Higler et al., 1998). The majority of the beds which were formed in the nineties were not stable enough to survive winterstorms and ice. In 1998 less than 200 ha of mussel beds were present. Outside the beds, there is a considerable population of mussels lying loose in small clumps on the sandflats (Stralen, 1998).

In Lower-Saxony, beds covered around 5000 ha in the period between 1950 and 1980 (Michaelis et al., 1995). According to Herlyn & Michaelis (1996), the area of mussel beds decreased to 2700 ha in 1989/1991, 1300 ha in 1994, and in spring 1996 only 170 ha were left (Herlyn et al., 1999). After a heavy spatfall in 1996, 1450 ha re-established of which 1280 ha survived the following severe winter (Dankers et al., 1999).

In Schleswig–Holstein, beds were destroyed in the middle of the 1980s by three consecutive winters. In 1987, extraordinary good spatfall occurred and this population has been followed since then. Ruth (1994) investigated the mussel population from 1989 to 1993, a study continued by Nehls (1999). The majority of the beds studied (63 out of 64) were present during 5 or more years. On average the beds covered 1830 ha. Savage (1956) mentions that extraordinary spatfall may determine the dominance of mussel beds in a given area within at least one decade.

The Danish section of the map of Dijkema et al. (1989) is based on aerial photographs from 1973 and 1974. No ground truth has been carried out in the late 1970s. According to Kristensen (1996) , few beds were present after the winter of 1979, but there was a large increase in 1980. In 1983 and 1984, 4000 ha were measured from aerial photographs. Fishery had a severe impact between 1984 and 1987 and also the severe winters of 1985/86 and 1986/87 destroyed many beds. Good spatfall occurred in 1987 and beds increased to 900 ha in 1991 and decreased to 620 ha in1993 (Kristensen, 1995). Fisheries impact was regulated and since then 1000 ha has recovered. In the Juvre Dyb, Manø and Knude Dyb areas, the beds have not returned after they were destroyed by the fisheries. In the Ho Bugt area (partly (66%) closed for fishery), almost 70% of the original beds has recovered.

Recovery and restoration

Policies

After the strong decline in distribution and abundance of intertidal mussel beds became apparent, the different countries took steps to reverse the situation. The aim is laid down in the earlier mentioned 'ecotarget', *'an increased area and a more natural distribution and development of natural mussel beds'* but the approach to reach this goal differs per country.

Denmark

In Denmark, mussel-culture is prohibited in the Waddensea, and therefore fishermen fish on wild mature mussels (shell length >5 cm). There is no relaying of seed from other areas, and no seed fishery takes place. The majority of the catch comes from the subtidal. Large intertidal areas are permanently closed for fisheries and in the open areas yearly TAC's are set depending on the production of the stock (Munch-Petersen & Kristensen, 2000). The ecotarget can be reached in the closed areas (50% of the intertidal) and possibly in the rest of the intertidal, as mature beds in the intertidal are rarely fished and only in years with low subtidal stock which has not been the case in the last 10 years.

Schleswig Holstein

The majority of the Schleswig Holstein Waddensea is designated as a National Park, with zones of different protection regimes. In the zones with the highest protection status no culture lots are allowed, and in the majority of the subtidal of these core zones seed fishery is also excluded. Landing of mature (>4 cm) wild mussels is prohibited. Seed fishery is allowed in the rest of the subtidal, but only for restocking the culture lots in Schleswig Holstein. All intertidal mussel beds are protected from fisheries.

Lower-Saxony

Mussel fishery and -culture is regulated by a blue mussel management plan (CWSS, 1999). On the basis of surveys between 1966 and 1997, three categories of beds were distinguished. When beds occurred in 7 or all of the 8 surveys they were included into categories I or II. In these categories, 37 beds were included. The remaining 150 beds were classified as category III.

Of the 37 category I and category II beds, 17 are closed for fisheries as a result of the management plan.

The Netherlands

As a result of the Esbjerg agreement (Esbjerg, 1991), 25% of the intertidal of the Dutch Waddensea was closed for shellfish fisheries. In the remainder, fishing was allowed, unless there was insufficient food for shellfish eating birds. In fact this was the situation for the period between 1991 and 1999, except for the autumn of 1994. On the basis of a sectoral management plan, the fisheries organisations indicated a small number (5) of potential beds in the open areas to be protected. There was also an agreement to first fish beds which were considered to be unstable.

In 1999, a new policy was developed. On the basis of a Potential Mussel Habitat Map (Brinkman & Stralen, 1999), 10% of the most suitable areas for mussel-bed development were closed for fisheries. The map was produced on the basis of inventories of (relatively young) mussel-beds between 1994 and 1998.

Fishing is allowed in the subtidal area and consists mainly of seed fisheries for restocking of culture lots. If less than 40 000 metric tons of mussels are present in the subtidal areas, then fishing is allowed in approximately 65% of the intertidal zone, but not on developing or mature mussel beds. If the intertidal is open for fishery, the total catch (subtidal + intertidal) should not exceed 40 000 tons. It is expected that the fisheries management plan will use the habitat map in order to fish the most unstable areas first.

In general, the management strategy in Germany and Denmark is a geographical separation of the fishery and conservation function by closure of (considerable) areas for fisheries. The Netherlands are aiming at co-existence by closing a smaller area and allow co-management in the majority of the area. For policy-makers it is important to have a good idea of the areas which might be important for future development of mature beds, either for restricting fisheries, allowing fisheries in areas unsuitable for stable beds or developing restoration programs.

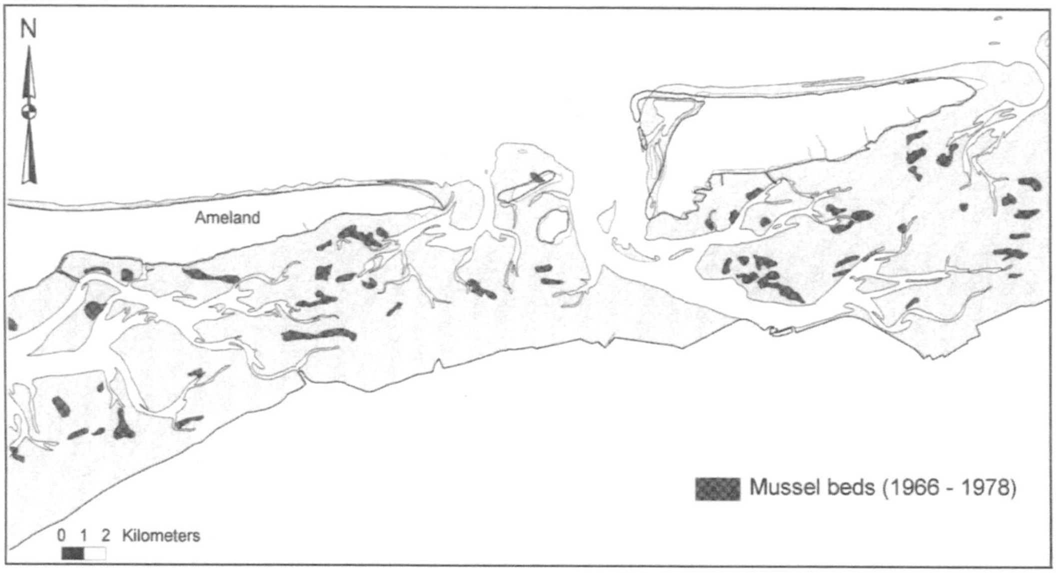

Figure 1. Mussel beds in the early seventies (Dijkema et al., 1989).

Use of potential mussel habitat map in The Netherlands

The ministry responsible for fisheries and nature conservation asked for a map on which management plans would be based. A first attempt was made in spring 1999.

The map used in The Netherlands for the management plan of 1999 (Brinkman & Van Stralen, 1999) can be considered as an interim map. The map was based on inventories of mussel beds over the preceding 5 years. Improvements to the map are under way.

In this paper, a mussel-habitat map will be described for the area between the mainland and the Islands of Terschelling, Ameland and Schiermonnikoog, based on long term data by Dijkema et al. (1989).

For the area under consideration, physical parameters which were available, and were considered to be relevant for the development and stability of mussel beds were stored in Arc-Info/GIS. These data have been collected for the description of Waddensea ecotopes (Wintermans et al., 1996; Jong et al., 1998). The parameters were submergence time, maximum current velocity after an extra high tide, wave related orbital velocities during a north-westerly storm, sediment characteristics (median grain size) and distance to the closest gully. The data were made available for a grid size of 50 * 50 m^2. Mussel distribution data from of Dijkema et al. (1989) were used for inclusion of the mussel data. The location of the beds is presented in

Table 1. The percentage of the observed beds in areas of suitability as calculated by the model

Suitability	Stability class of beds			
	I	II	III	IV
Lowest 70%	61%	57%	55%	17%
10–30%	10%	10%	18%	25%
5–10%	23%	33%	18%	33%
Best 5%	6%	0%	9%	25%
Total	52%	13%	27%	8%

Figure 1. With regression analysis, the suitability of each grid cell, which is characterised by five physical variable values, was calculated. The value thus computed was assumed to present the suitability of that cell for a musselbed of the Dijkema type. The suitability map (Fig. 2) presents the suitability of the intertidal in percentage of area in different shades of grey.

In the area under consideration, mussel bed surveys have been carried out since 1966. Only rarely was the areal extent of the beds was measured, but reasonably reliable position estimates were obtained. A total of 7 surveys in which the whole area was investigated were considered and 97 beds were recognised. When a bed was not recognised in a particular year, it does not always mean that it was not present, but it may have been neglected because it was not interesting for fisheries. Each bed was classified into one of four classes

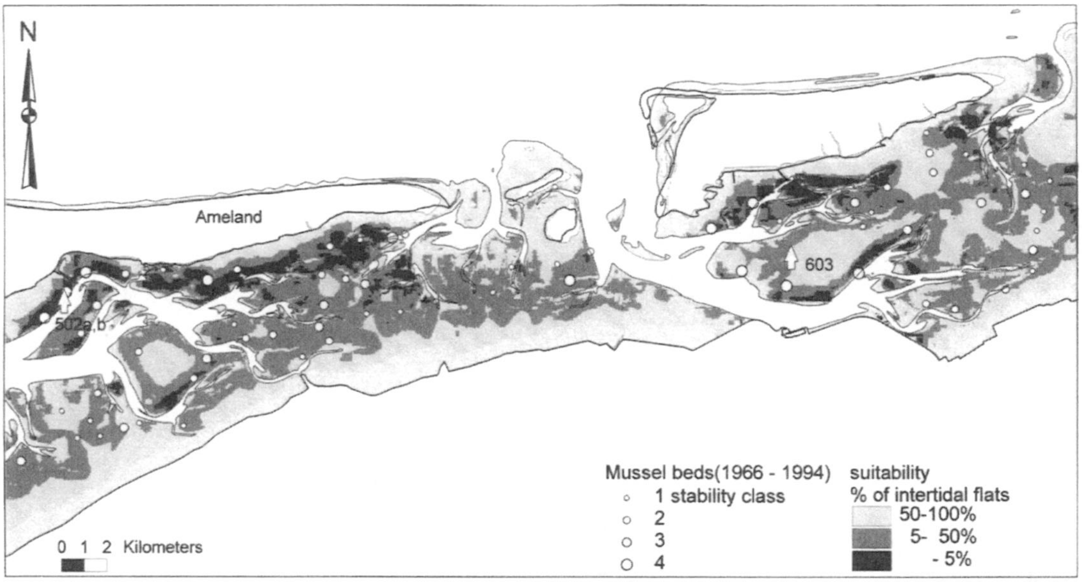

Figure 2. Habitat-suitability for mussel beds, and location of beds of different stability classes between 1966 and 1994. The darkest shade indicates the best 5% of the area and the lightest the least suitable 50% of the intertidal. Also indicated the location of beds 502 a,b and 603.

for stability:

I = present in 1 survey
II = present in 2 or 3 surveys
III = present in 4 surveys
IV = present in 5 or more surveys

The beds in the different classes are plotted in the potential map (Fig. 2) and it is clear that most beds fall within the area which is considered to be the best 50% of the intertidal for the occurrence of mussel beds. This is also demonstrated in Table 1, giving information for the whole Dutch Waddensea, which indicates that the beds which were classified as most stable, the majority (83%) occurred in the best 30% of the area. The beds which were considered least stable occurred mainly in the least suitable 70% of the area. These results indicate that the model is reliable and may be used in policy and management.

Recent development of some musselbeds in The Netherlands

In The Netherlands, in 1999, mussel fishery in the intertidal has not been allowed because sufficient seed was available in the subtidal. The preliminary map (Brinkman & Van Stralen, 1999) has been used for closure of areas for cockle fishery because there are indications that cockle beds may be an important substrate for mussel-spat. Several new mussel beds (1999 spatfall) have now developed on cockle beds.

Table 2. Development of mussel beds 502 a,b and 603 in the study area

Bed nr	Cover (%)				Area
	1997	1998	1999	2000	(ha)
502 a,b	19.5	15.7	24.3	28.7	18
603	15.1	23.7	45.8	46.4	7

Because insufficient mussels were available as food for the bird population, mussel fishery in the intertidal has been absent since 1994. Several beds have survived since then. Two beds will be treated in some detail. The locations are indicated in Figure 2. The beds are robust and contain mussels of different year classes (Fig. 3), also from years that spatfall in the intertidal was almost absent.

The beds developed from spatfall in 1994 and were severely damaged (the mussels on bed 603 almost completely disappeared) in the winter of 1995/1996. The beds recovered because new spat fell in the remnants, and as indicated in Table 2, the area within the bed covered by mussels increased.

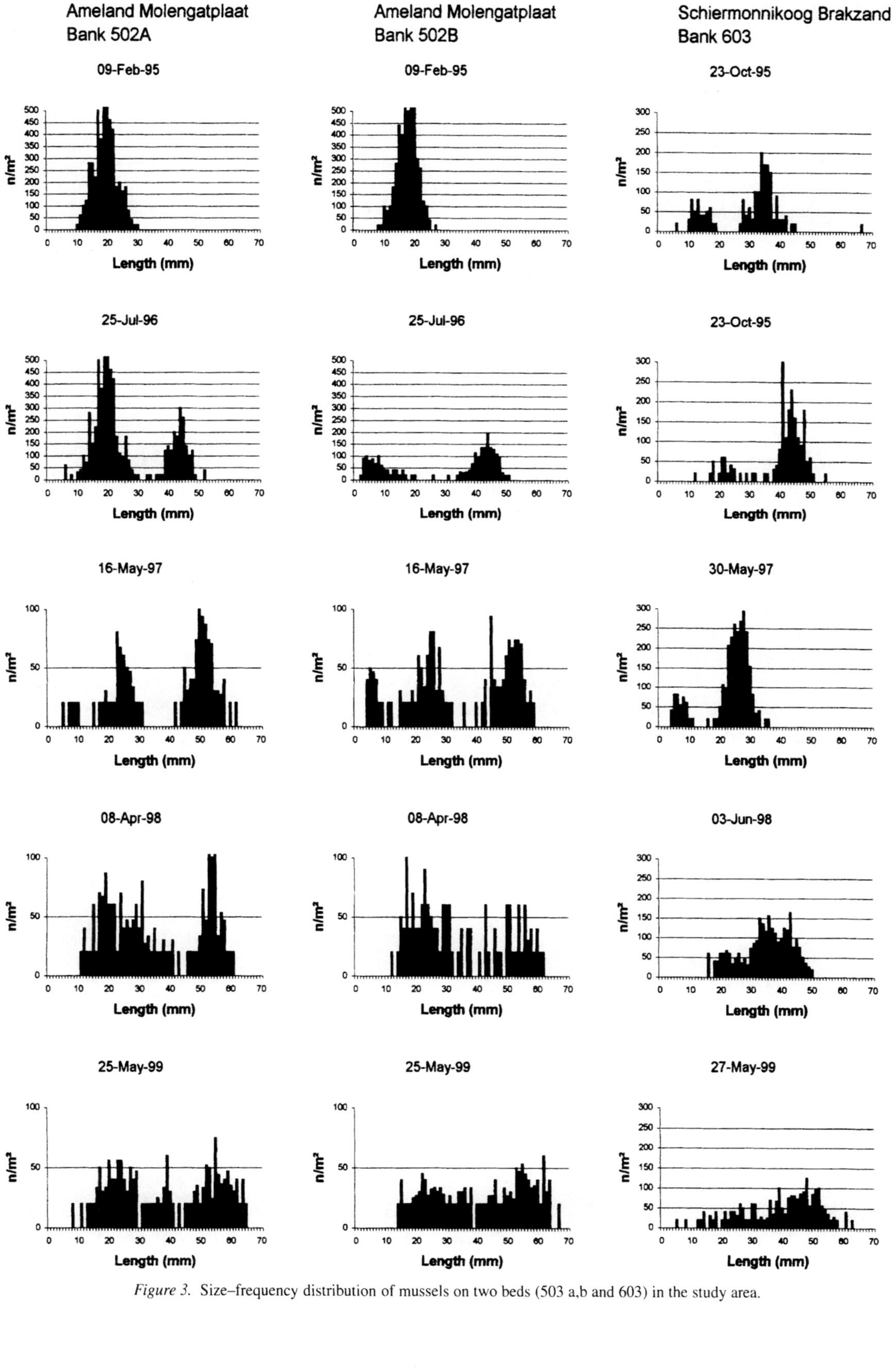

Figure 3. Size–frequency distribution of mussels on two beds (503 a,b and 603) in the study area.

In Denmark and Germany, beds have also started to redevelop. Spatfall often occurs on existing beds or remnants of beds after severe winters. However, the total area is still considerably less than in the early 1980s. In Denmark and Lower-Saxony, recent inventories have not been published yet. The only reliable estimates available are for Schleswig Holstein. Nehls (1999) found 67 beds. The beds were small (average size 8 ha) with a total area of 600 ha in 1998, which is considerably less than the area present in the late eighties. The average mussel covered area within the beds was 34%. In 1999, one bed had disappeared, and almost all investigated beds had increased in size. The total surface increased to 952 ha (Nehls et al., 2000).

Conclusions

The different approaches in the different countries provide an opportunity for a comparison of the management options in order to select the best way to reach the goal (more intertidal mature beds) and a sustainable fishery.

In Denmark, there is the possibility to compare developments in open and closed areas. This is also the case in Lower-Saxony and The Netherlands, but in these countries intertidal fisheries are restricted to seed fishery. Schleswig Holstein can potentially act as a control because all intertidal flats are closed irrespective of possibilities for development of stable beds. There is also a possibility to study the development of subtidal beds. In Lower-Saxony, an additional comparison is possible as beds with different levels of estimated stability are protected, while also beds within the different classes are open to fishery. Furthermore, in The Netherlands there is an additional possibility to compare developments in areas with different levels of suitability for development of stable beds.

The different management schemes have been in operation for too short a time period, and monitoring programs are not yet completely comparable for firm conclusions to be drawn, but in all areas there is recovery.

It is evident that recovery of mussel beds is slow. The same can be said for the Wash where intertidal beds are still absent (Vine, 1999), and for the Exe where beds did not re-establish after removal (McGrorty, pers. comm.). In the Waddensea, beds seem to firmly establish and maintain themselves if they can survive one or two winters, and are not fished. It is, therefore, essential that fisheries are banned from areas with good conditions for stable beds to develop. These areas can be predicted with habitat suitability maps as mentioned above.

The maps can also be used to regulate other forms of activities, such as cockle fishery, dredging, sand-, gravel- or shell-mining which might influence the substrate suitable for mussel-spatfall. It is uncertain if beds will develop in areas where they have long disappeared. In this case, active restoration (relays or seeding) may be necessary. Selection of the right place will be made easier by the availability of habitat suitability maps.

Acknowledgements

We thank Maarten Ruth, Per Sand Kristensen, Gert Millat, Marc Herlyn and Georg Nehls for their comments on an early version of the manuscript. Comments of two anonymous referees are highly appreciated, and improved the readability.

References

Asmus, H., 1987. Secondary production of an intertidal mussel bed community related to its storage and turnover compartments. Mar. Ecol. Progr. Ser. 39: 251–266.

Beukema, J. J., 1982. Annual variation in reproductive success and biomass of the major macrozoobenthic species living in a tidal flat area of the Waddensea. Neth. J. Sea Res. 16: 37–45.

Blok De, J. W. & H. J. F. M. Geelen, 1958. The substratum required for the settling of mussels (*Mytilus edulis* L) Extrait des archives Neerl Zool Tome XIII I Suppl 1958: 446–460.

Brinkman, A. G. & M. Van Stralen, 1999. Toelichting habitatkaart stabiele mosselbanken Waddenzee. IBN-DLO & RIVO-DLO. Bijlage bij brief van Staatssecretaris van LNV G.H. Faber aan Voorzitter Vaste Commissie voor LNV Tweede Kamer Staten Generaal dd 1 maart 1999, bij Beleidsbesluit Schelpdiervisserij Kustwateren 1999–2003.

Bryant, D., 1999. Longline mussel culture experiments in 'The Wash'. King's Lynn Fishing Industry Co-operative, King's Lynn: 22 pp.

CWSS, 1999. Lower Saxony blue mussel management plan. Wadden-Sea Newsletter 1999–1: 42–43.

Dame, R. & N. Dankers, 1988. Uptake and release of materials by a Waddensea mussel bed. J. exp. mar. Biol. Ecol. 118: 207–216.

Dame, R., N. Dankers, T. Prins, H. Jongsma & A. Smaal, 1991. The influence of mussel beds on nutrients in the western Waddensea and Eastern Scheldt estuaries. Estuaries 14: 130–138.

Dankers, N., 1993. Integrated estuarine management – Obtaining a sustainable yield of bivalve resources while maintaining environmental quality. In Dame (ed.), Bivalve Filter Feeders in Estuarine and Coastal Ecosystem Processes. NATO ASI Series Vol G 33. Springer, Berlin/Heidelberg: 479–511.

Dankers, N. & K. Koelemaij, 1989. Variations in the musselpopulation of the Dutch Waddensea in relation to monitoring. Helgwiss. Meeresunters. 43: 529–535.

Dankers, N., M. Herlyn, P. Sand, Kristensen, H. Michaelis, G. Millat, G. Nehls & M. Ruth, 1999. Blue mussels and Blue mussel beds in the littoral. In De Jong, F., J. Bakker, C. van Berkel, K. Dahl, N. Dankers, C. Gätje, H. Marencic & P. Potel (eds), Quality Status Report Waddensea Ecosystem No. 9. Common Waddensea Secretariat: 141–145.

Dare, P. J., 1976. Settlement growth and production of the mussel *Mytilus edulis* L in Morecambe Bay. England Fishery Invest Lond (Ser 2) 28: 1–25.

Dittmann, S., 1990 Mussel beds – amensalism or amelioration for intertidal fauna. Helgoländerwiss. Meeresunters 44: 335–352.

Dijkema, K. S., G. Van Tienen & J. J. Van Beek, 1989. Habitats of the Netherlands, German and Danish Wadden Sea 1:100 000. Research Institute for Nature Management, Texel/Veth Foundation, Leiden: 24 maps.

Dijkema, R., 1992. Spatfall and recruitment of mussels (*Mytilus edulis)* and cockles *(Cerastoderma edule)* on different locations along the european coast. ICES C.M. 1992/K:45 Shellfish Committee.

Dijkema, R., 1993. Spatfall and recruitment of mussels (*Mytilus edulis)* and cockles *(Cerastoderma edule)* on different locations along the European coast. ICES C.M. 1993/K:62 Shellfish Committee Ref. D.

Ens, B. J., G. J. M. Wintermans & C. J. Smit, 1993. Verspreiding van overwinterende wadvogels in de Nederlandse Waddenzee. Limosa 66: 137–144.

Esbjerg, 1991. Ministerial declaration of the Sixth Trilateral Governmental Conference on the protection of the Waddensea. CWSS, Wilhelmshaven.

Flemming, B. W. & M. T. Delafontaine, 1994. Biodeposition in a juvenile mussel bed of the East Frisian Waddensea. Neth. J. aquat. ecol. 28: 289–297.

Herlyn, M. & Michaelis, 1996. Untersuchungen zur Entwicklung von Miesmuschelbänken der niedersächsischen Watten, unter Berücksichtigung der Miesmuschelfischerei. Forscungsstelle Küste, Norderney, Forschungsbericht 108-02-085/21: 91 pp +figs.

Herlyn, M. & G. Millat, 2000. Decline of the intertidal blue mussel *(Mytilus edulis)* stock at the coast of Lower Saxony (Waddensea) and influence of mussel fishery on the development of young mussel beds. Hydrobiologia 426: 203–210.

Herlyn, M., G. Millat & H. Michaelis 1999. Einfluss der Besatzmuschelentnahme auf die Entwicklung eulitoraler Neuansiedlungen von *Mytilus edulis* L. im niedersächsischen Wattenmeer. NLÖ-Forschungsstelle Küste 9/1999: 27 pp.

Higler, B., N. Dankers, A. Smaal & V. De Jonge, 1998. Evaluatie van de ecologische effecten van het reguleren van schelpdiervisserij in Waddenzee en Delta op bodemorganismen en vogels. Appendix 5, 17 pgs, 14 figs In Van Dijk, J. J. & R. Heling (eds), Structuurnota Zee- en Kustvisserij; Evaluatie van de maatregelen in de kustvisserij gedurende de eerste fase (1993–1997). Ministerie van LNV, Den Haag.

Hilgerloh, G., M. Herlyn & H. Michaelis, 1997. The influence of predation by herring gulls *Larus argentatus* and oystercatchers *Haematopus ostralegus* on a newly established mussel *Mytilus edulis* bed in autumn and winter. Helg. Meeresunters. 51: 173–189.

Jong, D. J. de, N. Dankers & R. J. Leewis, 1998. Naar ecologische kaarten van de Waddenzee. BEON-rapport 98-13. ISSN 0924-6576: 25 pp.

Kam, J. Van De, B. Ens, T. Piersma & L. Zwarts, 1999. Ecologische atlas van de Nederlandse wadvogels. Schuyt & Co. Haarlem: 368 pp.

Kamps, L. F., 1962. Mud distribution and land reclamation in the eastern Wadden shallows. Rijkswaterstaat Comm. 4, Den Haag: 73 pp.

Kristensen, P. S., 1995. Aerial surveys, biomass estimates, and elimination of the mussel *(Mytilus edulis L.)* in the Danish Waddensea, 1991–1994. ICES. Shellfish Committee CM/K: 44: 22 pp.

Kristensen, P. S., 1996. Thirty years of blue mussel (Mytilus edulis L.) stock, and mussel beds in the Danish Wadden Sea. Dan. Inst Fish. Res: 15 pp.

Kristensen, P. S., 1997. Blamuslingebestanden I det danske Vadehav august 1996. DFU-rapport nr 36–97: 27 pp.

Kröncke, I., 1996. Impacts of biodeposition on macrofaunal communities in intertidal sandflats. In Dworschak, P. C., M. Stachowittsch & J. A. Ott (eds), Influence of Organisms on their Environment. The Role of Episodic Events. Proc. 29th EMBS Vienna. P.S.Z.N.I: Mar. Ecol. 17: 159– 174.

Kuenen, D. J., 1942. On the distribution of mussels on the intertidal sand flats near Den Helder. Arch. Neerl. Zool. 6: 117–160.

Linke, O., 1954. Die Bedeutung der Miesmuscheln fur die Landgewinnung im Wattenmeer. Natur. und. Volk. 84: 253–261.

Maas-Geesteranus, R. A., 1942. On the formation of banks by *Mytilus edulis* L. Arch. Neerl. Zool. 6: 283–326.

McGrorty, S., R. T. Clarke, C. J. Reading & J. D. Goss-Custard, 1990. Population dynamics of the mussel *Mytilus edulis*: density changes and regulation of the population in the Exe estuary Devon. Mar. Ecol. Progr. Ser. 67: 157–169.

Millat, G., 1996. Entwicklung eines methodisch-inhaltlichen Konceptes zum Einsatz von Fernerkundungsdaten für ein Umweltmonitoring im niedersächsischen Wattenmeer. Schriftenreihe der Nationalparkverwaltung Niederssächsisches Wattenmeer, Wilhelmshaven, Band 1: 125 pp.

Michaelis, H., B. Obert, I. Schultenkötter & L. Böcker, 1995. Die Miesmuschelbestände der Niedersächsischen Watten, 1989–1991. Ber. Forschungsstelle Küste Norderney 40: 55–71.

Möbius, K., 1877. Die Auster und die Austernwirtschaft. Wiegund Hempel and Parey Berlin: 1–126.

Munch-Petersen, S. & P. S. Kristensen, 2001. On the dynamics of the stocks of blue mussels (*Mytilus edulis L.)* in the Danish Wadden Sea. Hydrobiologia 465 (Dev. Hydrobiol. 160): 31–43.

Nehls, G., 1999. Miesmuschelmonitoring im Nationalpark Schleswig-Holsteinisches Wattenmeer 1998. Landesamtes für den Nationalpark Schleswig-Holsteinisches Wattenmeer, Tönning: 56 pp.

Nehls, G., G. Sheiffahrt & G. Sach, 2000. Miesmuschelmonitoring im Nationalpark Schleswig-Holsteinisches Wattenmeer 1999. Landesamtes für den Nationalpark Schleswig-Holsteinisches Wattenmeer, Tönning: 58 pp.

Nehls, G. & M. Thiel, 1993. Large scale distribution patterns of the mussel *Mytilus edulis* in the Waddensea of Schleswig Holstein: do storms structure the ecosystem? Neth. J. Sea Res. 31: 181–187.

Obert, B & H. Michaelis, 1991. History and ecology of the mussel beds (*Mytilus edulis* L.) in the catchment area of a Waddensea tidal inlet. In Elliot, M. & J. P. Ducrotoy (eds), Estuaries and Ccoasts; Spatial and Temporal Intercomparisons. Olsen & Olsen: 185–194.

Oost, A. P., 1995. Dynamics and sedimentary development of the Dutch Waddensea with emphasis on the Frisian Inlet. Geologica Ultraiectina 126: 455 pp.

Prins, T. C. & A. C. Smaal, 1990 Benthic–pelagic coupling: The release of inorganic nutrients by an intertidal bed of *Mytilus edulis* In Barnes, M. & R. M. Gibson RM (eds), Trophic Relationships in the Marine Environment. Proc. 24th Eur. Mar. Bio. Symp., Aberdeen Univ. Press: 89–103.

Pulfrich, A., 1995. Reproduction and recruitment in Schleswig Holstein Waddensea edible mussel (*Mytilus edulis* L.) populations. Thesis Inst. Für Meereskunde. Cristian-Albrechts Uni. Kiel.

Ruth, M., 1994. Untersuchungen zur Biologie und Fischerei von Miesmuscheln im Nationalpark "Schleswig – Holsteinisches Wattenmeer". Inst. f. Meeresforschung, Uni. Kiel: 327 pp.

Savage, M., 1956. The great spatfall of mussels (*Mytilus edulis* L.) in the river Conway estuary in spring 1940. In Fisheries Investigations Ser.II, Vol. XX (7): 22 pp.

Seed, R. & T. H. Suchanek, 1992. Population and community ecology of *Mytilus:* In Gosling, E. M. (ed.), The Mussel *Mytilus:* Ecology, Physiology, Genetics and Culture. Elsevier, Amsterdam: 87–169

Stade, 1997. Ministerial declaration of the Eighth Trilateral Governmental Conference on the protection of the Waddensea. (English version), Common Waddensea Secretariat, Wilhelmshaven: 100 pp.

Stralen, M. R. Van, 1998. De ontwikkeling van het mosselbestand in de Waddenzee en Oosterschelde na 1992. RIVO-rapport C.006.98: 31 pp.

Svane, I. & M. Ompi, 1993. Patch dynamics in beds of the blue mussel *Mytilus edulis* L.: Effects of patch size, and position within a patch. Ophelia 37: 187–202.

Tsuchyia, M. & M. Nishihira, 1985. Islands of *Mytilus* as habitat for small intertidal animals: effect of island size on community structure. Mar. Ecol. Prog. Ser. 25: 71–81.

Tsuchyia, M. & M. Nishihira, 1986. Islands of *Mytilus* as habitat for small intertidal animals: effects of *Mytilus* age structure on the species composition of the associated fauna and community organisation. Mar. Ecol. Prog. Ser. 31: 171–178.

Verwey, J., 1952. On the ecology and distribution of cockle and mussel in the Dutch Waddensea. Their role in sedimentation and the source of their food supply. With a short review of the feeding behaviour of bivalve molluscs. Arch. Neerl. Zool. 10: 171–239.

Vine, H., 1999. Development of Blue Mussels on the Wash/North Norfolk coast. Waddensea Newsletter 1999. 1: 36–38.

Vooys, C. De, 1999. Numbers of larvae and primary plantigrades of the mussel *Mytilus edulis* in the western Dutch Waddensea. Neth. J. Sea Res. 41: 189–201.

Wintermans, G., N. Dankers, R. Leewis, P. Molegraaf, D. de Nooyer, S. Reents, F. Steyaert & R. Wegman, 1996. Ecotopes in the Waddensea. In Wintermans, G. (ed.), Habitatkartering en – Beschrijving van Nederlandse Kustwateren. BEON-Rapport 96-5 ISSN 0924-6576.

Zwarts, L., 1991. Mosselbanken: wadvogels op een kluitje. Vogels 66: 8–12.

Hydrobiologia **465:** 31–43, 2001.
G. Burnell (ed.), Coastal Shellfish – A Sustainable Resource.

On the dynamics of the stocks of blue mussels (*Mytilus edulis* L.) in the Danish Wadden Sea

Sten Munch-Petersen & Per Sand Kristensen
The Danish Institute for Fisheries Research (DIFRES), Charlottenlund Castle, DK-2920 Charlottenlund, Denmark
E-mail: smp@dfu.min.dk

Key words: Mytilus edulis, growth, mortality, production, yield, assessment methodology

Abstract

As biological basis for the monitoring programme for the commercially exploited stock(s) of mussels (*Mytilus edulis* L.) in the Danish Wadden Sea, samples of mussels have been collected regularly since 1986, both from sub-tidal and inter-tidal mussel beds. These samples are the basis for the estimation of total biomass. They also provide data on size frequency distributions, which have been analysed for cohort identification resulting in length at age data, which again have been used for estimating parameters (L_∞ and K) for the von Bertalanffy growth equation (VBGE) as well as mortality parameters. By applying these in the Beverton & Holt model, estimates of average biomass and annual production (P) of the mussels have been obtained together with possible fisheries yields from the beds. The growth and mortality parameters and the figures for annual production and P/B are compared with figures from other investigations. These analyses have been the basis for annual assessments of the mussel stocks, which again are used in the current management of mussel fishery in the Danish Wadden Sea.

Introduction

The Danish Wadden Sea is the northernmost part of a mostly continuous Wadden Sea area along North Sea coast of Holland, Germany and the southwestern part of Denmark. In the Danish part of the Wadden Sea, the stocks of mussels (*Mytilus edulis* L.) have been commercially exploited regularly since the 1980s. The annual landings have fluctuated between a few hundred tonnes (1979–1981, 1997) up to more than 27 000 tonnes in 1985, see Table 1. Since 1990, the commercial exploitation of mussels in the Danish Wadden Sea has been concentrated in very few areas, mainly in the Ho Bight area in the northernmost part, see Figure 1.

The sudden expansion of this mussel fishery and the destruction of a large number of inter-tidal mussel beds during the ice winters 1985/1986 and 1986/1987 resulted in a special focus on this fishery by the bodies responsible for nature conservation policies for the Wadden Sea as a RAMSAR[1] and bird protection area. Since 1987 management of this mussel fishery has been based on annual TACs ('Total Allowable Catch') and these, in turn, are set at a level, such that the commercial exploitation should not limit the amount of mussels as prey for birds in the Wadden Sea.

[1] Convention on Wetlands of International Importance especially as Waterfowl Habitat (Ramsar, Iran 1971).

Thus a main problem concerning the fishery is how much can be harvested annually without limiting the sea birds' access to their prey? To elucidate this problem, one has to consider two major aspects:

The current size of standing stock (biomass). An estimate of the annual average biomass requires:

- an estimate of the total area covered by mussel beds.
- representative samples of mussels from the mussel beds providing figures of density in numbers and weight.

For the assessment, the mussel beds in the Danish Wadden Sea are identified from aerial photographs. The photographs are used both for estimates of total area of mussel beds as well as for stratifying the biological sampling. The procedures have been described in detail by Munch-Petersen & Kristensen (1987), Munksgaard, (1989), Kristensen (1995). The possibility

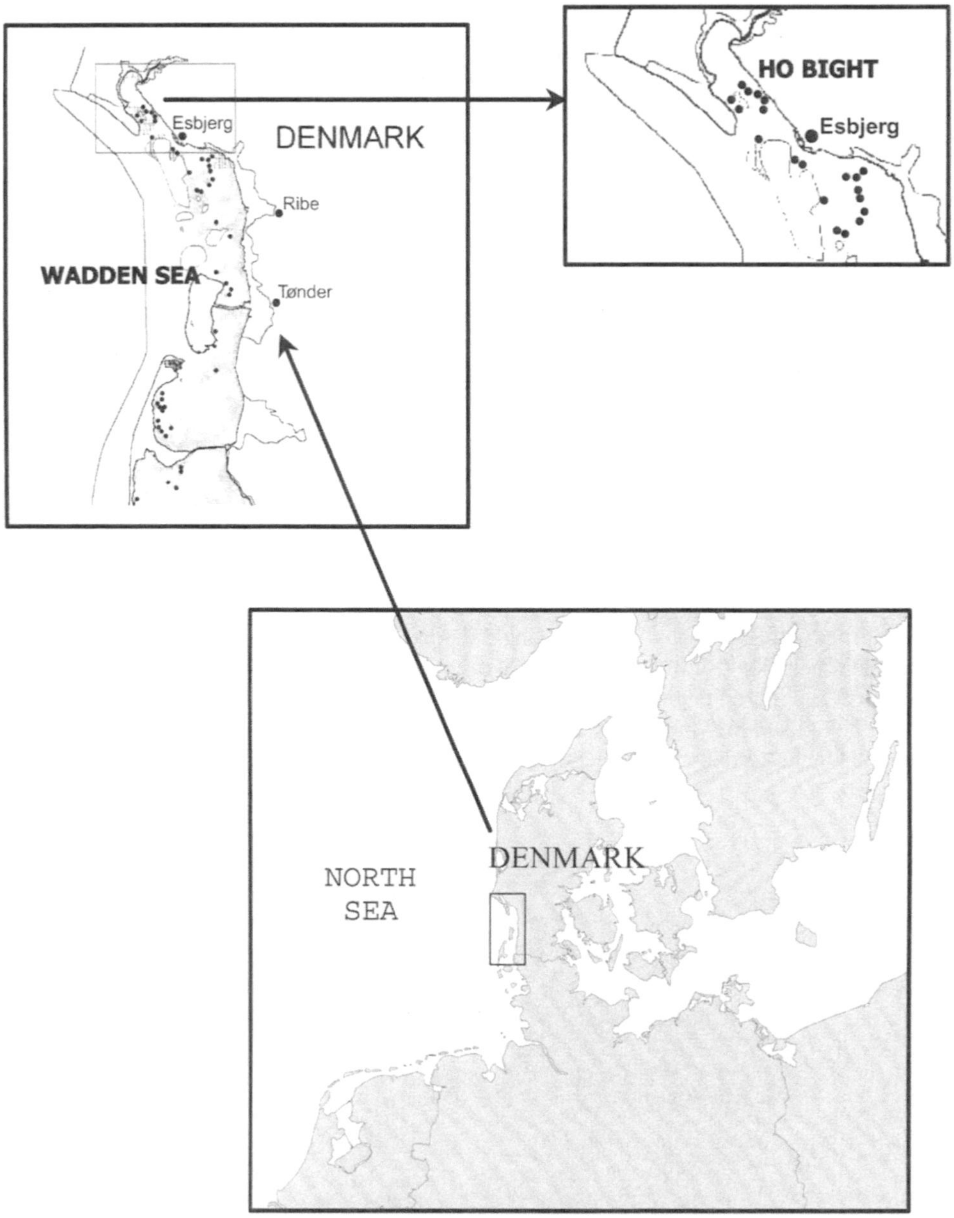

Figure 1. Location of investigated area in the Wadden Sea.

of reducing the sampling costs by introducing special techniques for analysis (scanning) of digitised (aerial) photos for density estimation was described by Munch-Petersen & Kristensen (1989):

The biological dynamics of the mussels, i.e. growth, mortality and, if possible, recruitment on the mussel beds. The commercial fishery accounts for part of the total mortality, and it is likely that predation of mussels by sea birds is responsible for the main part of the natural mortality.

In a paper presenting a first assessment of the mussel stock using the Beverton & Holt model (Munch-Petersen & Kristensen, 1987) the growth and mortality parameters used were taken from earlier investigations (Theisen, 1968). However, the data collected regularly since 1987 have provided sufficient information for

Table 1. Danish landings of mussels (*Mytilus edulis*) in the Danish Wadden Sea, 1979–1998

Year	Landings, *t*
1979	305
1980	293
1981	131
1982	1144
1983	2147
1984	14533
1985	27099
1986	17564
1987	17384
1988	1161
1989	1403
1990	1759
1991	5539
1992	5030
1993	3153
1994	4635
1995	1382
1996	9850
1997	263
1998	3100
Average, 1989–1998	3611

new estimates of growth and mortality parameters and subsequent figures for the annual biomass and production. The main purpose of this paper is to present the new data together with the main features of the analyses as well as to emphasise the applicability of this assessment model regarding natural mussel production both as prey to for instance sea birds and as yield in commercial fisheries. Furthermore, to evaluate the results such as estimates of annual production, these figures are compared with production figures from other similar investigations.

Materials and methodology

Collection of data

Samples of mussels have been collected regularly since 1986, according to a stratified scheme based on aerial photographs combined with the practical possibilities for field sampling. A significant part of the sub-tidal mussels samples have been obtained from the commercial catch of one of the few mussel fishing vessels, which has license to operate in the Wadden Sea. In this case, the mussels were caught with a commercial mussel dredge. The tow time and tow speed (approximately 2 knots) together with the width of the dredge give the area of the sample ('swept area'). The total catch in each tow was assessed by the skipper. The inter-tidal samples were collected randomly at low tide by placing a 50*50 cm frame on the beds and removing all mussels within the frame. All the inter-tidal mussel beds were reached by the institute's own vessel. Calculating the total biomass of mussels on an inter-tidal bed also involves an estimate of mussel coverage on the particular bed, i.e. the fraction of the bed actually covered by mussels at the observed density, see Munch-Petersen et al. (1987). As mentioned above, in recent years the commercial exploitation of the mussels has mainly taken place in the Ho Bight, see Figure 1. Here the total area subject to fishing amounts to around 11 km^2.

In both cases, depending of the amount of mussels, sub-samples of around 5 kg including empty shells were brought ashore for further analysis. Normally the amount of live mussels measured and weighed was from 1–2 kg. The size of the mussels was measured with a vernier calliper to the nearest mm below, or in some cases measured on a measuring board to the nearest 5 mm below. In selected samples the tissue of individual mussels was removed from the shells dried and weighed as AFDW (ash free dry weight).

As fieldwork is costly and one of the main purposes of the biological sampling has been to obtain estimates of standing stock, the sample scheme has unfortunately not been optimal regarding cohort identification through time. In particular, the 0 group specimens were under represented in the dredged commercial samples.

Data compilation and methods of analysis

As mentioned above, most of the mussels in the samples were measured in mm except from 1996 to 1998, where all specimens were measured in 5 mm units. In order to eliminate some of the 'noise' likely to be found in the 1 mm groupings, all the length data were grouped in 5 mm units for the cohort separation analyses. However, in order to make the results comparable with other investigations, all lengths have been expressed in mm.

Prior to cohort identification, the sample data were grouped into the two categories: sub-tidal and inter-tidal. Since the data are also used for estimation of

Table 2. Average density of mussels in Ho Bight in 1998. Numbers and weight per m^2

Shell length (mm)	Sub-tidal & inter-tidal pooled		
	Observed	Calculated weights	
	Numbers/ m^2	Fresh weight grams/m^2	AFDW grams/ m^2
0	3	0.00	0.00
5	13	0.46	0.02
10	18	3.13	0.13
15	15	6.99	0.30
20	13	12.62	0.54
25	27	49.76	2.12
30	78	234.10	9.99
35	97	446.26	19.04
40	71	474.05	20.23
45	58	543.51	23.19
50	48	604.63	25.80
55	28	459.85	19.62
60	12	245.60	10.48
65	4	97.34	4.15
70	1	40.20	1.72
75	0	12.28	0.52
80	0	2.12	0.09
85	0	2.52	0.11
90	0	0.00	0.00
95	0	0.00	0.00
100	0	0.00	0.00
Total number:	**486**		
Total (grams/m^2):	**3019**	**3235**	**138**

total biomass, empty samples were also included. The averages of the samples (numbers and weights) are expressed per m^2 (sq m), unless otherwise specified.

Figure 2 shows examples of the size composition in the samples from inter-tidal as well as sub-tidal mussel beds. These particular samples were all collected in Ho Bight in August 1998. Figure 2 also shows the distribution in all samples combined, and Table 2 gives the average density (numbers/m^2) behind the figures as well as the observed densities (grams/m^2). It is seen that there is a considerable variation in the pattern of the size distributions within the same area. It is a common feature for most of the samples, that there is some variation among the size (length) distributions between samples taken within the same area at the same time, both inter-tidal and sub-tidal mussel beds. One of the main reasons for this variability is probably the large local variations in spat fall (larval settlement), but local differences in growth may also have some influence on this, see also Section Cohort identification.

Cohort identification

The partitioning of the length frequency data into cohorts was done by applying the method of Bhattacharya assuming normal distributed individual sizes in each cohort (Bhattacharya, 1967). The same method was applied by Bayne & Worrall (1980) for mussels in waters near Plymouth in U.K. This standard method is available in the FiSAT Stock Assessment Tools software package (Gayanilo et al., 1994). In Bhattacharya's method, each distribution is determined in turn, beginning on the 'left side' of the composite distribution. It is the user that has to subjectively decide which modes to select and then allocate age to them. However, due to a decreasing growth rate and larger spread in size among the older individuals, composite size distributions of the older age groups will usually not show distinct modes and such age groups can therefore not be properly separated by this method.

In order to test for any differences in growth between sub-tidal and inter-tidal mussels, the data were separated into two categories prior to the analyses, cf. above:

Size distributions of mussels from inter-tidal beds, where exposure is common at low tide.
Size distributions of mussels from sub-tidal beds, which are permanently covered by seawater.

For mussels in the Wadden Sea it is known from the samples that there are (1) variations in time of spat fall within the same area, (2) sometimes two larval settlements within a season, and (3) perhaps local variation in growth conditions. These variations add to difficulties in allocation of the correct age to a mode. The actual criteria used are outlined in Section on Size distributions and cohort identification.

Estimation of growth and mortality

The model used for growth description is the von Bertalanffy growth equation (VBGE):

$$l_t = L_\infty(1 - e^{-K(t-t_0)}) \quad \text{(growth in length).}$$

The VBGE is widely used because it normally fits reasonably well to observed growth patterns of a wide range of marine bivalve and fish species. Furthermore, its parameters (K, L_∞ and t_0) can easily be estimated from either size at age data or size increment data. The actual method for estimation of the

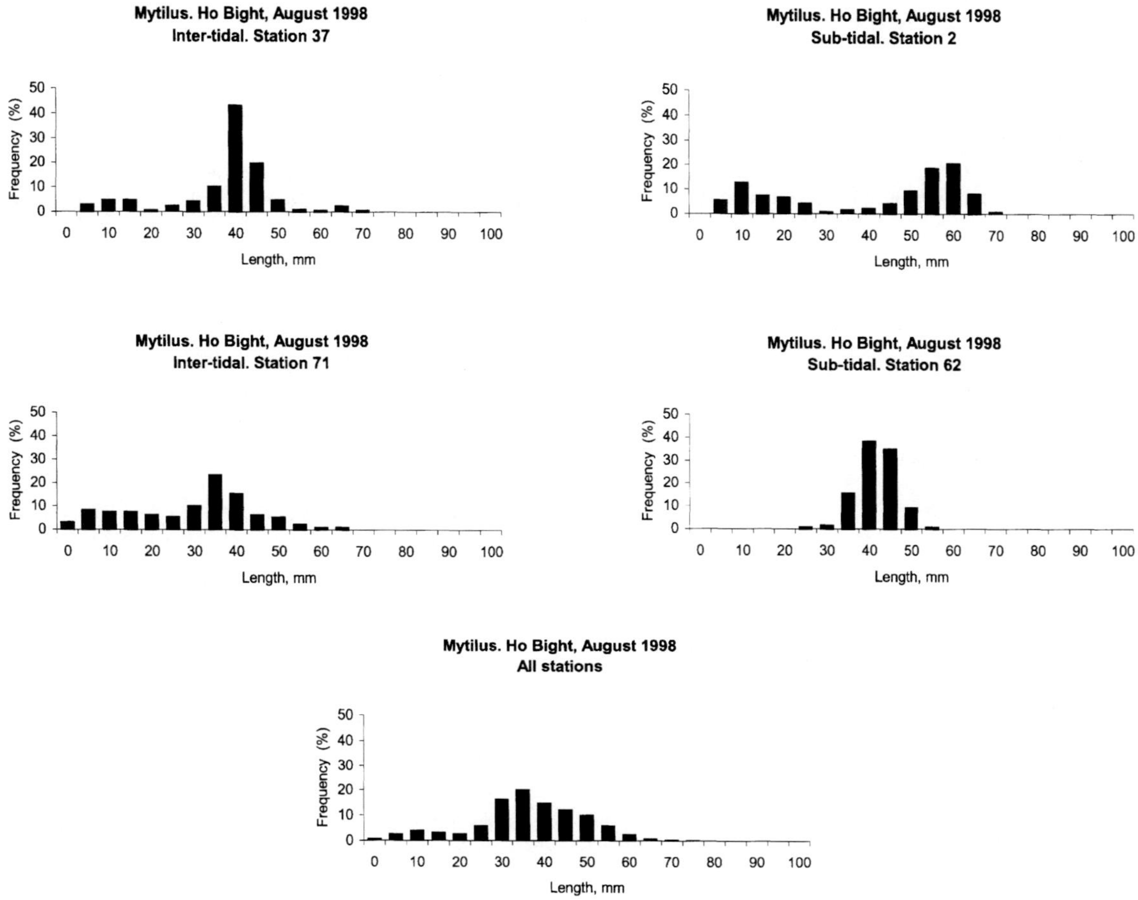

Figure 2. Ho Bight, 1998: Length distributions in selected samples (stations) and the average for all stations.

parameters was a least square routine available in the above-mentioned FiSAT package. In order to convert growth in length to growth in weight, an isometric length-weight relationship, $w = q * l^3$, was assumed.

The estimated growth curves, i.e. the parameters K and L_∞ are compared by using the so-called Φ' (phi prime) test for overall growth performance:

$$\Phi' = \ln(K) + 2\ln(L_\infty).$$

This index of growth performance (Munro & Pauly, 1983) is empirically based on the fact that calculated index values for different stocks of the same species are very similar. This index of growth performance also implies that growth curves based on different age-length observations from the same stock can have the same Φ' value and yet be different in shape (Sparre & Venema, 1992). In such cases one finds a negative correlation between L_∞ and K.

The mortality is expressed by the exponential decay model: $N_t = N_0 {*} \mathrm{e}^{-Zt}$, where Z is a coefficient of instant total mortality. Z is partitioned in fishing mortality, F, and natural mortality, M ($Z = F+M$).

Estimation of annual production and yield

One of the advantages of the models for growth and mortality given above is that they easily can be incorporated in the more complex models by Beverton & Holt (1957) for calculating production and fisheries yield. For instance, annual (average) biomass (B) and the annual production (P) for a single individual belonging to age group (a) are given by:

$$B(a) = \int_{a}^{a+1} \mathrm{e}^{(-(M(a)+F(a))t} w(t)\mathrm{d}t, \qquad (1)$$

$$P(a) = \int_a^{a+1} e^{(-(M(a)+F(a))t}(dw/dt)dt, \quad (2)$$

where d*w*/d*t* is the growth rate (in weight) expressed by the VBGE. Similarly, the annual amount of natural deaths (*D*) and annual yield (*Y*) (deaths due to fishing) are given by:

$$D(a) = M(a) \int_a^{a+1} e^{(-(M(a)+F(a))t} w(t)dt, \quad (3)$$

$$Y(a) = F(a) \int_a^{a+1} e^{(-(M(a)+F(a))t} w(t)dt, \quad (4)$$

where $w(t)$ is the weight expressed by the VBGE and $M(a)$ and $F(a)$ are the coefficients of natural and fishing mortality at age a. Thus, the expressions (2), (3) and (4) can provide relative figures for how annual changes in standing crop are distributed in the components: natural deaths (predation), fisheries yield and net production by age group.

The applicability of the predictions of annual production, *P*; natural deaths, *D*; fisheries yield, *Y*; and stock biomass, *B*, by the Beverton & Holt model is somewhat restricted by the rigorous assumptions of a steady state population. For instance, it is unrealistic to assume constant growth and mortality rates for all individuals in the stock for as many years as there are age groups. Also, the assumption of constant recruitment is unrealistic. However, even if the prediction figures may not be very exact, their magnitudes will still reflect how annual production and thus yield by age group is affected when for instance the mortality parameter values change. Since the mussel populations in the Wadden Sea is of importance both for the commercial fisheries and for instance as prey for mussel eating birds, this model is well suited as a tool for the management of this resource. An example of application of the model has been presented earlier by Munch-Petersen & Kristensen (1987).

Results

Size distributions and cohort identification

Individual cohorts are generally easy to identify and follow during their first season. Later, however, the size of the individuals in the particular cohorts will vary to an extent such that the various cohort-specific size distributions seemingly merge into one. Only in one case was it possible to follow a particular cohort through 5 years. It was the 1987 cohort on an inter-tidal mussel bed, which was not commercially exploited. Except for this case, it has not been possible to follow a specific cohort from one season to the next in the length frequency samples. Bhattacharya's method facilitates the partitioning of a composite distribution into separate distributions. However, in the subsequent allocation of age to the identified cohorts subjectivity in assessing the age cannot be avoided.

In distinguishing between age 0 and age 1, the size in relation to the month of sampling was considered. Spat fall may occur from early May to October, and field observations have shown that in some cases with early spat fall 0-group individuals may reach a size of up to around 30 mm in their first season. However, from the data collected it appears that 'on average' 0-group mussels will attain a size of up to around 15 mm after spat fall in the period before winter. A 'birthday' of 1 July was set as the 'average' time of spat fall. In allocating age 0 or age 1 to modes $\leq$15 mm the time (month) of sampling in relation to 1 July was considered as well as the size of adjacent modes. Even though components with modes $\geq$40 mm can be distinguished in the analysed length data, there is a very large degree of uncertainty in allocating age to these. Thus mainly age groups $\leq$3 are used in the subsequent estimation of growth parameters.

The variation in size-at-age for the identified age groups appears from Figure 3A ('observed lengths'). The estimated mean sizes of the age groups for sub-tidal and inter-tidal beds also seem to be similar in their range. In fact, a statistical comparison (by *t*-tests) of the estimated mean sizes of sub-tidal and inter-tidal age groups 0–2 did not show any significant differences at the 95% level.

Growth

Growth in length

Parameters of the von Bertalanffy growth equation (VBGE) have been estimated on basis of the mean lengths-at-age of the cohorts identified in the length frequency data and are shown in Table 3. As mentioned above, on one particular inter-tidal bed a single 1987 spat fall could be followed for 5 years, see Figure 3B. It is evident, that the mussels in this cohort during their growth on the average attain a lesser size

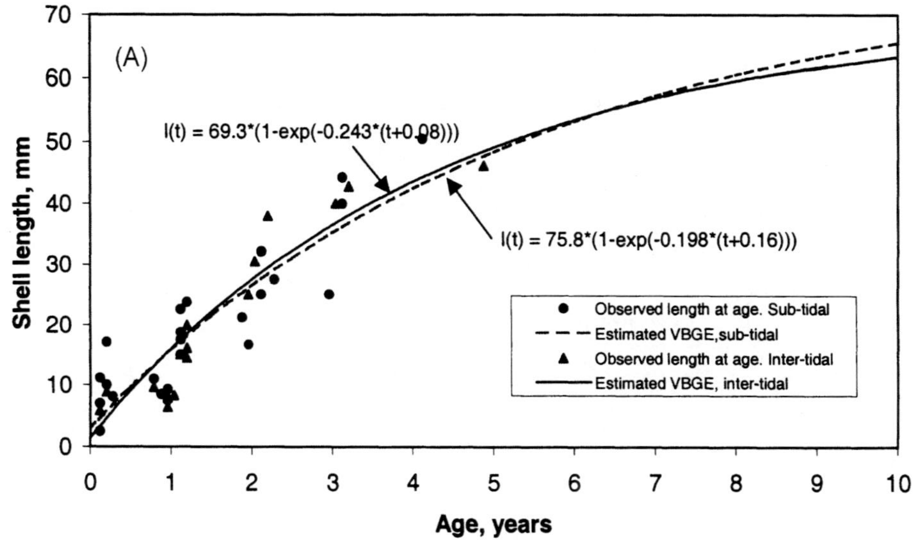

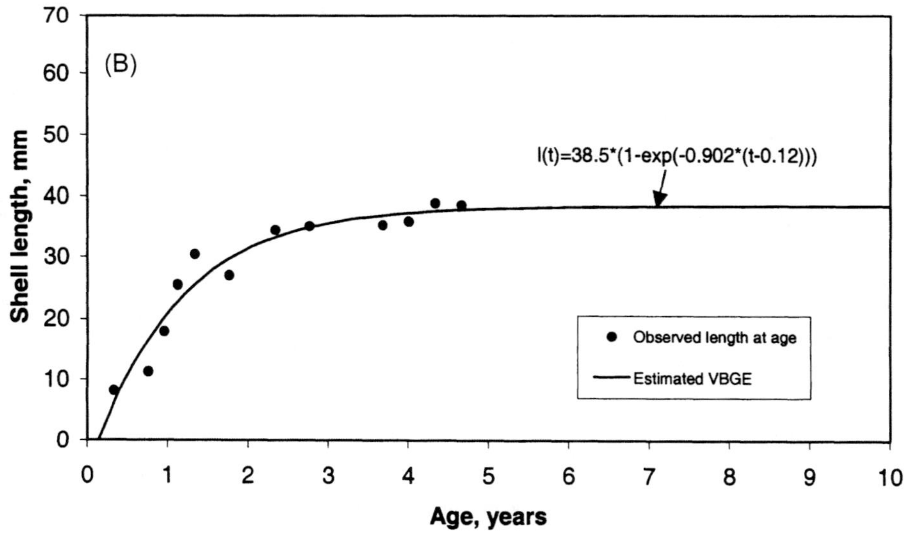

Figure 3. 3A: Estimated von Bertalanffy growth curves for *Mytilus* in Ho Bight, 1990–1998. Sub-tidal and inter-tidal beds. 3B: Estimated von Bertalanffy growth curves for *Mytilus* in Ho Bight, 1990–1998. A particular inter-tidal 1987 cohort (see text).

at ages ≥3 than observed on the other inter-tidal beds sampled. However, regarding the observed lengths as age groups 0, 1 and 2 and comparing with Figure 3A, there seems to be no great difference in size between this cohort and the ones from the other beds. This seemingly retarded growth for older individuals may be explained by some selective predation by for instance by birds, which tend to take only fast growing and therefore larger mussels. The resulting growth pattern observed for this particular mussel bed must therefore be considered somewhat extreme.

The calculated VBGE curves for sub-tidal and inter-tidal mussels are shown in Figure 3A and B shows the curve the 1987-cohort. The large variances (given as S.E.) of the estimated VBGE parameters for the sub-tidal and inter-tidal mussels (Table 3) are illustrated by a large scatter of the particular length at age observations around the estimated growth curves (Fig. 3A). The growth parameters for the 1987 cohort

Table 3. Estimated VBGE parameters, see text

Parameter	Sub-tidal	Inter-tidal	Sub- & Inter-tidal pooled	1987-cohort (Inter-tidal)
L_∞ (mm)	75.8	69.3	75.8	38.5
s.e. (L_∞)	48.8	24.8	26.6	1.989
K	0.196	0.243	0.203	0.902
s.e.(K)	0.181	0.14	0.106	0.202
t_0	−0.198	−0.08	−0.16	0.12
Φ' (Phi prime):	7.0266	7.0622	7.0616	7.2034
W_∞ (g AFDW)	1.62	1.24	1.62	0.21

Table 4. Estimates for 1998 of annual production, deaths (predation) and fisheries yield for current exploitation level ($F = 0.15$). By age group
St. crop (1) = Standing crop at the beginning of the year
St. crop (2) = Standing crop at the end of the year

(A) Ash free dry weight (AFDW) per m^2

Age group	Nos/ m^2	Mortality M	Mortality F	AFDW (grams) St.crop (1)	*P*	*D*	*Y*	*B*	St.crop (2)	*P*/*B*
0	37	1.74	0	0.00	0.11	0.04	0.00	0.03	0.06	4.24
1	36	0.35	0	0.36	1.46	0.31	0.00	0.89	1.51	1.64
2	112	0.35	0	6.76	8.75	3.35	0.00	9.57	12.16	0.91
3	129	0.35	0.15	19.88	12.60	7.44	3.19	21.26	21.84	0.59
4	72	0.35	0.15	19.93	7.94	6.75	2.89	19.29	18.22	0.41
5+	99	0.35	0.15	54.11	10.81	16.65	7.13	47.56	41.14	0.23
Totals:	485			101.02	41.67	34.54	13.22	98.60	94.93	**0.42**

(B) Raised to total fresh weight (tonnes) in the Ho Bight (total area = 11 km^2)
(fresh weight = 23.44 * AFDW)

Age group	Tonnes St.crop (1)	*P*	*D*	*Y*	*B*	St.crop (2)
0	0	28	12	0	7	17
1	92	377	81	0	230	388
2	1742	2256	863	0	2467	3135
3	5124	3247	1918	822	5481	5630
4	5137	2047	1740	746	4971	4698
5+	13948	2787	4291	1839	12260	10605
Totals:	26042	10742	8905	3407	25416	24472

are based on a wide range of ages, and the variances are much smaller here.

Since the length-at-age data for sub-tidal and inter-tidal mussels are rather similar it was to be expected that the estimated parameters values for the two data sets also would be similar (Table 4). This is not a proof of no difference in growth between inter-tidal and sub-tidal mussels, but the available data do not suggest any significant difference. Taking one further step by applying the Φ' (phi prime) test, see Section Estimation of growth and mortality, for all 3 parameter sets, one obtains rather similar values, see Table 4. This suggests that even the growth of the 1987 cohort complies with the growth performance of the two other, although the older age groups showed very little growth.

Since there is little difference between the inter-tidal and sub-tidal parameters, it was decided to use the growth parameters based on pooled data (excluding the 1987 cohort) in the assessments. The results for the particular 1987 cohort indicate, however, that extreme deviations may not be infrequent.

Growth in weight

The VBGE parameter-value W_∞ for growth in weight, see Table 3, has been calculated using an isometric length–weight relationship $w = q * l^3$, derived from a series of length measurements and corresponding weights. Fresh weights as well as AFDW (ash free dry weight) were measured, but for comparison with other investigations the weights are expressed in AFDW, see Figure 4. In Table 2, the calculated fresh weights from the estimated length weight relationship are compared with the observed average weight per m^2 (inter-tidal and sub-tidal mussel beds pooled). It is seen that the deviation from the calculated weights lies around 7% above the observed weights. However, we consider this discrepancy insignificant in relation to the magnitude of the total biomass, yield and production.

Estimates of mortality

Mortality parameters have been estimated as coefficients of instantaneous mortality (unit: $year^{-1}$) assuming the exponential decay model for a cohort, see Section Estimation of growth and mortality. As mentioned in Section Size distributions and cohort identification, for most samples it was not possible to follow cohorts from one year to the next in the samples. However, in one area a specific 1987 cohort be identified over a longer time range, and the estimates of mortality presented here are based on this inter-tidal cohort which resulted from a high larval settlement in 1987. Regular sampling between 1987 and 1992 provided data on abundance for estimation of survival. Combining all these data in a regression analysis gives an overall value for instant annual mortality of 0.84 for the whole period (around 1700 days). However, it appears from Figure 5, that there is a much higher mortality during the first year of this cohort (0-group individuals) than for the following years. Such a trend in mortality coefficients seems natural, and when splitting the entire period into two where the first period roughly covers the first year one obtain coefficients of 1.7 for the 0-group and 0.35 for all the remaining ages, see Figure 5.

Since this particular area is not subject to commercial exploitation, these mortality estimates refer to natural mortality, M, only. In fact, it is likely that the major part of M is due to predation by various sea birds (Faldborg et al., 1994), see Section Biomass, production and yield. These mortality figures confirm the assumption of a rather low mortality for older mussels on inter-tidal beds, where no fishery takes place. For comparison, the annual natural mortality figures for transplanted mussels in the Wadden Sea estimated by Theisen (1968) varied between around 0.6 for mussels of sizes 25–30 mm to around 0.33 for mussels of sizes around 50 mm. The data by Faldborg et al. (1994) on mussel densities (also for the 1987 year class) during 1989–1990 in another sub-area in Ho Bight provide an M-value of around 0.5. Mortality estimates from other localities lead to parameter values of similar magnitudes, e.g. Kristensen & Lassen (1997) estimated an annual natural mortality of around 0.32 for mussels >30 mm in Limfjorden in areas temporarily closed to commercial fishery.

Biomass, production and yield

The estimated figures for standing crop, annual production, fishery yields, biomass are based on:

- The average density of mussels (sub-tidal and inter-tidal pooled) for 1998 as shown in Table 2, and a total area with mussel beds estimated to around 11 km^2.
- The estimated growth parameters for sub-tidal and inter-tidal beds pooled given in Table 3. For simplicity in the calculations t_0 has been set = 0. The mortality parameters are described in Section Estimates of mortality and shown in Figure 5.

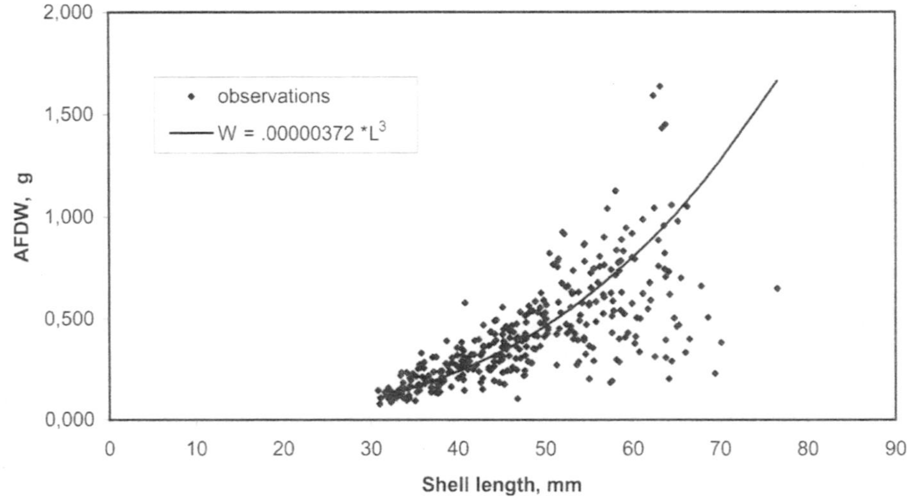

Figure 4. Mytilus in the Ho Bight: Isometric length–weight relationship.

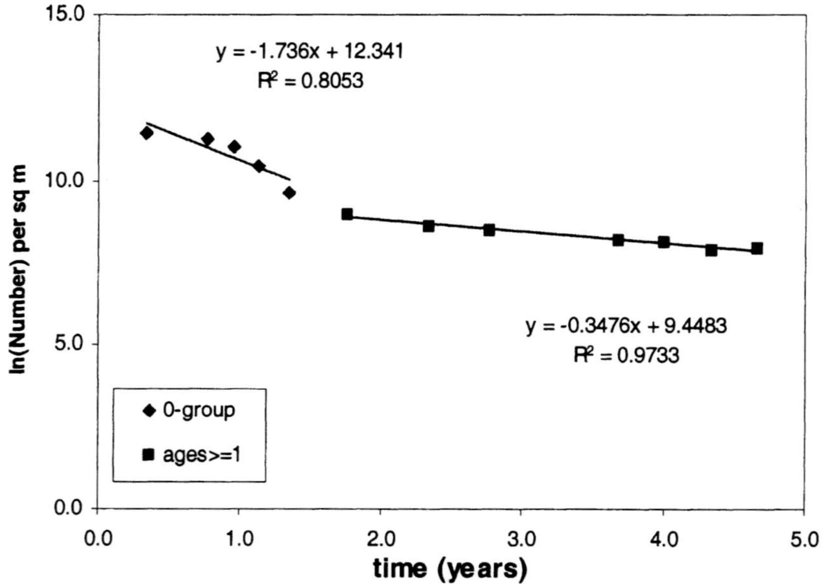

Figure 5. Mytilus in the Ho Bight: Estimation of mortality for the 1987 cohort. M = 1.74 for the 0-group. M = 0.35 for the ages ≥ 1.

- An average age distribution for 1998 obtained by converting the observed length distribution (numbers by length group/m^2) to numbers by age group/m^2 by the VBGE.

The calculations are given both as AFDW per m^2 and as total fresh weight for the whole area in consideration. The current level of fishing mortality (for 1998) has been determined from the total landings in 1998, which amounted to 3100 tonnes. This is below the 1989–98 average of around 3600 tonnes, see Table 1. In recent years almost all landings has come from the Ho Bight area. The 3100 tonnes in 1998 correspond approximately to a removal of 10–15% of the estimated total. Using this figure to set a level for fishing mortality gives a figure for F of around 0.15 for age groups ≥ 3, and in the calculations shown in Table 4A & B $F = 0.15$ has been used.

Table 4A shows the standing crop at the beginning and end of the year, annual production, deaths and average biomass during the year as grams AFDW / m^2

Table 5. Estimates of food demand for the three main species of sea birds in the Danish Wadden Sea

Species	Annual number of bird days[a]	Daily food demand per indivdual AFDW (grams)[b]	Annual food demand AFDW (tonnes)	Demand for mussels (40%) AFDW (tonnes)	Demand for mussels (40%) fresh weight (tonnes)
Oyster catcher	7 117 500	40	285	114	2669
Herring gull	4 901 220	90	441	176	4135
Eider	5 490 695	124	681	272	6382
All species	17 509 415	80	1407	563	13186

[a]Laursen et al. (1997).
[b]Goss-Custard et al. (1987), Clausen (1988), Swennen (1976).

by age group. The standing crop (by age group) at the start of the year (St. crop(1)) is simply the weight of that age group taken from the VBGE multiplied by the number belonging to that age group. The changes in biomass during the year are distributed in the components: production (P), natural deaths (D) and fisheries yield (Y). Also shown is the average biomass during one year (B). The standing crop at the end of the year, (St. crop (2)) is = St. crop(1) + $P - D - Y$. Table 4B gives the totals (in tonnes) for Ho Bight based on the above mentioned total area of 11 km^2 for Ho Bight.

When evaluating the figures in Tables 4A & B, it is the magnitude and relative contribution of the particular age groups to the totals, which is important when assessing the dynamics of this particular combination of density and size (age groups) of the mussels. Note that the P/B ratios decrease with increasing age. The calculated age specific P/B depends entirely on K and the mortality parameters. The estimates of P are especially sensitive to K in the growth equation. This is a consequence of the growth rate, dw/dt in the expression for production (1) in Section Estimation of annual production and yield, being proportional to K. It is seen, that P for each of the age groups 0–4 exceeds or almost balances the deaths, even with a fishery for age groups 3 and 4.

Turning to the totals it appears that for this particular (1998) age (size) distribution approximately 50% of the annual production comes from age groups 2 and 3. It also appears that the decrease in total standing crop during the year is rather small with this level of fishery. Figure 6 shows how increasing levels of fishing (exploitation) will influence production, P and natural deaths, D. It is of particular interest to compare our figures for B, P and D with those of Faldborg et al. (1994). Their figures (AFDW per m^2) refer to a particular sub-area in Ho Bight with mussel densities of around 10 times our average and were obtained on a monthly basis (see Table 1 in Faldborg et al., 1994), and their figures are also correspondingly around 10 times ours.

Mussel predation by sea birds

The model also provides information on the amount of mussels predated (natural deaths) compared to the fishery, see Section Estimates of mortality. Tables 4A & B show a magnitude of around 9000 tonnes of natural deaths with a fishery yield of around 3400 tonnes. Figure 6 shows, how this amount would be for different levels of exploitation. Such information is of special relevance to the management of the mussel fishery in the Wadden Sea, where leaving sufficient food availabile for the common sea birds (e.g. eiders, oyster catchers a.o.) has been a main objective in recent years management strategy.

The basis for assessing whether there is sufficient mussels for the birds has been:

- estimates of the daily food demand for three main bird species;
- estimates of annual number of bird days.

From such information one can estimate total annual food demand by the important bird species in the area. One must also have some data for how much the mussels on the average cover this food demand. From such information one can come up with rough estimates of the birds' demand for mussels and compare these with calculations like the ones given in Table 4B.

Following the figures of Goss-Custard et al. (1987), Clausen (1988) and Swennen (1976), see Table 5 the average daily food demand is 80 grams AFDW, and according to Swennen (1976) 40% of this is covered by mussels. An estimate by Laursen (1997) of the annual number of bird days for the total Dan-

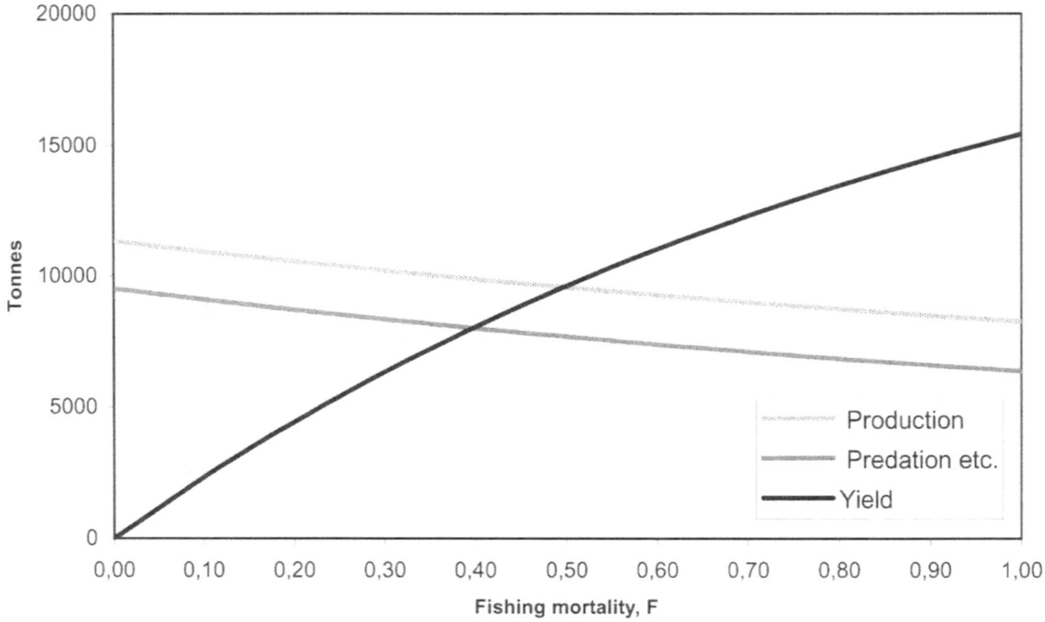

Figure 6. Mytilus in the Ho Bight: Annual production (*P*) natural deaths (*D*) and fisheries yield (*Y*) as function of fishing mortality (*F*). 1998 scenario.

ish Wadden Sea is also given in Table 5 This leads to an estimated annual demand of around 13 000 tonnes mussels (fresh weight) for the entire Danish Wadden Sea. Since the estimates in Table 4 B only cover a minor part of the Danish Wadden Sea, a reasonable conclusion on basis of these data would be that there is sufficient availability of mussels for these sea birds with the given exploitation level.

Discussion and conclusions

The estimates of annual production, *P*, per m^2 presented here can be compared with other production estimates for *Mytilus*. Table 6 summarises production figures for various *Mytilus* populations expressed as average *P*/*B*, the ratio annual production over average biomass taken over all size or age groups. To compare with other Wadden Sea mussel populations, the investigations by Faldborg et al. (1994), Craeymeersch et al. (1986) and Asmus (1987) are relevant. Their estimates, based on inter-tidal mussel population, are based on *in situ* measurements of changes in biomass per m^2 for particular time periods for the total range of size classes, and can be regarded as observed, actual, figures for particular mussel beds for a particular period of time. Conversely, the figures presented in Tables 4A & B do not refer to a specific mussel bed, but refer to inter-tidal and sub-tidal mussel beds combined, with an average size distribution in 1998 for a particular exploitation level. In addition, however, it is possible to see the contribution of each age group to the totals.

Note that our figure for P/B of 0.42 is of the same magnitude as the figures of 0.36 by Asmus (1987) and 0.5 by Faldborg et al. (1997) and Craeymeersch et al. (1986), see Table 6. Thus, there is good agreement between these 3 estimates of productivity figures for mussels in the Wadden Sea. As further examples of *P*/*B* values, the estimates the by Bayne & Worrall (1980) for two mussel populations near Plymouth mentioned in Section Cohort identification have also been included. However, their estimates do not refer to population values, but are averages over age groups and are therefore not comparable with the other figures. Thus, it appears the overall range of the average *P*/*B* ratio for mussel populations in the Wadden Sea is 0.4–0.5. But the actual value of the average will depend on the size distribution in the population.

The main purpose of this paper has been to describe the methods applied for assessment of the mussel resources in the Danish Wadden Sea and to demonstrate their applicability as a basis for management of this resource in a situation where the total cost of the sampling programme is limiting the sampling

Table 6. Comparison of estimates of annual *P/B* for *Mytilus edulis*

	P/B
Craeymeersh et al. (1986):	0.50
Asmus (1987):	0.36
Bayne & Worrall (1980), site 1	0.92
Bayne & Worrall (1980), site 2	0.51
Faldborg et al. (1994)	0.50
These investigations (Table 4A):	0.42

intensity. As in all cases when fitting biological data to simple models, which are to cover large areas as well as large variation in dynamics, one has to sacrifice some detail. However, we think that for assessment purposes the advantages of fitting variable growth and mortality patterns to the simple models and then combining them into a simple model like the Beverton & Holt model, balances the loss of detail and possibilities of coping with special cases and local variation.

The 'point estimates' of biomass and production presented by Faldborg et al. (1994), Craeymeersch et al. (1986) and Asmus (1987) may provide more precise estimates of particular situations sampled. However, a major advantage by using models like the one used here is that once the basic data (e.g. length and density data) have been collected and the parameters for the model determined, the model can, besides simulate the current situation, readily provide estimates based on alternative input parameters and thus provide scenarios to elucidate problems of the type: 'How would biomass, production etc. change if, say, the fishing effort is doubled, fishing is banned etc.'? Indeed, the methodology used here, based on the Beverton & Holt model is certainly not new on the market, and the problem of getting adequate recruitment data has not been solved. However, when sampling resources are strongly limited we think that the method presented here still is well suited for providing management advice for the mussel resources in the Danish Wadden Sea.

References

Asmus, H., 1987. Secondary production of an intertidal mussel bed community related to its storage and turnover compartments. Mar. Ecol. Prog. Ser. Vol. 39: 251–266.

Bayne, B. L. & C. M. Worrall, 1980. Growth and Production of Mussels *Mytilus edulis* from Two Populations. Mar. Ecol. Prog. Ser. Vol. 3: 317–328.

Beverton, R. J. H. & S. J. Holt, 1957. On the dynamics of exploited fish populations. Fish. Invest., Ser. II, 19. 533 pp.

Bhattacharya, C. G., 1967. A simple method of resolution of a distribution into Gaussian components. Biometrics, 23: 115–135.

Clausen, C. V., 1988. Spatial distribution and foraging behaviour of gulls (*Larus* spp.) in the Danish Wadden Sea. M.Sc. thesis. University of Aarhus, Denmark.

Craeymeersch, J. A., P. M. J. Herman & P. M. Meire, 1986. Secondary production of an intertidal mussel (*Mytilus edulis* L.) population in the Eastern Schelte (S.W. Netherlands). Hydrobiologia 133: 107–115.

Faldborg, K., K. T. Jensen & L. Maagaard, 1994. Dynamics, growth, secondary production and elimination by water fowl of an intertidal population of *Mytilus edulis* L. Ophelia Suppl. 6: 187–200.

Gayanilo, F. C., P. Sparre & D. Pauly, 1994. FAO-ICLARM Fish stock Assessment (FiSAT) Users Guide. FAO Comp. Inf. Ser. (Fish.) 7.

Goss-Custard, J. D. & S. E. A. Durrell, 1987 Age-related effects in Oyster catchers, *Haematopus ostralegus*, feeding on mussels, *Mytilus edulis*. I. Foraging efficiency and interference. J. Anim. Ecol. 56: 521–536.

Kristensen, P. S., 1995. Aerial surveys, biomass estimates, and elimination of the mussel population (*Mytilus edulis* L.), in the Danish Wadden Sea, 1991–1994. ICES. Shellfish Committee. C.M. 1995/K: 44. 22 pp.

Kristensen, P. S. & H. Lassen, 1997. The production of relaid blue mussels (*Mytilus edulis* L.) in a Danish fjord. ICES Jour. Mar. Res., 54: 854–865.

Laursen, K., J. Frikke & J. C. Salvig, 1997. Vandfugle i relation til menneskelig aktivitet i Vadehavet 1980–1995, med en vurdering af reservatbestemmelser. Danmarks Miljø undersøgelser. 73 s.- Faglig rapport fra DMU, nr. 187.

Munch-Petersen, S. & P. S. Kristensen, 1987. Assessment of the stocks of mussels in the Danish Wadden Sea. ICES. Shellfish Committee. C.M. 1987/K: 13. 21 pp.

Munch-Petersen, S. & P. S. Kristensen, 1989. On the applicability of aerial survey techniques for recording and estimating densities of mussel beds . ICES. Shellfish Committee. C.M. 1989/K: 24. 16 pp.

Munksgaard, C., 1989. Investigations of the mussel stocks in the Danish Wadden Sea. Rept. to the Forest and Nature Agency. Inst.of Genetics & Ecol., Aarhus University. 39 pp. (In Danish).

Munro, J. L. & D. Pauly, 1983. A simple method for comparing growth of fishes and invertebrates. ICLARM Fishbyte, 1: 5–6.

Sparre, P. & S. C. Venema, 1992. Introduction to tropical fish stock assessment. FAO Fisheries Technical Paper 306/1.

Swennen, C., 1976. Populatiestructuur en voedsel van de Eiderend (*Somateria mollissima*) in de Nederlandse Waddenzee. Ardea, 311–371. 64 pp.

Theisen, B. F., 1968. Growth and mortality of culture mussels in the Danish Wadden Sea. Medd. Danm. Fiskeri og Havunders. N.S. 6: 47–78.

Hydrobiologia **465:** 45–48, 2001.
G. Burnell (ed.), Coastal Shellfish – A Sustainable Resource.

The role of lobster (*Homarus* spp.) hatcheries in ranching, restoration and remediation programmes

C. A. Burton
Sea Fish Industry Authority, Seafish Aquaculture, Marine Farming Unit, Ardtoe, Acharacle, Argyll PH36 4LD, U.K.

Key words: European lobster, fisheries management, hatcheries, *Homarus americanus*, *Homarus gammarus*, lobster, ranching, remediation, restocking, restoration, stocking, stock enhancement, stock management

Abstract

Juvenile lobsters can be reared in hatcheries. Their use in ranching programmes has been demonstrated successfully and exploitation of the idea by fisheries is commencing. Studies are underway to investigate the use of these juveniles in restoring damaged or threatened populations. However, the concept of using them in remediation projects remains theoretically possible but untested.

Introduction

Hatchery techniques for the production of juvenile clawed lobsters (*Homarus* spp) are well established and described (van Olst et al., 1977; Waddy, 1988; Beard & Wickins, 1992; Burton, 1992). The output from the hatcheries has many potential uses; intensive 'farming', where they are reared from cradle to table in controlled conditions, is one of them, but within the field of resource management, the most likely uses are in ranching, restoration or remediation programmes.

Ranching

Releasing hatchery-reared juvenile lobsters into the wild to enhance fisheries has a history going back over 100 years (e.g., Fullerton, 1896; Herrick, 1909). In the US, Hughes began a release programme with the American Lobster (*Homarus americanus* Milne-Edwards) at Martha's Vineyard in 1951, which is still running today (Aiken & Waddy, 1995). Previous European attempts, such as those in France (Henocque & Lorec, 1979), lasted only a short time and were inconclusive. However, despite the resources committed to them, none of these projects could demonstrate that the released lobsters were surviving in the wild or had any effect on fishery landings.

In 1982, the UK embarked upon a series of collaborative stocking experiments which demonstrated conclusively, for the first time, that hatchery-reared juvenile European Lobsters (*Homarus gammarus* L.) released to the wild would survive to market size and would contribute to fishery returns (Burton, 1993, 1997; Bannister et al., 1994; MAFF, 1995; Cook, 1995). To achieve this, each juvenile released was tagged internally with a binary-coded micro-wire tag (Wickins et al., 1986). Consequently, each animal recaptured by the commercial fishery could be identified positively together with its year and time of release.

Initially, for the scientific trials, the cost of each juvenile released was high, in the order of £1.12–£1.25, but recent efforts by a commercial hatchery operator have resulted in a dramatic reduction in price, to approximately £0.36 (Coghill, 1997; Burton & Adamson, in press).

The positive results obtained from the trials (MAFF, 1995) encouraged the UK inshore fishing industry to consider the possibility of local stocking initiatives. Orkney began the process and these investigations received a further boost in 1997 when UK legislation, enabling local fishery control (Regulated Fishery Order) or conferring property rights to a designated species in a defined area (Several Fishery Order), was extended to encompass crustacea. Shetland became the first area to take advantage of these provisions in October, 1999 when a local fishery-led consortium was granted a Regulated Fishery Order for their waters (Johnson, 1999). As part of the development plans, their lobster stocking programme will be expanded and improved (Napier, 2000).

The economic viability of stocking programmes is controversial. Many consider the costs to be too high for the returns generated when examined from a strictly monetary view-point. However, when considered from a fishery management perspective an immediate benefit becomes apparent; once stocking programmes begin, the fishers actively seek other fishery controls, in order to protect their investment, and they are willing to enter into constructive dialogue with fishery managers (Burton, 2000). In the UK, the impetus provided by the stocking programmes has made the inshore fishing industry more receptive when considering or demanding other, more conventional, management measures, such as V-notching or a ban on landing berried females. The socio-economic impact of stocking programmes on often fragile rural fishing economies remains unquantified as does the potential long-term biological impact on stocks. Preliminary modelling of the latter suggests that this could be far longer than previously anticipated, extending over several generations (Burton, 2000).

The UK was able to benefit from the hatchery and stocking experience of those who went before, from Fullerton to Hughes, when designing its own experimental programme and was able, by using the micro-tag system, to demonstrate conclusively that hatchery-reared juveniles were surviving and contributing to commercial fishery landings (Bannister et al., 1994; MAFF, 1995). In turn, Ireland (Mercer, pers. commun.), Norway (Tilreth & Uglem, pers. commun.), Spain (Carrasco, pers. commun.) and even the US (Bayer, pers. commun.), were able to draw upon the UK results to support their own programmes and many report outcomes which are considered a success from their point of view (Van der Meeren & Naess, 1993; Carrasco & Barros, 1996; Browne & Mercer, 2000; Van der Meeren & Agnalt, 2000). Other countries remain interested in the prospects of lobster ranching and may begin their own programmes when conditions are favourable. Italian researchers were able to use the data to begin a project (Scovacricchi & Cervelli, pers. commun.).

Restoration

Natural or anthropogenically induced environmental events can bring lobster stocks in an area to the brink of extinction. One place where this has occurred is the northern Adriatic Sea. Here, a number of factors, from industrial input to transient anoxic events and illegal fishing, have combined to reduce European Lobster populations to extremely low levels. In addition to studying the remnant wild population, Italian researchers began rearing juvenile lobsters from locally caught broodstock, using hatchery techniques similar to those utilised in the UK and Ireland (Scovacricchi & Burton, 1998). These juveniles were released on to natural habitats in a protected marine reserve at Miramare near Trieste and their fate studied. Initial results were reported as favourable, with the juveniles surviving and growing, and it is hoped to embark upon a larger restoration trial in the future (Scovacricchi, 1999).

Initially, the objective will be to boost numbers in selected locations in an attempt to create breeding centres. Over time and with the addition of the natural, wild recruitment, it may be possible to slowly rebuild the population. In the extreme long-term, it may even be possible to recommence economic exploitation of the stocks, although this remains highly speculative at this stage. Some support can, however, be drawn from the fact that many molluscan fisheries have been successfully restored and managed in a similar way (see other papers in this volume).

Recent studies are beginning to demonstrate the degree of genetic variation within and between lobster stocks (Jorstad & Farestveit, 1999). Whilst the concept is theoretical and untested, it is possible to foresee that where genetically distinct or phenotypically unique populations exist, which may be vulnerable to human activity, such as that in northern Norway, a combination of good management and captive breeding for release could be used to safeguard the future of the stock, protect the gene-pool and assist in regenerating numbers. Such efforts have direct parallels in terrestrial and avian conservation and management, where captive breeding and release programmes are routine.

Remediation

Considered very simply, any development activity in the coastal zone can directly affect fisheries in the area. Nearly all the impacts are detrimental in one way or another. In most cases, it is the sessile or habitat-dependent species, such as molluscs and crustacea, which are disadvantaged before the more mobile fish species. When high-value animals, such as lobsters, are affected, fishers consider it important to lessen the economic impact.

In the UK, wherever major marine civil engineering projects are envisaged, for example coastal protection defences, harbour extensions, bridge building or tunnel construction, it has become almost standard practice to negotiate a compensation package with fishery interests in order to ameliorate any detrimental effects they might experience. As part of this package, stocking new or alternative habitats, which could be specially created, with hatchery-derived juveniles may be considered. Unfortunately, these negotiations are considered commercially sensitive, which makes it difficult to give specific examples, and few are recorded or reported in the literature, so most knowledge of these matters has to remain anecdotal. Similar situations may pertain in other countries.

At its simplest, the proposal may be to lay hatchery-reared stock on to existing natural habitats close to the work area. Ideally, these areas should be far enough away so as to be unaffected by the work, but close enough to the fishing ports so that the fleet does not incur greatly increased costs. The seeded animals are allowed to grow on naturally and enter the wild-stock fishery after a number of years. This is, in effect, another form of ranching, but without any additional intervention. Alternatively, where appropriate, it may be possible to modify the design of the coastal structure to provide, upon completion, suitable habitat for economically important fishery species, such as lobsters and crabs or molluscan shellfish. However, on many occasions, there appears to be a gap between what an engineer and a biologist consider 'appropriate' modifications to create 'suitable habitat'. In time, this may be overcome through increased dialogue and better understanding on both parts. Finally, when a habitat is being totally destroyed (e.g. by dumping or dredging) and few natural alternatives exist, a purpose-designed artificial structure, specifically created for the target fishery, may be considered. The potential of this approach has been demonstrated (Jensen et al., 1994; Wickins, 1994), but it tends to be the most expensive option.

To date, none of the proposals mentioned above have progressed past the discussion and feasibility study stage. Economic criteria are often cited for the non-progression of the idea. Initially, the fishery may be receptive to and favour stocking proposals, but as time progresses, the short-term cashflow advantage offered by monetary payment may become more attractive. On paper, this option may also appear cheaper for the contractor, who often opt for the short-term expedient measure of paying monetary compensation to the fishers, rather than pursuing a programme which may be better for the longer-term health of the stocks. However, as environmental awareness increases and the socio-economic future of fishing-based communities becomes more important politically, this policy may change. The decrease in the unit cost of hatchery-reared juveniles, particularly lobsters, may also assist in the reassessment of the priorities in such cases.

Conclusion

Because initial investigation of lobster hatchery techniques was driven by the desire to culture these valuable species, there has been a tendency to focus on the prospects for either intensive or extensive aquaculture. Intensive culture is unlikely to be economically viable in the near future, but extensive techniques or ranching appear promising. However, the output of juveniles from a hatchery can also be used for the alternative, but related, purposes of stock restoration or damage remediation. These can be important and should be considered when planning and approving hatchery applications in the future.

References

Aiken, D. E. & S. L. Waddy, 1995. Chapter 8: Aquaculture. In Factor, J. R. (ed.), Biology of the Lobster *Homarus americanus*. Academic Press, San Diego, CA: 528 pp.

Bannister, R. C. A., J. T. Addison & S. R. Lovewell, 1994. Growth, movement, recapture rate and survival of hatchery-reared lobsters (*Homarus gammarus* Linnaeus 1758) released into the wild on the English east coast. Crustaceana 67 (2): 156–172.

Beard, T. W. & J. F. Wickins, 1992. Techniques for the production of juvenile lobsters ((*Homarus gammarus* (L.)). Fisheries Research Technical Report 92. MAFF Directorate of Fisheries Research, Lowestoft. 22 pp.

Browne, R. & J. P. Mercer, 2000. A review of innovations concerning Ireland's Lobster (*Homarus gammarus*) fishery (1992–1998). In Rance, P. (ed.), Lobsters: The Next Generation. Developing ideas in Stock Enhancement and Preservation. Orkney Fisheries Association, Kirkwall: 18–19.

Burton, C. A., 1992. The techniques of lobster stock enhancement. Seafish Report 396. Sea Fish Industry Authority, Edinburgh. 55 pp.

Burton, C. A., 1993. The United Kingdom lobster stock enhancement experiments. In Baine, M. (ed.), Proceedings of the 1st British Conference on Artificial Reefs and Restocking. International Centre for Island Technology, Heriot-Watt University, Edinburgh: 22–35.

Burton, C. A., 1997. Lobster stock enhancement; Phase II. Shellfish News 4: 10–11.

Burton, C. A., 2000. The management and enhancement of lobster stocks: The Scottish experience. In Rance, P. (ed.), Lobsters: The Next Generation. Developing ideas in Stock Enhancement and Preservation. Orkney Fisheries Association, Kirkwall: 31–37.

Burton, C. A. & K. Adamson, in press. A preliminary study of the cost of operating a lobster hatchery in Orkney and the development of a model for future hatchery and stocking programmes. Sea Fish Industry Authority, Edinburgh.

Carrasco, J. F. & A. R. Barros, 1996. Primos datos sobre marcado de juveniles de bogavante europeo *Homarus gammarus* (Linnaeus 1758) llevados a cabo en el Principado de Asturias. Bol. Inst. Esp. Oceanogr. 12 (2): 145–149.

Coghill, A., 1997. Fishermen back lobster conservation. West Highland Free Press, Friday 29 August 1997: 9.

Cook, W., 1995. A lobster stock enhancement experiment in Cardigan Bay. Final report. North Western & North Wales Sea Fisheries Committee, Lancaster. 33 pp.

Fullerton, J. H., 1896. The European Lobster. Breeding and development. Fourteenth Annual Report of the Fishery Board for Scotland. Part III: 187–222.

Henocque, Y. & J. Lorec, 1979. General policy on lobster restocking in France. Aquaculture Extensive et Repeuplement, Mai 1979. Centre Oceanologique de Bretagne, Brest: 29–30.

Herrick, F. H., 1909. Natural history of the American Lobster. Bull. US Bur. Fish. 29: 148–428.

Jensen, A. C., K. J. Collins, E. K. Free & R. C. A. Bannister, 1994. Lobster (*Homarus gammarus*) movement on an artificial reef: the potential use of artificial reefs for stock enhancement. Crustaceana 67 (2): 198–211.

Johnson, P., 1999. Shetland controls its shellfish stocks. Fishing News 4467: 1.

Jorstad, K. E. & E. Farestveit, 1999. Population genetic structure of lobster (*Homarus gammarus*) in Norway, and implications for enhancement and sea-ranching operation. Aquaculture 173: 447–457.

MAFF, 1995. Lobster Stocking: Progress and Potential. Significant results from UK lobster restocking studies 1982–1995. MAFF Directorate of Fisheries Research, Lowestoft. 11pp.

Napier I., 2000. Development of a shellfish management system in Shetland. In Rance, P. (ed.), Lobsters: The Next Generation. Developing ideas in Stock Enhancement and Preservation. Orkney Fisheries Association, Kirkwall: 38–39.

Scovacricchi, T., 1999. Tentativi di produzione di giovanili di astice europeo *Homarus gammarus* (Linnaeus 1758) (Decapda, Nephropidae) per il ripoplamento di tegnue (beachrock outcrops) alto-adriatiche. Relazione Finale. CNR Istituto di Biologia del Mare, Venezia. 86 pp.

Scovacricchi, T. & C. A. Burton, 1998. Lobster (*Homarus gammarus* L.) research in the northern Adriatic Sea. The Lobster Newsletter 11 (1): 11.

Van der Meeren, G. I. & A. Agnalt, 2000. Fishery, management and restocking of the European Lobster (*Homarus gammarus*); the Norwegian experience. In Rance, P. (ed.), Lobsters: The Next Generation. Developing ideas in Stock Enhancement and Preservation. Orkney Fisheries Association, Kirkwall: 24–26.

Van der Meeren, G. I. & H. Naess, 1993. Lobster (*Homarus gammarus*) catches in southwestern Norway, including the first recaptures of previously released juveniles. ICESCM1993/K:29 Sess R. 9 pp (mimeo).

Van Olst, J. C., J. M. Carlberg & R. F. Ford, 1977. A description of intensive culture systems for the American Lobster (*Homarus americanus*) and other cannibalistic crustaceans. Proceedings 8th Annual Meeting World Aquaculture Society, 9–13 January 1977, San Jose, Costa Rica. 271–292.

Waddy, S. L., 1988. Farming the Homarid lobsters: State of the art. World Aquaculture 19 (4): 63–71.

Wickins, J. F., 1994. Do artificial reefs have 'lobster appeal'? Fishing News 4191: 6.

Wickins, J. F., T. W. Beard & E. Jones, 1986. Microtagging cultured lobsters, *Homarus gammarus* (L), for stock enhancement trials. Aquaculture Fish. Man. 17: 259–265.

Hydrobiologia **465**: 49–62, 2001.
G. Burnell (ed.), Coastal Shellfish – A Sustainable Resource.

An historical overview of the Republic of Ireland's lobster (*Homarus gammarus* Linnaeus) fishery, with reference to European and North American (*Homarus americanus* Milne Edwards) lobster landings

R. M. Browne[1], J. P. Mercer[2] & M. J. Duncan[2]
[1]*Taighde Mara Teo., Carna, Co. Galway, Ireland*
Zoology Department, National University of Ireland, Galway, Ireland
[2]*Shellfish Research Lab, Carna, Co. Galway, Ireland*

Key words: lobster, *Homarus gammarus*, Republic of Ireland, inshore-fishery, Europe, *Homarus americanus*, North America, Canada, USA

Abstract

Historical developments of the Republic of Ireland's *Homarus gammarus* fishery are reviewed and compared in the context of landings in Europe and the North American *H. americanus* fishery. *H. gammarus* fishing has provided essential income for over a century to many Irish coastal communities. There have been significant fluctuations (greater than 460%) in recorded landings over this time and the relative importance of *H. gammarus* fishing has changed with the establishment of ancillary crustacean fisheries. In 1994, *H. gammarus* landings reached an all-time high of 714 tonnes, with a landed value of IR £6.3 million; in 1999 landings were 597 tonnes valued at IR £6.15 million. Current Republic of Ireland lobster fishery regulations are: (a) minimum carapace length of 85 mm, (b) ban on landing lobsters that have been 'V-notched' or have a mutilated tail fan, and (c) prohibition of capturing lobsters by SCUBA diving. Chronological details on recent legislation, regulations and landings relating to Ireland's lobster fishery sector are reviewed. In the last 40 years landings of *H. gammarus* in the Republic of Ireland exhibit signs of an overexploited stock characterised by a reduction in catch per unit effort and there has been a general decline of total landings for *H. gammarus* throughout Europe. This contrasts with North America, where many regulations govern the *H. americanus* fishery and landings appear to be in a healthier state.

Introduction

The fishery for crustaceans in the Republic of Ireland's waters is part of a complex interrelated inshore fishery involving lobster (*Homarus gammarus* Linnaeus), spiny lobsters (*Palinurus elephas* Fabricius, *Palinurus mauritanicas* Gruvel), edible crab (*Cancer pagurus* Linnaeus), velvet swimming crab (*Macropipus puber* Linnaeus), spider crab (*Maia squinado* Herbst), prawn (*Palaemon serratus* Pennant) and other species. Traps and tangle nets are generally used to catch these commercially important decapods.

The most recent available data from the Department of Marine and Natural Resources (DoM&NR) valued the Republic of Irelands total inshore trap and fixed net crustacean catch at IR£18.02 million in 1999 (excluding Dublin Bay prawn or Norway lobster *Nephrops norvegicus* Linnaeus, as they are generally concentrated in deep water and caught by larger boats, using trawl nets). Landings of *H. gammarus* in 1999 were 597.6 tonnes valued at IR£6.152 million. This represents approximately 6% of the total live weight of the inshore crustacean tonnage and 34% of its value. However, between 1994 and 1997 lobster landings declined from 715 to 513 tonnes respectively, a 28% reduction in volume and a 29.2% (IR£1.88 million) decline in monetary value (DoM&NR, 1994, 1997). This instability has prompted fishermen's concern about lobster stocks.

For over a century, lobster (*H. gammarus*) fishing has provided essential income for many Irish coastal communities; however, the fishery has experienced large fluctuations in landings (Fig. 1) that have also been reflected in other European countries. In this

Table 1. Number of boats and fishers in Irish waters in 1876 (Buckland & Walpole, 1877)

Status	Boats	Men	Boys
Full-time	1283	5731	374
Part-time	644	2162	86
Occasional	4038	14 880	460

Table 2. Landings and values of *H. gammarus* in Ireland, England /Wales, and Scotland in the year 1912 (Selbie, 1914)

Ireland		England/Wales		Scotland	
Number	Value (£)	Number	Value (£)	Number	Value (£)
508 986	17 490	640 860	30 576	624 703	32 173

review, the development of Ireland's *H. gammarus* fishery is roughly divided into three periods, as Dow (1967) used for the Maine lobster (*H. americanus*) fishery: (1) prior to 1914; (2) between 1914 and 1939; (3) post 1939.

Lobster fishery prior to 1914

Andrews (1847) (in Edwards, 1971) examined Ireland's coastal fisheries during the famine and found that an excellent supply of lobsters and crabs were available along the Clare and Galway coasts. Brabazon (1848) (in Buckland & Walpole, 1877) reported that traps, called 'pots' or 'creels', made of wickerwork or of a wicker frame covered with netting, were used to catch lobsters. The Annual Reports of the Inspectors of Irish Fisheries for the years 1880–1885, refer to lobsters and crabs being taken in traps of the ink-well type and made of willow. Even in the late 1800s traps were regarded as a recent invention and prior to their introduction 'hoop' nets were used. Hoop nets required constant attention by fishermen to check if any lobsters were on the surface of the net. The substitution of pots or creels for hoops enabled a greater fishing effort as they retained the catch within and required only occasional attention from the fishermen.

Initially, lobster fishing was a part-time occupation of fishermen-farmers (Gibson, 1967). The number of traps per boat varied with location. In some places fishing boats carried as many as 100 traps; in others as few as 20 were carried (Buckland & Walpole, 1877). In 1876 the number of boats engaged in fishing was 5965, crewed by 22 773 men and 920 boys (Buckland & Walpole, 1877; see Table 1). Blake et al. (in Buckland & Walpole, 1877) found that a lobster fishery existed in many parts of the country with no general evidence of declining stocks. They held inquiries at 62 locations around the coast, only 19 of which reported diminishing catches. Nevertheless, in the majority of locations many fishermen were in favour of prohibiting the landing of lobsters less than 9 inches (23 cm) in total length (TL). In the 1870s, methods for the transportation of shellfish improved, new markets became available and it was predicted that an increase in the lobster fishery was likely (Buckland & Walpole, 1877). This prediction proved to be true and the trend in the development of lobster fishing owed much to the expansion of the English and continental markets (Gibson, 1967).

Ireland's first annual records of national lobster landings were found in the Scientific Investigations of 'The Decapoda Reptantia of the Coast of Ireland' (Selbie, 1914), in which the number of animals landed were recorded for the years 1900–1912. Ireland's DoM&NR and Dow (1980) provided data on the quantity and value of lobsters landed annually from 1926 to 1973. After 1973, landings were recorded by weight.

Over the years, fishermen and merchants have raised questions about the validity of landing figures in Ireland (pers. commun.). In the absence of reliable data to the contrary, it is reasonable to employ these landing figures as representitive of the trends of lobster landings.

The total number of lobster landings recorded between 1900 and 1912 in Ireland showed a steady increase from around 128–228 tonnes per annum. The numbers of animals landed are compared with those in England/Wales and Scotland for one year (1912) in Table 2 (Selbie, 1914) to illustrate the similarities in landings and discrepancies in value of the fishery between Britain and Ireland. For example, one pound sterling in 1912 could purchase 21 lobsters in Britain and up to 29 in Ireland. The landing statistics showed a slight but steady decline before World War I for England, Wales and Scotland (Allen et al., 1926), but unfortunately no data have been found for the fishery in Ireland for the period from 1912 to 1926.

Between 1919 to 1939

Figure 1 gives a representation of the trend in Republic of Ireland lobster landings from 1930 to 1999. After World War I there was an increase in the Republic of Ireland's lobster landings up to 1927. In that year 956 686 *H. gammarus* (approximately 430 tonnes) were landed representing a 300% increase from the start of the century. In the same period, Allen et al. (1926) reported a considerable increase in lobster landings recorded for the neighbouring fisheries of England/Wales and Scotland. They attributed this in large part to the 'protection' afforded the fishery due to a reduction in fishing effort because of World War I. After 1935, around the Republic of Ireland an ever-increasing proportion of the lobster catch was taken mainly during the summer by boats adapted specifically for lobster fishing to the exclusion of other activities (Gibson, 1967). Between the years 1927 and 1940/1941 there was a pronounced and sustained decline in catches to 223 917 lobsters (approximately 135 tonnes) in the Republic of Ireland.

Landings post 1939

In the Republic of Ireland, an increase in lobster landings occurred between 1940 and 1947. In 1951, an Irish minimum total length (TL) of 9 inches (23 cm) (measured from the tip of the rostrum to the end of the tail when flat) for *H. gammarus* was prescribed by the Undersized Sea-fish Order (S.I. No. 359 of 1951). Over the years further changes were made in size and other regulations to ensure a continuing supply of lobsters.

Gibson (1967) noted there was an increase in the number of motorised boats used to fish around the Irish coast from 1954. Though from approximately 1947 to 1967 there was no change in the number of pots fished (Gibson, 1967). He also reported that the weather in 1959 was the best in 20 years and Irish lobster landings reached a peak for the decade of nearly 350 tonnes in that year (Gibson, 1967). However, a dramatic decline (44%) in landings followed until 1963.

Subsequently there has been an apparent increase in Irish lobster landings, with intermittent troughs until 1984, as shown in Figure 1. Recorded landings after this period have been somewhat erratic and in 1994 they reached an all time high of 715 tonnes. In 1997, landings fell to 513 tonnes.

Irish legislation and control measures

Currently fishery management in the Republic of Ireland is operated through a centralised system with the DoM&NR acting as the responsible enforcement authority. Ireland's legislative framework governing regulation and management is provided by the Fisheries Acts 1959–1997. Part XIII of the 1959 Act regulates access to fishing grounds, registration and licensing of fishing vessels, conservation, fisheries control and monitoring. The EU's Multi-annual Guidance Programme (MAGP) objective was established to achieve sustainable fisheries (BIM, 1999). The chronological development of fishery regulations relating to the Republic of Irelands lobster fishery are given in Table 3:

Until 1992, the only management of Irish lobster stocks was a minimum size 240 mm (total length) or 85 mm (carapace length) and a ban on the capture of lobsters by SCUBA diving.

The catalysts in the Republic of Ireland leading to the national adoption of the 'V-notch' conservation measure were complex and a brief overview is presented here. Concern by fishermen about declining catch per unit effort (CPUE) led to the establishment of lobster hatcheries in 1992 (Shellfish Research Laboratory, Carna) and 1994 (South Wexford Lobster Fishermen's Co-op., Carne). These hatcheries acted as focal points for seminars, conferences and information exchange. Education and debate were also provided at local meetings around the coast by industry and scientific representatives from USA (Maine) and Ireland. These workshops promoted the idea of using conservation measures to enhance lobster fisheries.

The voluntary conservation practice of protecting selected mature lobsters by 'V-notching' their tail fan began on a limited scale in 1993. In 1994 legislation to protect 'V-notched' lobsters was introduced in Ireland at the bequest of fishermen, primarily those belonging to the South Wexford Lobster Fishermen's Co-op. That regulation required that all lobsters with a 'V-notch', or any form of mutilation to their tail fan must be returned to the sea. Predominantly, ovigerous female lobsters have been notched, with non-ovigerous females and some males to a lesser extent. EU and governmental financial support served to give impetus to 'V' notch programs. Also in that year, regional lobster fishing associations and Co-operatives were formed around Ireland and the Irish Lobster Association (ILA) was established to represent these organisations. Later in 1994 the Maine Lobster Association

Table 3. Legislation and control measures regulating the Republic of Ireland's *H. gammarus* fishery

Year	Regulation	Description
1951	Undersized sea-fish (crabs and lobsters) Order	This statutory instrument introduced a minimum total length for lobsters of nine inches (230 mm).
Fisheries Acts, 1959 to 1964.	By-law No. 533 (S.I. No. 533 of 1964).	This act prohibits the taking or attempting to take any shellfish by means of "skin diving".
1963	Undersized sea-fish (lobsters) Order, 1963 (S.I. No. 52 of 1963).	Concern about the taking of undersized berried females, led to the conclusion that a minimum size (83 mm, carapace length) was the best method of conservation.
1980		In 1980 an 85mm minimum carapace length (CL) size regulation was introduced. The main practical advantage in measuring the carapace is that it may be done with great accuracy and considerable speed" (Gibson, 1961). CL is measured from the rear edge of the eye socket to the midpoint of the posterior edge of the carapace.
1994	Lobster (conservation of stocks) Order, 1994 (S.I. 304).	This regulation requires the return to the sea of all lobsters that have a "V" notch in one or more of their uropods or telson.
1995	Sea Fisheries (minimum size of lobster) Order, 1995 (S.I. No. 162 of 1995).	The effect of this order is to prescribe and adopt Commission Regulation (EEC) No. 345/92 of 27th January 1992 and in particular Article 5, Annex IV except insofar as it relates to the minimum size of lobsters by overall length.
Commercial fishing for lobsters in Ireland is also subject to EU regulation as outlined below:		
2002	EU Regulation 3094/86- technical conservation regulation which applies to the landings of crustaceans in European waters.	Minimum landing sizes are applied to lobster (*H. gammarus*) of 85 mm carapace length and 240 mm overall length (except for Skagerrak and Kattegat). A new EU regulation creates a legal minimum size of 87 mm CL on 1st January, 2002 (Shellfish News, 1999).
National control mechanisms with regard to fishing vessels.		
1983	Fisheries (Amendment Act 1983).	An Act to make further provisions as regards sea-fishing and for that purpose to amend and extend the Fisheries Acts, 1959 to 1980, and to make certain provisions regarding registration under the mercantile Marine Act, 1955 and entry in the fishing boat register under part IV of the Merchant Fishing Act, 1984.
1989	Merchant Shipping (registry, lettering and numbering of fishing boats) Regulations, 1989. SI. No. 344 of 1989.	This order established a register of fishing vessels. In order to register under the 1989 Regulations, vessels are required to hold a Sea Fishing Boat Licence under Section 222B of the Fisheries (Consolidation) Act, 1959 (BIM, 1999).
	EU Council Regulation 3760/92 provides for the licensing of all fishing vessels.	

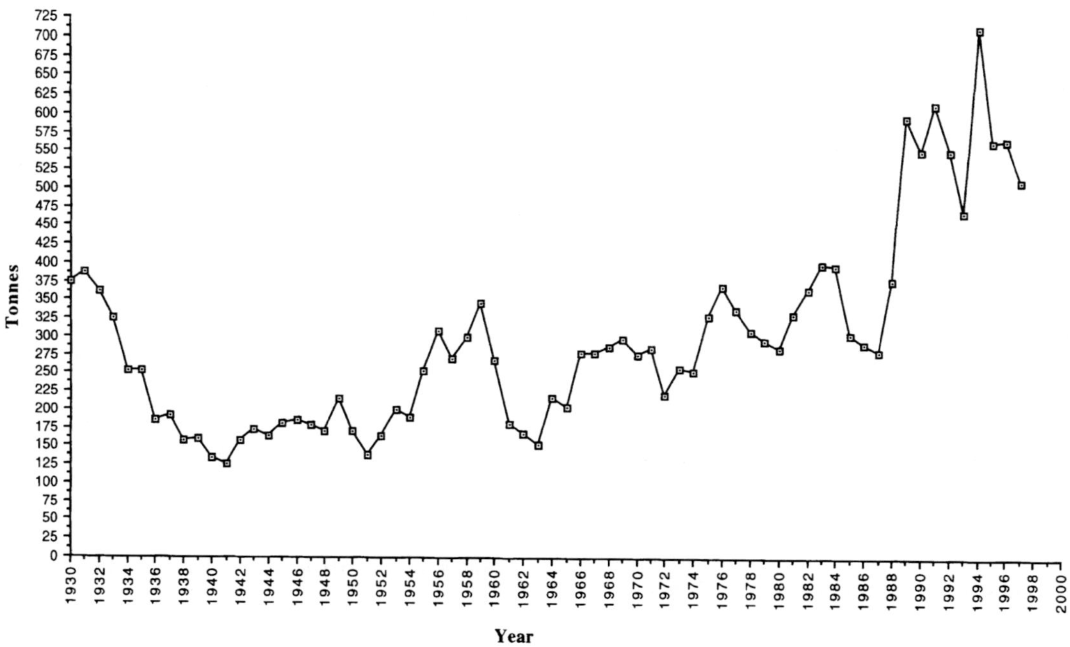

Figure 1. Landings (1930–1999) of *H. gammarus* in Ireland (Dow 1980 and DoM&NR data).

(MLA) invited Irish lobster fishing representatives to observe the Maine fishery and its regulations in operation. On their return to Ireland these representatives further strengthened the national impetus for conservation in combination with a lecture circuit of Ireland by the director of the Maine Lobster Institute (University of Maine, Orono).

The 1995 Sea Fisheries Order adopted 85 mm CL as the sole measure for minimum size abandoning the dual measurement of total length. Part of the reasoning behind this move could be ascribed to Watson (1974), who found that the flexible abdomen of a lobster may be stretched, adding up to 6mm to the total length.

More recently in Ireland there has been interest in establishing a forum for debate on management strategies. In 1999 Bord Iascaigh Mara (BIM, Ireland's sea fishery board) published a report that proposed a national Inshore Fisheries Advisory Committee (IFAC) be established. The committee includes representatives of BIM, DoM&NR, Marine Institute (National body responsible for undertaking, co-ordinating and promoting marine research in Ireland), Central and Regional Fisheries Boards and the Fishing Industry. The brief of this committee is to advise the Minister for the Marine on "policy and strategies for the inshore sector and to act as a consultative/liaison forum". In addition to establishing the IFAC, Local Inshore Fisheries Development Committees (LIFDC) would bring together the many groups that participate in the inshore sector. These include existing fishing Co-ops, and local management groups such as the proposed Co-ordinated Local Aquaculture Management System (CLAMS), which all would be important participants of these local structures (BIM, 1999). Control and enforcement of inshore management measures is provided by DoM&NR and the Central and Regional Fisheries Boards with assistance from the Naval service and Air Corps (BIM, 1999).

In summary, the legislation governing lobster (*H. gammarus*) fishing in the Republic of Ireland to date is comprised of:

- a minimum carapace length of 85 mm;
- a ban on landing 'V-notched' lobsters or those with mutilated tails;
- a prohibition on taking shellfish by SCUBA;
- all commercial fishing vessels around Ireland must be licensed (including boats used for lobster fishing).

For *H. gammarus* there is, at present no legislation in Ireland:

- to prohibit the taking of berried females;
- for an upper size limit;

- for escape vents in traps for juveniles or undersized lobsters;
- to limit the number of traps and or type of fishing gear;
- to limit the number of fishers;
- to limit the seasons (although nature partially enforces this with weather conditions);
- to provide for biodegradable panels in traps to reduce ghost-fishing;
- to provide log books detailing fishing effort and returns.

Value of Irish landings for *H. gammarus*

The clawed lobster is a recognised high value seafood resource and Irish *H. gammarus* are marketed live with the majority of the catch sold for export. The total annual landed values (IR£) of the Irish *H. gammarus* catch from 1926 to 1999, shown in Figure 2 indicates that returns began to increase in the late 1950s and early 1960s. During the 1990s there has been a considerable variation in the total landed value (e.g. 1993 IR£million 4.2 and 1994 IR£million 6.3).

Catch per unit effort (CPUE)

Data relating to CPUE over the last 100 years are difficult to find. However fishermen believe they can detect the impact of their activities through the progressive reduction in number of lobsters per trap hauled and decrease in average size of lobster landed. Larger animals have become rarer and the size distribution of the catch has become clustered around the minimum legal size (Gibson, 1967; Watson, 1974).

Gibson (1965, 1967) found, in the 1960s, a linear relationship between time spent fishing and yield of lobsters in Ireland. Suggesting a maximum sustainable yield had not been reached. Nolan et al. (1997) made a comparison of the 1995 catch per trap in Co. Wexford with that reported by Gibson (1970) and found that there had been a 3-fold reduction over the 36-year period. O'Driscoll (1998) also indicated that trials carried out in the mid-1960s showed that the Irish fishery yielded one lobster for every three trap hauls while in 1997 one lobster was produced for every 10 trap hauls. This decline in CPUE may be explained by increased lobster fishing effort due to the decline in salmon drift-netting (BIM, 1999), the technological innovations in fishing practices, and the catastrophic decline in spiny lobster (*P. elephas*) landings resulting in increased economic pressure on fishermen to bolster their incomes. In the 'Irish Inshore Fisheries Sector Review and Recommendations May 1999', respondents to a survey indicated a 15.4% decrease in CPUE per annum for the inshore fishing sector over 11 years (BIM, 1999).

Table 4. Gross registered tonnage (GRT) of Irish fishing vessels in 1995 (FAO, 1998)

GRT	0–4.9	5–24.9	25–49.9	50–99.9	100+
Number	347	559	146	175	126
Tonnage	955	4861	5178	12 792	34 098

Number of lobster vessels and fishermen

Since its inception, the Irish lobster fishery has largely been a small vessel fishery. The total number of vessels of all kinds engaged in lobster fishing in 1969 was 671 (Gibson, 1970). In 1973, 675 lobster fishing vessels were recorded (Bhatnagar, 1974). These vessels fished 42 259 traps of which 51% were for lobster and the remainder for crawfish. After 1973, data on the number of vessels and fishermen actively fishing for lobsters is less clear. In 1981, it was reported that there were a total of 3095 fishing vessels in Ireland and 8740 fishermen, of which 3464 were permanently employed in fishing while 5276 fished part time. It was estimated that 1500 of these vessels were less than 15 m in length. These figures suggest that the majority of fishing vessels operate in the inshore area. Details of the Irish decked fishing fleet are presented in Table 4.

In 1999 it was estimated that there were over 1744 vessels (779 registered and a further 965 applications for registration) less than 15 m in length in the inshore fishery sector (inshore is defined as within 19 km of shore). It includes demersal, pelagic, shellfish, salmon and sea angling fisheries (BIM, 1999). Of this number 1314 were <10 m. The results of the 1999 BIM survey indicate that there were approximately 3700 jobs in the inshore sector of which approximately 2700 were full time.

Irish lobster fishery in European context

In Norway, Sweden and Denmark there have been dramatic declines in the lobster fisheries this century

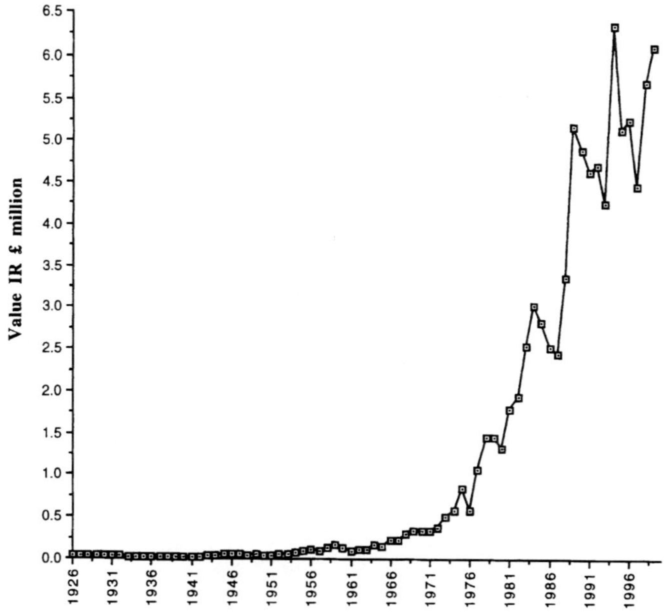

Figure 2. Total landings value of *H. gammarus* (1926–1999) for the Republic of Ireland (DoM&NR).

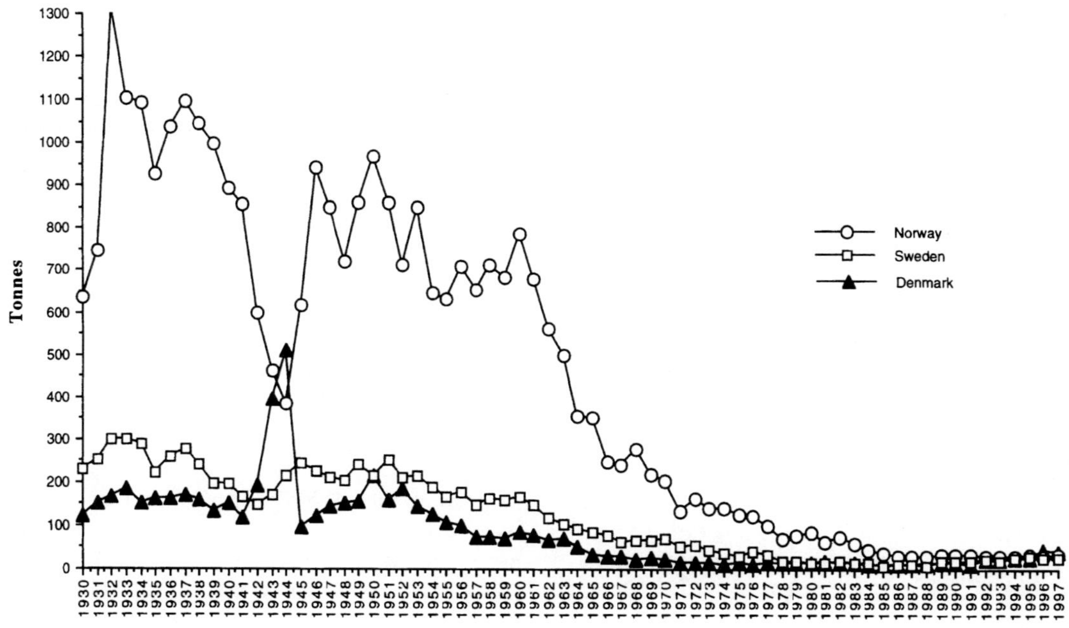

Figure 3. Annual *H. gammarus* landings (tonnes) for Norway, Sweden and Denmark (1930–1996) (compiled from: Dow 1980, Bulletin Statistique, and FAO 1995 & 1998).

Table 5. Maximum and minimum landings (tonnes) of *H. gammarus* by year, in selected European countries (compiled from: Dow 1980, 1945–70 Bulletin Statistique, 1971–95 FAO Yearbook [Table B43])

Country	Year	Max. tonnes	Year	Min. tonnes
Norway	1932	1313	1992	28
Sweden	1933	299	1985	9
F. R. Germany	1996	17	(0 tonnes frequently recorded)	
Denmark	1944	511	1991	10
Netherlands	1934	32	1994	0
Scotland	1948	997	1941	297
N. Ireland	1949	30	1997	8
Ireland	1994	715	1941	126
Isle of Man	1982	34	1997	0
England/Wales	1997	836	1980	311
Channel Isles	1996	260	1974	30
France	1958	625	1985	176
Spain	1980	206	1981	4
Portugal	1968	229	1994	2

NB: when compiling lobster catch data some variation was found in landing figures between sources. Generally these differences do not significantly affect the trend in landings except for the following, when discrepancies greater than 50 tonnes per annum were noted; Denmark, 1945–1948, England & Wales, 1945–1950; Scotland, 1945–1946 and 1950–1957; and Ireland 1949. For these years, data used to compile figures and tables are from Bulletin Statistique.

(see Fig. 3). The peaks and troughs of *H. gammarus* landings in selected European countries are given in Table 5, below and total annual landings in Appendix 1.

In Norway, yearly catches ranged from 1313 tonnes maximum (1932) to less than 30 tonnes in 1990 (van der Meeren, 1990), though van der Meeren and Tveite (1998) regard the landing statistics as unreliable. They estimate that there are currently more than 400 000 small boats (i.e., not registered as fishing boats) along the Norwegian coast and a large proportion of Norwegians spend their vacations fishing (van der Meeren & Tveite, 1998). In addition to this unrecorded fishing pressure, they state that the summer sea temperature along much of the coast of Norway may be below 15°C, which is marginally low to support lobster larval and juvenile development (van der Meeren & Tveite, 1998).

Other northern European countries and regions have had declines in recorded lobster landings similar to those of Norway. For example Sweden's landings peaked in 1933 at 299 tonnes and Denmark's in 1944 at 511 tonnes. In 1997 landings of 27 and 39 tonnes were recorded for each country, respectively. Landings of lobsters around the Shetland Islands reached a peak of 179 tonnes in 1962 and declined to 12 tonnes in 1997 (G. Arthur, North Atlantic Fisheries College, Shetland pers. commun.).

In the Mediterranean, the decline in *H. gammarus* catches has also been equally catastrophic. Dow (1980) reported that landings of *H. gammarus* in Italy fell from a peak of nearly 1000 tonnes in 1962 to zero recorded landings after 1965. Yugoslavian and Algerian catches were each around 100 tonnes from which they declined to zero (Dow, 1980). The most phenomenal rise and fall of landings of *H. gammarus* recorded are those of Turkey where there was an increase in landings from 500 tonnes in 1957 to 2200 tonnes in 1964, and a subsequent decline to zero (Dow, 1980).

When compared to those reported for both Scandinavian and Mediterranean countries landings from England/Wales, Scotland and France appear relatively static, though exhibiting the characteristics of fully exploited stocks (Fig. 4).

H. gammarus compared to *H. americanus* landings

Figure 5 illustrates the overall trend of landings of *H. gammarus* in Europe, (Norway, Sweden, Denmark, Germany, Belgium, Netherlands, France, Spain, Portugal, England and Wales, Channel Isles, Scotland, Isle of Man, Northern Ireland and the Republic of Ireland).

Total European lobster landings average approximately 2400 tonnes and they may be significantly under-recorded or under-reported (Bannister, 1999). A steady decline of the *H. gammarus* stocks can be seen in Figure 5, from the 1030s until the 1990s, whereas Figure 6 demonstrates the contrasting increase in Canadian and USA landings since the 1970s (Appendix 1). In 1994 total *H. gammarus* landings were 2362 tonnes compared to a total of 71 663 tonnes for *H. americanus*. In the North American fishery up to 1992 there was also a concurrent and dramatic increase in effort. For example, Addison & Fogarty (1992) found that landings in the State of Maine were slightly higher than those of a century ago and the number of traps fished had increased by a factor of 20. They speculated that expansion of the fishery and improved technology were contributing factors to improved landings. Miller (1994) felt that the increase in landings of *H.*

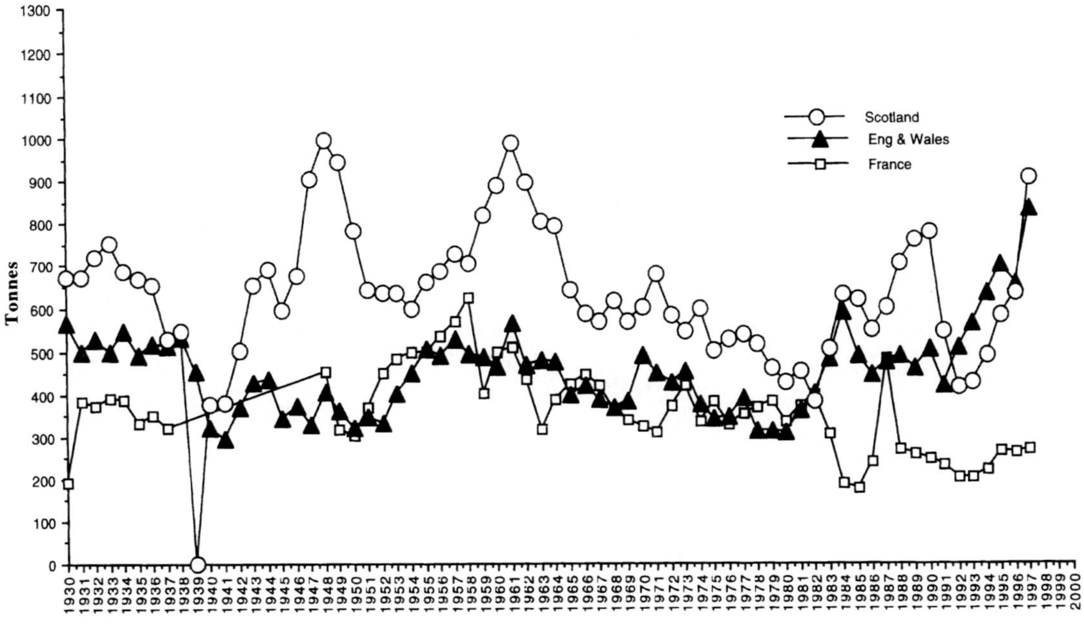

Figure 4. Annual *H. gammarus* landings (tonnes) for Scotland, England & Wales, and France (1930–1997) (Dow (1980); 1945–70 Bulletin Statistique (1995); HMSO 1998; 1971–95 FAO 1995 & 1998).

Figure 5. Annual total landings (tonnes) of *H. gammarus* in Europe (1930–1994) (Dow (1980); 1945–70 Bulletin Statistique (1995); HMSO 1998; 1971–95 FAO 1995 & 1998). NB: this figure excludes historical Mediterranean (Italy, Yugoslavia, Turkey and Algerian) landings reported by Dow (1980).

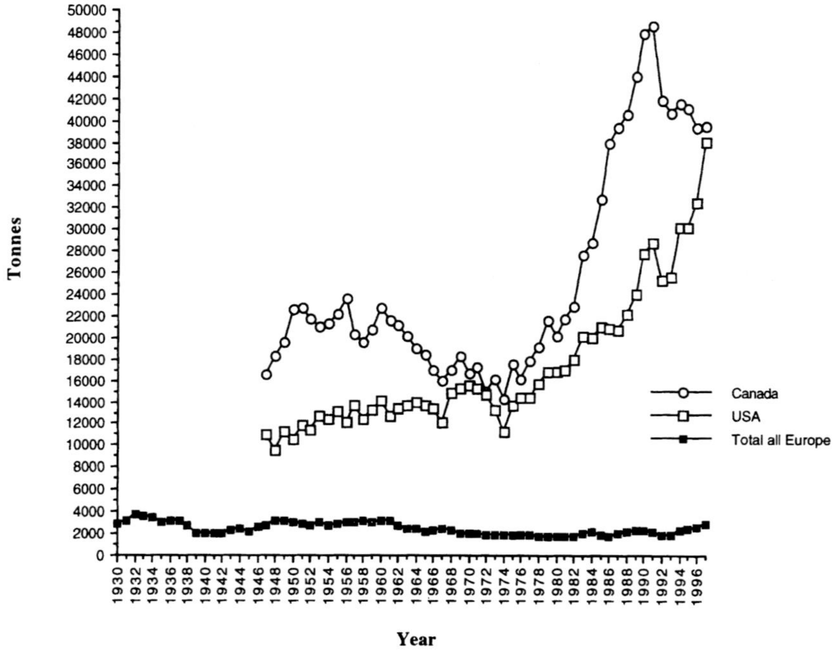

Figure 6. Annual landings (tonnes) of lobster (*H. gammarus* and *H. americanus*) in Europe, Canada and USA (Dow (1980); 1945–70 Bulletin Statistique (1995); HMSO 1998; 1971–95 FAO 1995 & 1998).

americanus in the 1980s and early 1990s was a result of enhanced abundance rather than purely increased fishing effort. His explanation is that the cause was a widespread, but unidentified, environmental factor.

Discussion

Management practices of the lobster fishery in the Republic of Ireland are basically limited to a minimum size regulation, the recent (1994) protection of 'V-notched' lobsters, and ban on their capture by SCUBA. A minimum size policy is, at present, the main method of managing lobster fishing in the rest of Europe. Currently the minimum carapace length (CL) for *H. gammarus* landed in European countries is 85 mm, but in 2002 a minimum CL of 87 mm will be introduced under EU fisheries legislation.

Ireland's lobster landings appear robust as seen in the data from 1940 up to 1997. However, this is not to imply that these landing figures are sustainable. The increases in tonnage landed over this period may be due to any one or all of the following:

(1) A decrease in landing of undersized animals due to increased compliance and enforcement of minimum size regulations. Also, improved conservation probably has increased egg production, which may result in an increase in recruitment.

(2) A minimum size of 83 mm CL was introduced in 1963 (before that it was 23 cm TL (ICES, 1974)) and the subsequent increase to 85 mm CL in 1980 has contributed to increased egg production.

(3) Increase in mean seawater temperature could possibly promote growth and recruitment of lobsters (ICES, 1974). In addition, Addison & Fogarty, (1992, 1993) suggest that an increase in catchability of *H. americanus* with increased temperatures could have resulted in improved catch rates fuelling further expansion of the fishery leading to a cascading effect of increasing catch and effort.

(4) Increase of fishing effort in terms of pot number, new offshore lobster fishing areas are now being exploited of the north-west of Ireland (O. Tully, Trinity College Dublin, pers. commun.) and technology improvements (the adoption of synthetic materials used in trapping and electronic fishing aids).

(5) Alteration in the methodology used to assess landing figures by the DoM&NR, e.g., up to 1973 landings of lobsters were recorded by animal number. Since then they have been recorded by weight,

although this is an unlikely explanation for such a sustained increase.

Conclusion

Landings of *H. gammarus* in the Republic of Ireland exhibit signs of a fully or overexploited stock with reported declining catch per unit effort. In Europe there has been an obvious decline of total landings. Historically, management practices have not proved successful in countering declining catches. In contrast, North America with many more rules and regulations governing the fisheries has dramatically increasing landings, which does not imply that the fisheries there are secure from changes in the longer term. Appropriate management must be adopted in Ireland/Europe if stocks of *H. gammarus* are to be stabilised.

The value of *H. gammarus* landings in Europe may appear to be small when compared to the value of pelagic fish catches but they are a source of revenue and employment throughout their distribution. Lobster fisheries are representative of crustacean fisheries in general and offer the potential for the introduction of enlightened management strategies. In short, a well-managed lobster fishery may be used as a role model for sustainable crustacean fisheries management on a broad scale. However, numerous regulations are of little value unless sensibly constructed on relevant biological data and adequately enforced. It is hoped that the lobster fishery will receive appropriate consideration from national bodies in order not only to maintain it but also to improve and enhance it.

Acknowledgements

We would like to thank all those that helped in compiling the lobster landing statistics, these include those in Irelands DoM&NR, Ann-Lisbeth Agnalt (Inst. Marine Research Nordnesgt), Daniel Latrouite (IFREMER), Steve Lovewell (CEFAS)) and acknowledge the financial support from Taighde Mara Teo. and FAIR CT-96-1775 during this study. In addition we would like to thank Brian Beal for his helpful comments on this paper.

References

Addison, J. & M. Fogarty, 1992. Juvenile lobster habitat limitation: What can landings tell us? Lobster Newslett. 5 (2): 10.

Addison, J. & M. Fogarty, 1993. Lobster landings reprise. Lobster Newslett. 6 (1): 10.

Allen, E. J., A. Bowman, R. Elnherst, J. Travis Jenkins, & E. S. Russel, 1926. Interdepartmental Committee on crabs and lobsters 1926. Report on lobsters. Berrill 1974, Interdepartmental Committee appointed by the Minister of Agriculture and Fisheries and the Secretary for Scotland to enquire into the Crab and Lobster Fisheries. Published by His Majesty's Stationary Office. 34 pp.

Bannister, R. C. A., 1999. A review of shellfish resources and their management. Dr. Walne Memorial Lecture. Thirtieth Annual Shellfish Conference 1999. The Shellfish Association of Great Britain, Fishmonger's Hall, London Bridge, London EC4R 9EL, 30 pp.

Bhatnagar, K. M., 1974. Lobster trap census, 1973. Department of Agriculture and Fisheries, Fisheries Division, Dublin 1. Fishery Leaflet Number 65. 13 pp.

BIM, 1999. Irish Inshore Fisheries Sector. Review and Recommendations. May 1999. 74 pp.

Buckland, F. & S. Walpole, 1877. Reports on the crab and lobster fisheries of England and Wales, Scotland and Ireland. also Aloysius Blake, J. Hayes and Brady, T.F. Presented to both Houses at Parliament by command of Her Majesty. London. Printed by G. E. Eyre and W. Spottiswoode.

Bulletin Statistique, 1971. Lobster (*Homarus gammarus*) landings (1945–1970) from the North East Atlantic (FAO area 27)-tonnes.

Dow, R. L., 1967. The influence of temperature on Maine lobster supply. Maine Sea and Shore Fisheries. Published by Maine and Sea Shore Fisheries under Appropriation 7745. 22 pp.

Dow, R. L., 1980. The clawed lobster fisheries. In: The Biology and Management of Lobsters, Vol. II: pp 265–315. Ecology and Management, Academic Press, 390 pp.

DoM&NR., 1994, 1997. Seafish landing records. Department of the Marine and Natural Resources. Leeson Lane, Dublin 2, Ireland.

Edwards, E., 1971. A contribution to the bionomics of the edible crab (*Cancer pagurus* L.) in English and Irish Waters. Submitted to the National University of Ireland for the Degree of Doctor of Philosophy, Zoology Department, University College, Galway, 123 pp.

FAO., 1995. The State of the World Fisheries and Aquaculture. FAO Fisheries Department of the UN, Rome, Italy.

FAO., 1998. Bulletin of Fishery Statistics. 1970, 1975, 1980, 1985, 1989–95. 35. ISSN 1014-1189.

Gibson, F. A., 1965. Catch and effort in lobster and crawfish fishery. Irelands, Department of Agriculture and Fisheries. Reprinted from the Department Report on the Sea and Inland Fisheries for the year 1965. pp. 3–7.

Gibson, F. A., 1967. Irish Investigations on the lobster (*Homarus vulgaris* Edw) 1967. Irish Fisheries Investigations Series B (Marine) No.1 1967. An Roinn Talmhaiochta agus Iascaigh, (Department of Agriculture and Fisheries). Fo-Roinn Iascaigh (Fisheries Division) Dublin. Government Publications Sales Office, G.P.O. Arcade, Dublin. 44 pp.

Gibson, F. A., 1970. Lobster trap census 1969. Fishery Leaflet No. 23. An Roinn Talmhaíochta agus Iascaigh, (Department of Agriculture and Fisheries), Fisheries Division 1970. 8 pp.

HMSO, 1998. Fish and shellfish landings in the United Kingdom (1970–1997). File srl-s:
maffstat
workbook
ewukqty.xlw. United Kingdom Sea Fisheries Statistics.

ICES, 1974. Report of the meeting of the *ad hoc* Group for the Study of the European Lobster (*Homarus gammarus*). Dublin, 12–14 March 1974. International Council for the Exploration of the Sea. C.M. 1974/K:3. Shellfish and Benthos Committee. 10 pp.

Miller, R. J., 1994. Why are there so many American lobsters? Lobster Newslett. 7 (1): 14, 15.

Nolan, F., O. Tully, S. Lysaght, & R. C. A. Bannister, 1997. A study of technical conservation measures on lobster stocks in N.W. Europe. Project no. 94/040. Final report-issued June 1997. Funded by the European Commission DG XIV. Research projects in support of the Common Fisheries Policy. 27 pp.

O'Driscoll, M., 1998. Lobster conservation & enhancement. Principal Fisheries Officer. Department of the Marine & Natural Resources. Irish Lobster Assoc. News Views. 4 (2).

Selbie, C. M., 1914. The decapoda reptantia of the coast of Ireland. Part 1. Palinura, Astacura, and Anomura (except Paguridea). Dept. of Agric. and Technical Instruction for Ireland. Fisheries Ireland. Sci. Invest. I.

Shellfish News, 1999. New National Shellfish Conservation Measures. Shellfish News 7 pp. 46. CEFAS Publication. ISSN 1363–4720.

van der Meeren, G. I., 1990. Acclimatization techniques for lobster experiences from experimental releases of juvenile lobsters to the sea. Sea ranching – Scientific Experiences and Challenges. Norwegian Soc. Aquacult. Res. pp. 95–97.

van der Meeren, G. I. & S. Tveite, 1998. Fishery and management of the lobster (*Homarus gammarus*) in Norway. Proceedings from the Seminar at Kvitsoy 1995. Havorskningstittet Institute of Marine Research: 72–76.

Watson, P. S., 1974. Investigations on the lobster (*Homarus gammarus*) in Northern Ireland – A progress report 1972–1973. International Council for the Exploration of the Sea. C.M. 1974/K:20. Shellfish and Benthos Committee. 21 pp.

Appendix 1. Annual lobster (*H. gammarus*) and (*H. americanus*) landings by country

Year	Den.	Nor.	Swe.	Ger.	Fra.	Spa.	Por.	Eng. Wal	Sco.	R.O. Ire.	N. Irel.	Isle of Man	Chan. Isle	Bel.	Neth.	Other Europe	Canada	USA
1930	122	636	229		193			564	673	374					23			
1931	150	745	250		383			499	673	388					17			
1932	167	1313	298		374			527	721	361					19			
1933	183	1103	299		393			498	753	325					32			
1934	150	1093	287		389			546	688	253				4	32			
1935	163	926	223		334			546	688	253				5	22			
1936	162	1037	259		352			517	653	185					23			
1937	170	1098	278		322			515	529	191				5	20			
1938	160	1046	239					533	546	157					15			
1939	134	998	197					456		159					13			
1940	150	892	194			40		323	376	135								
1941	118	856	167					297	379	126								
1942	191	597	148					368	502	157								
1943	394	463	171					430	655	173								
1944	511	384	213			18		434	691	164	18			0	2			
1945	95	615	245	0		12		343	595	180	15			0	5			
1946	122	941	224	0		20		374	676	185	20			0	5			
1947	145	850	211	0		20		330	903	179	15			0	6		16500	10900
1948	150	720	203	9	455	30		408	997	171	25			1	7		18300	9500
1949	156	861	239	0	316	48		361	947	214	30			1	7	45	19600	11200
1950	216	969	215	7	304	19		320	784	170	14			1	13	37	22600	10500
1951	157	862	252	9	368	29		346	643	139	10			2	17	32	22700	11800
1952	186	712	210	3	499	32		331	635	164	11			3	12	37	21700	11400
1953	145	848	216	11	485	37		403	635	200	18			3	10	36	21000	12700
1954	124	648	188	4	499	34		450	597	189	17			3	9	30	21200	12300
1955	108	632	167	1	497	34		506	662	253	19			3	7	30	22100	13100
1956	101	708	178	1	537	32		492	688	308	21			3	7	35	23500	12000
1957	74	655	148	4	568	53		528	728	270	17			2	6	29	20200	13700
1958	75	714	164	4	625	68		495	704	300	28			4	8	1130	19500	12400
1959	72	684	160	6	401	57		489	819	347	12			4	7	30	20700	13200
1960	85	787	168	7	497	37		465	890	267	18			4	8	36	22700	14100
1961	76	681	147	6	509	26		565	991	180	13			8	5	34	21600	12700
1962	67	561	120	8	437	24		469	898	167	15			9	1	35	21100	13400
1963	71	498	105	0	318	5		480	805	153	17			10	19	50	20100	13700
1964	50	353	92	3	388	23		477	793	217	18			7	5	31	19000	14000
1965	35	350	86	1	426	20		398	643	205	15			3	12	219	18400	13700
1966	30	248	78	1	446	20		420	589	278	11			4	16	262	17000	13400

Continued on p. 62

Appendix 1. Continued

Year	Den.	Nor.	Swe.	Ger.	Fra.	Spa.	Por.	Eng. Wal	Sco.	R.O. Ire.	N. Irel.	Isle of Man	Chan. Isle	Bel.	Neth.	Other Europe	Canada	USA
1967	30	239	64	1	422	161	190	387	568	279	12			2	14	258	15900	12100
1968	24	276	66	1	361	99	229	371	616	287	12			1	23	39	17000	14800
1969	25	218	66	1	340	18	229	383	568	298	13			0	32	72	18200	15300
1970	22	202	72	1	324	47		491	602	277	13			0	15	10	16600	15500
1971	15	133	51	1	310	20		451	678	285	23			0	14	49	17300	15300
1972	16	161	54	1	373	16		429	585	221	22			0	22	82	15000	14600
1973	13	142	45	1	420	10		455	545	258	28	7	30	0	0	77	16100	13200
1974	11	140	38	0	336	12	23	377	600	253	22	9	40	0	0	77	14251	11152
1975	14	127	31	0	385	14	4	342	503	330	21	6	35	0	0		17487	13698
1976	12	121	41	1	328	29	13	346	528	369	12	5	50	0	0		16028	14398
1977	14	100	33	1	353	69	1	393	541	339	15	4	79	0	0		17802	14382
1978	16	70	19	1	368	62		313	517	309	18	3	75	1	0		19177	15612
1979	16	76	19	1	383	21	2	315	461	295	16	2	70	0	0		21579	16867
1980	15	84	15	0	337	206	26	311	428	285	20	1	65	0	0		20089	16761
1981	17	63	15	0	372	4	17	361	453	331	15	34	45	16	0		21696	17007
1982	16	74	17	1	398	13	10	403	383	366	13	3	171	0	0		22805	17892
1983	14	58	16	0	306	10	9	483	506	400	17	8	191	2	0		27655	20052
1984	10	43	16	0	189	8	8	591	630	398	14	2	143	4	0		28694	19943
1985	10	38	9	0	176	33	5	491	619	305	12	2	107	0	0		32797	20934
1986	10	31	11	0	241	6	6	446	550	292	14	1	132	0	0		37970	20890
1987	10	30	10	0	485	5	6	475	603	281	17	1	160	0	0		39430	20665
1988	15	28	9	1	271	7	5	492	706	379	15	3	125	0	0		40512	22064
1989	14	34	19	0	259	10	6	460	760	595	17	5	129	0	0		43957	24007
1990	14	33	18	1	247	13	6	507	781	553	11	4	120	0	0		47857	27677
1991	10	35	19	0	234	10	5	421	546	615	24	1	128	0	0		48493	28729
1992	21	28	22	0	202	14	5	510	416	552	16	3	129	0	0		41827	25307
1993	22	28	18	2	202	22	4	564	430	470	14	12	123	0	0		40688	25634
1994	26	30	26	2	223	17	3	636	491	715	22	14	114	0	0		41537	30126
1995	27	34	29	9	266	25	2	701	583	565	14	0	260	3	0		41059	30122
1996	44	30	26	17	264	20	3	656	634	567	13	26	221	0	0		39369	32496
1997	39	35	27	0	269	30	2	836	907	513	8				0		39500	38066
1998								800		611								
1999										598								

Data compiled from Ireland's DoM&NR, Dow 1980, Bulletin Statistique 1971, HMSO 1998, HMSO 1998, and FAO yearbook 1998. However, some variation was found in landings figures between sources. For these yeas HMSO 1998 and Bulletin Statistique were used.

Hydrobiologia **465:** 63–72, 2001.
G. Burnell (ed.), Coastal Shellfish – A Sustainable Resource.

Searching for the early benthic phase (EBP) of the European lobster: a trans-European study of cobble fauna

Adrian Linnane[1], Brendan Ball[1], John P. Mercer[2], Ronan Browne[2], Gro van der Meeren[3], Halldis Ringvold[3], Colin Bannister[4], David Mazzoni[5] & Brian Munday[1]
[1]*The Martin Ryan Marine Science Institute, National University of Ireland, Galway, Ireland*
[2]*National University of Ireland, Galway, Shellfish Research Laboratory, Carna, County Galway, Ireland*
[3]*Institute of Marine Research, Austevoll Aquaculture Research Station, N-5392 Storebo, Norway*
[4]*CEFAS, Lowestoft Laboratory, Pakefield Road, Lowestoft, Suffolf NR33 OHT, U.K*
[5]*DIPROVAL - via del Guasto 5/B, Universita Degli Studi di Bologna, 40126 Bologna, Italy*

Key words: European lobster (*Homarus gammarus*), recruitment, cobble fauna, scientific diving

Abstract

The re-stocking or enhancement of Homarid populations has been the focus of several research projects in recent years. However, despite being able to successfully rear lobsters in captivity, there are few reports of newly settled European lobsters (*Homarus gammarus*) in the wild. This is of serious concern for fishery managers when trying to ascertain if restocking or enhancement programmes are successful at the fishery level. Based on published work from the western Atlantic seaboard, cobble and boulder substrata have been identified as an important nursery habitat for early benthic phase (EBP) American lobsters (*Homarus americanus*). This paper, based on an E.U. funded research project (LEAR), describes the results of quantitative airlift suction sampling from cobble habitat in Norway, Ireland the U.K., and Italy. In total, 67 m^2 of cobble was sampled, but no EBP lobsters were found. Overall, crustaceans and molluscs were the most abundant species in all countries. While the collective densities of animals per $m^2$of cobble were similar to that of the United States, the species diversity in Europe was noticeably higher. Among the crustacea, the reptant decapods dominated at all sites. These findings tend to support the modern day 'exclusion hypothesis' that competition and predation from other species limits the successful recruitment of EBP European lobster to the benthos. The results highlight the need to go beyond the single species approach to understanding the factors that influence the recruitment of *Homarus gammarus*.

Introduction

The European lobster *Homarus gammarus* (Linnaeus) has a broad geographical distribution which extends from the Arctic Circle to Morocco. In its northern range, it occurs from the Lofoten Islands in northern Norway to southeastern Sweden and Denmark, but is apparently blocked from inhabiting the Baltic Sea by lowered salinity and temperature extremes. Its distribution southwards extends along the mainland European coast and around Ireland, the United Kingdom and the Azores, to a southern limit of about 30° north latitude on the Atlantic coast of Morocco. The species also extends, though less abundantly, throughout the coastal and island areas of the Mediterranean Sea and has been reported from the westernmost end of the Black Sea in the Straits of Bosporus (Dybern, 1973; Holthuis, 1974).

Despite being relatively common in coastal waters, however, the early life history and recruitment processes of the European lobster are not well understood. While the larval and adult stages have been identified in the wild (Tully & Ó Céidigh, 1987), data on the distribution and abundance of newly settled or early benthic phase (EBP) individuals are sparse. This paucity of information is of major concern to researchers as it is essential for both current lobster re-stocking programmes (Addison & Bannister, 1994; Gendron, 1998; Agnalt, et al. 1999) and long-term stock and fisheries management strategies.

On the western Atlantic seaboard, the spatial and temporal patterns of benthic recruitment within the American lobster *Homarus americanus* (Milne-Edwards) have recently been identified (Wahle & Incze, 1997). Advances in benthic sampling, such as the airlift suction sampler (Incze & Wahle, 1991), have allowed researchers to ascertain that EBP American lobsters have a strong association with shelter-providing habitat, in particular, cobble-boulder substrata (Hudon, 1987; Wahle & Steneck, 1992; Wahle & Incze, 1997).

Building on the success of the airlift sampler at identifying EBP American lobsters, suction sampling of cobble-boulder habitat in European waters was undertaken in 1997. The aim of the census was to locate EBP European lobsters while at the same time determine the effects of competition, predation and habitat carrying capacity on these stages. It should be noted that this is the first known quantified census of cobble habitat in European waters.

Sites

Four European countries were involved in the sampling programme; Ireland, Norway, Italy and the U.K. Within each country, four cobble-boulder locations suitable for suction sampling were identified by means of reconnaissance dives (Table 1).

Norway

Sampling was undertaken on the west coast of Norway from August to October 1997, in an area south of Bergen at Bjoernafjord. The four sites were located in depths ranging from 6.5 to 14 m. The substratum consisted of cobble/boulder type habitat overlying sand or bedrock. The temperature ranged from 12.5 to 19 °C during the sampling period.

Ireland

Field sampling was carried out on the west coast of Ireland in September 1997, at an area known as Aughinish Shoal in Inner Galway Bay. The site was situated along the base of a reef where the water depth ranged from 9 to 14 m. Each of the four sampling locations were comparable in their cobble size and composition, with the underlying substratum consisting of either sand or bedrock.

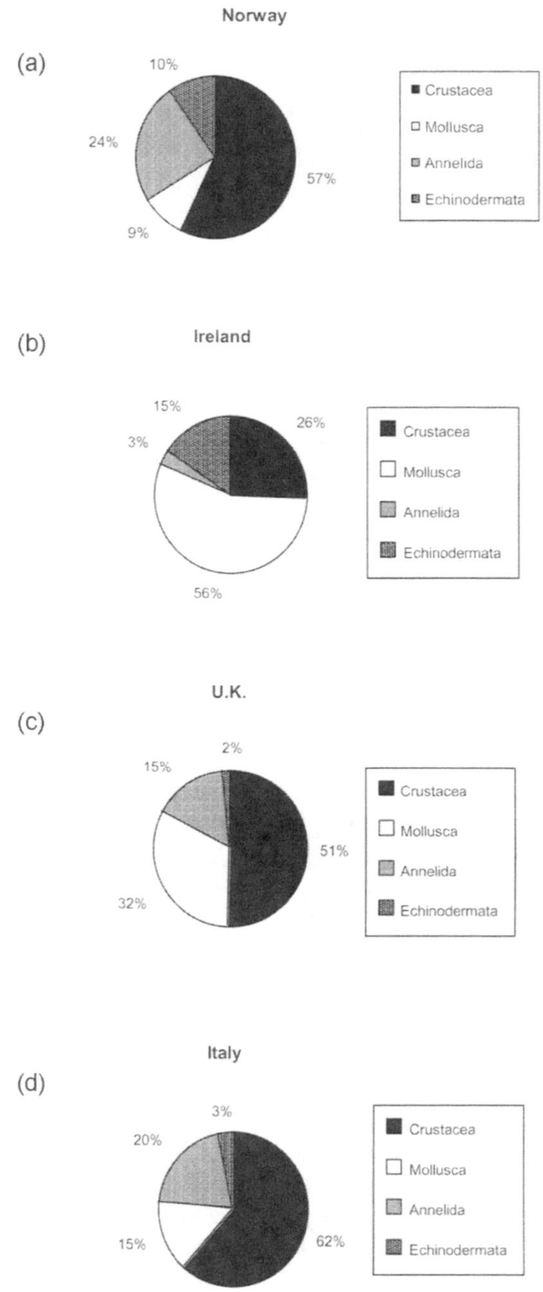

Figure 1. (a–d). The percentage breakdown of the major animal phyla in each European country as calculated from density per m^2.

U.K

Field work took place in Bridlington Bay on the English east coast, north of the River Humber in October,

Table 1. Details of the four site locations within each European country

	Ireland (Aughinish Shoal)	Norway (Bjoernafjorden)	U.K. (Bridlington Bay)	Italy (Elba Island)
Depth (m)	9–14	6.5–14	12–17	9–13
Substratum	cobble on sand or bedrock	cobble/boulder on sand or bedrock	cobble/boulder on muddy clay or shellsand	cobble on sand or shingle
Time of year	September	August–October	October	June–July
Location				
A	53° 10′ · 62 N 09° 07′ · 8 W	60° 09′ · 01 N 05° 34′ · 09 E	53° 58′ · 20 N 0° 10′ · 55 W	42° 44′ · 10 N 10° 19′ · 30 E
B	53° 10′ · 57 N 09° 06 · 88 W	60° 13′ · 01 N 05° 18′ · 08 E	53° 55′ · 30 N 0° 07′ · 10 W	42° 42′ · 40 N 10° 25′ · 23 E
C	53° 10′ · 62 N 09° 06′ · 88 W	60° 06′ · 02 N 05° 16′ · 02 E	53° 55′ · 65 N 0° 05′ · 75 W	42° 42′ · 10 N 10° 23′ · 60 E
D	53° 10′ · 56 N 09° 07′ · 28 W	60° 08′ · 06 N 05° 10′ · 03 E	53° 53′ · 00 N 0° 06′ · 60 W	42° 44′ · 30 N 10° 19′ · 45 E

1997. Due to strong tidal flows in the area, diving was restricted to little more than 60–90 min on either side of slack water at depths of 12–17 m. The substratum consisted of cobble and boulders overlying muddy clay or shellsand.

Italy

Suction samples were taken during the period June to July 1997, at a site north east of Elba Island on the west coast of Italy. The water depth ranged from 9 to 13 m with the substratum consisted of cobble on a sand or shingle bottom.

Materials and methods

The sampling protocol used in the study was broadly similar across all four countries. Within Ireland, Norway and Italy, 38, 42 and 40, 0.5 m^2quadrats were sampled respectively. Due to logistic problems within the U.K., the number of quadrats sampled was 14. To ensure standardisation, dive teams in each country utilised identical sampling gear. In addition, a sampling standardisation exercise was carried out in Ireland by dive teams from each participating country, prior to commencement of the sampling programme.

Benthic sampling

Benthic sampling of cobble habitat was conducted using a diver-operated airlift suction sampler (Incze & Wahle, 1991; Wahle & Steneck, 1991). The airlift consisted of a 170 cm length of 7.0 cm diameter black PVC tube, with a SCUBA cylinder supplying air fitted 10 cm above the mouth. The other end of the tube was affixed to a 1 mm mesh nylon bag that could be removed, closed and replaced underwater.

The suction sampler was operated by 2 divers and was used in conjunction with a 0.5 m^2quadrat constructed from buoyant PVC piping, a 30 cm high sheet of 1 mm mesh curtain and a heavy chain base. This design prevented the escape of mobile decapods from the quadrat.On sampling, the quadrat was placed haphazardly on a suitable cobble patch. The cobble stones were then removed by hand by one diver while the second operated the airlift, ensuring that the mouth of the tube reached into the interstitial spaces of the substratum. Divers worked slowly until all of the interstitial spaces were exposed and sampled. Periodically, sample bags were returned to the surface where they were analysed immediately or preserved for future examination.

Faunal preservation and analyses

Prior to identification, samples were preserved by freezing or storage in 70% alcohol. All animals were identified at least to class level. While very little is known about the species interactions within cobble habitat, mobile reptant decapods are considered to be the primary competitors of juvenile *H. gammarus* for this resource. Therefore, the data analysis focuses on the crustaceans and all such animals were identified to species level.

Numerical data of decapod abundance were analysed using a clustering index to find affinities among sampling locations within each of the European countries. This multivariate analysis was carried out using the PRIMER statistical software package. Multidimensional scaling (MDS) was the chosen method of ordination and was applied using log (x+1) transformed data.

Results

Figure 1a–d show the principle animal phyla present within cobble substratum in each of the four European countries. Despite extensive sampling, no EBP lobster was found by dive teams in any of the sites. Apart from Ireland, where molluscs dominated, crustacea were the most numerically abundant phylum within each country.

Norway

Overall, 17 species of crustaceans (Fig. 1a) were sampled at the Norwegian sites. The galatheidae were the dominant family (Fig. 1) with *Galathea dispersa* and *G. squamifera* having the highest densities (Table 2). The pandalid shrimp *Pandalina brevirostris* and the portunid swimming crab *Liocarcinus arcuatus* were also recorded.

The annelids were the second most dominant phylum with sedentary polychaeta species prominent in all locations. The polyplacophora, bivalvia and gastropoda constituted the molluscs while the asteroidea, echinoidea and ophiuridae were the principle echinoderm classes.

Ireland

The most abundant of the molluscs (Fig. 1b) were the gastropoda, which in turn were entirely dominated by a single animal, the thick lipped dog whelk *Hinia incrassata*. Densities of this species ranged from 149–192 individuals/m^2.

A total of 32 crustacean species were recorded (Table 2). The porcellanidae clearly dominated (Fig. 2) with *Pisidia longicornis* having the highest density. Also present, but at markedly lower densities, were the alpheidae (*Athanas nitescens*), xantidae (*Xantho pilipes*) and galatheidae (*Galathea strigosa*).

Echinoderms (Fig. 1a) were common at all sites and while the ophiuroidea dominated, the echinoidea and holothuroidea were also present. The errantia were the dominant polychaetes.

U.K.

A total of 15 crustacean species were recorded at the U.K. cobble plots (Table 2). The species composition was similar to that recorded in Ireland (Fig. 2) with the porcellanid crab *P. longicornis* dominating. Both galatheids (*G. strigosa*) and portunids (*Liocarcinus pusillus*) were also present but at markedly lower densities.

The molluscan species sampled consisted almost entirely of the bivalves *Mytilus edulis* and *Mya truncata* while the annelids were numerically dominated by the polychaetes, *Scoloplos armiger* and *Nephtys caeca*. Ophiuridea were the dominant echinoderm.

Italy

While crustaceans dominated the Italian sites with 20 decapod species, the overall densities of families were noticeably lower than other European countries sampled (Fig. 2). The alpheidae were dominant in all locations with *Alpheus macrocheles*, *A. dentipes* and *Athanas nithescens* being the prominent species. Of the xanthidae present, *Xantho pilipes* and *X. incisus* were the main species while *Pisidia blutei* and *P. longicornis* made up the porcellanidae sampled.

The eunicidae were the dominant polychaeta, while the polyplacophora, gastropoda and bivalvia constituted the molluscs censused. The ophiuroidea and echinoidea were the principle echinoderms.

Trans-European Comparisons

Cluster analysis (Bray–Curtis Index) of decapod species densities in each sampling station produced dendrograms separating groups of localities in each country (Fig. 3). Italy, Norway and Ireland are presented as 3 distinct groups each having a high level of in-site similarity in terms of species composition. Such

Table 2. List of decapod species and their density (individuals/m^2 ± SD) as sampled within each European country

Family and species	Ireland	Norway	U.K.	Italy
Porcellanidae				
Porcellana platycheles	0	0	0	0.2 ± 0.4
Pisidia longicornis	52 ± 23	6 ± 3	70 ± 64	0
P. bluteli	0	0	0	2 ± 2
Galatheidae				
Galathea strigosa	5 ± 3	0.2 ± 0.2	3 ± 2	0
G. intermedia	0	6 ± 7	0	0
G. squamifera	0	11 ± 7	0	0.4 ± 0.3
G. dispersa	0	78 ± 56	0	0
G. bolivari	0	0	0	0.8 ± 0.5
Paguridae				
Paguridae sp.	5 ± 2	3 ± 2	0.75 ± 1.2	1.5 ± 0.8
Xanthidae				
Xantho pilipes	6 ± 0.3	0.6 ± 0.6	0	3 ± 3
X. incisus	0	0	0	15 ± 16
X. poressa	0	0	0	0.9 ± 0.7
Pilumnus hirtellus	0.3 ± 0.4	0	0	0
P. villosissimus	0	0	0	0.7 ± 0.4
Portunidae				
Pirimela denticulata	0.05 ± 0.1	000.05 ± 0.1		
Necora puber	0.7 ± 0.7	0	0	0
Carcinus maenas	0.05 ± 0.1	0	0	0
Liocarcinus pusillus	0.05 ± 0.1	0.8 ± 0.2	3 ± 2	0
L. corrugatus	0.05 ± 0.1	0	0	0
L. arcuatus	0	14 ± 20	0.1 ± 0.2	0
Majidae				
Inachus phalangium	0.9 ± 0.5	0	0	0
Macropodia linaresi	1 ± 0.6	0	1.2 ± 0.50	
M. deflexa	0.5 ± 0.4	0	0	0
M. rostrata	0.05 ± 0.1	1 ± 1.3	0	0
M. czerniavskii	0	0	0	0.1 ± 0.2
M. tenuirostris	0.05 ± 0.1	0	0	0
Eurynome spinosa	0.1 ± 0.1	0	0	0
Achaeus gracilis	0	0.2 ± 0.2	0	0
Pisa armata	0.05 ± 0.1	0	0	0
Hyas araneus	0	0	0.05 ± 0.1	0
Alpheidae				
Athanas nitescens	6.2 ± 1.6	7 ± 6	02 ± 0.5	
Alpheus macrocheles	1.9 ± 0.9	0	0	4 ± 1
A. dentipes	0	0	0	2 ± 0.7

Continued on p. 68

Table 2. Continued

Family and species	Ireland	Norway	U.K.	Italy
Hippolytidae				
Hippolyte varians	1.8 ± 0.6	0	1.2 ± 10	
Eualus occultus	2.2 ± 1.2	0	0	0
E. pusiolus	0.05 ± 0.1	0	0	0
Caridion steveni	0.05 ± 0.1	0	0	0
C. gordoni	0	0.6 ± 0.7	0	0
Thoralus cranchii	0	0	0.1 ± 0.2	0
Leucosiidae		0	0	0
Ebalia tuberosa	0.05 ± 0.1	0	0	0
E. tumefacta	0.1 ± 0.2	0.05 ± 0.1	0	0
Ilia nucleus	0	0	0	0.05 ± 0.1
Palaemonidae				
Palaemonetes varians	0.1 ± 0.2	0	0	0
Crangonidae				
Crangon crangon	2 ± 1	0	0	0
Philocheras sculptus	0.05 ± 0.1	0	0	0.05 ± 0.1
P. fasciatus	0.16 ± 0.2	0	0	0
Pontophilus trispinosus	0	0	1.5 ± 1.7	0
Processidae				
Processa edulis crassipes	0.4 ± 0.2	0	0	0
Pasiphaeidae				
Pasiphaea sp.	0.05 ± 0.1			
Pandalidae				
Pandalina brevirostris	0.2 ± 0.3	16 ± 110.1 ± 0.2	0	
Pandalina sp.	0	0	0	0.4 ± 0.3
Upogebiidae				
Upogebia deltaura	0	2 ± 1	0	0.5 ± 1
Atelecylidae				
Atelecycles rotundatus	0	0	1.5 ± 0.9	0
Caridae				
Caridae sp.	0	0	3 ± 2	0
Cancridae				
Cancer pagurus	0	0.05 ± 0.1	0.1 ± 0.2	0
Diastylidae				
Diastylis bradyi	0	0	0.1 ± 0.2	0
Diogenidae				
Calcinus tubularis	0	0	0	0.7 ± 0.3
Pirimelidae				
Sirpus zariquiey	0	0	0	0.05 ± 0.1

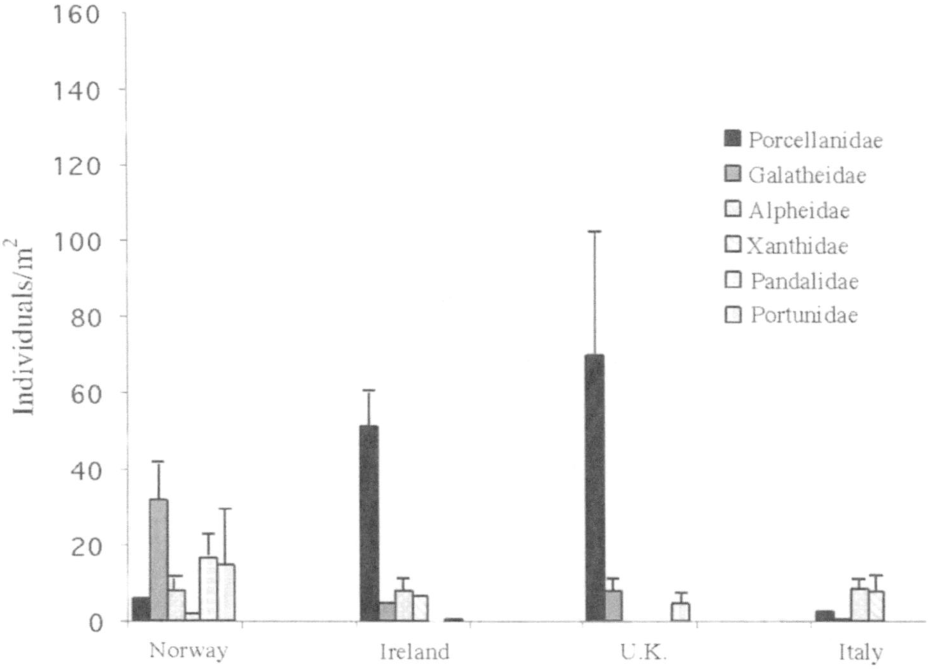

Figure 2. Average population densities (+ 1 SE) of decapod fauna in cobble habitats of four European countries.

similarity was not evident in the U.K. Italy appears to differ somewhat from Norway and Ireland in terms of species composition. These results are confirmed by MDS plots of the fauna in each country (Fig. 4).

Discussion

The present study has demonstrated the efficiency of suction sampling for investigating cobble substrata. While unsuccessful at detecting EBP European lobsters, suction sampling has opened a previously unavailable window into the possible interactions of lobsters with coexisting fauna. Apart from Ireland, the crustacea were the most numerically abundant group in cobble sites and thus appear to have a close association with this shelter-providing habitat. This association is well documented for many species such as American lobster (Wahle & Steneck, 1991), freshwater crayfish (Stein & Magnuson, 1976) and spiny lobster (Childress & Herrnkind, 1994). The relationship is closely linked to protection from predation (Barshaw & Lavalli, 1988; Wahle, 1992; Wahle & Steneck, 1992) and food supply (Barshaw, 1989; Lavalli & Barshaw, 1989).

Wahle (1998) proposed a number of hypotheses as to why the EBP European lobster is rare in nature despite being so ecologically similar to its American counterpart. Based on a preliminary number of suction samples from the west coast of Ireland and the Channel Islands, he suggests that species interactions might inhibit successful recruitment to the benthos.

Wahles' initial trials, combined with the current more comprehensive study, reveal differences in species diversity between European and U.S. cobble sites. Taxonomic diversity is low in the coastal Gulf of Maine with *H. americanus* and *Cancer irroratus* being the two most abundant subtidal species in cobble. Within European waters, it is clear that taxonomic diversity is much greater. This current study identified 21 families of decapods, half of which contained more than one species. It is reasonable to speculate, therefore, that high species diversity may well be the causative factor affecting lobster recruitment.

It is unknown if this interaction is predatory or competitive. Wahle & Incze (1997) have experimented in the wild to ascertain whether shelter may be a limiting resource for the American lobster. Within 24 h of seeding cleared cobble plots, average lobster densities dropped from 24 to 5.3, ind/m^2. This density remained constant for a 2 month period before a gradual attrition

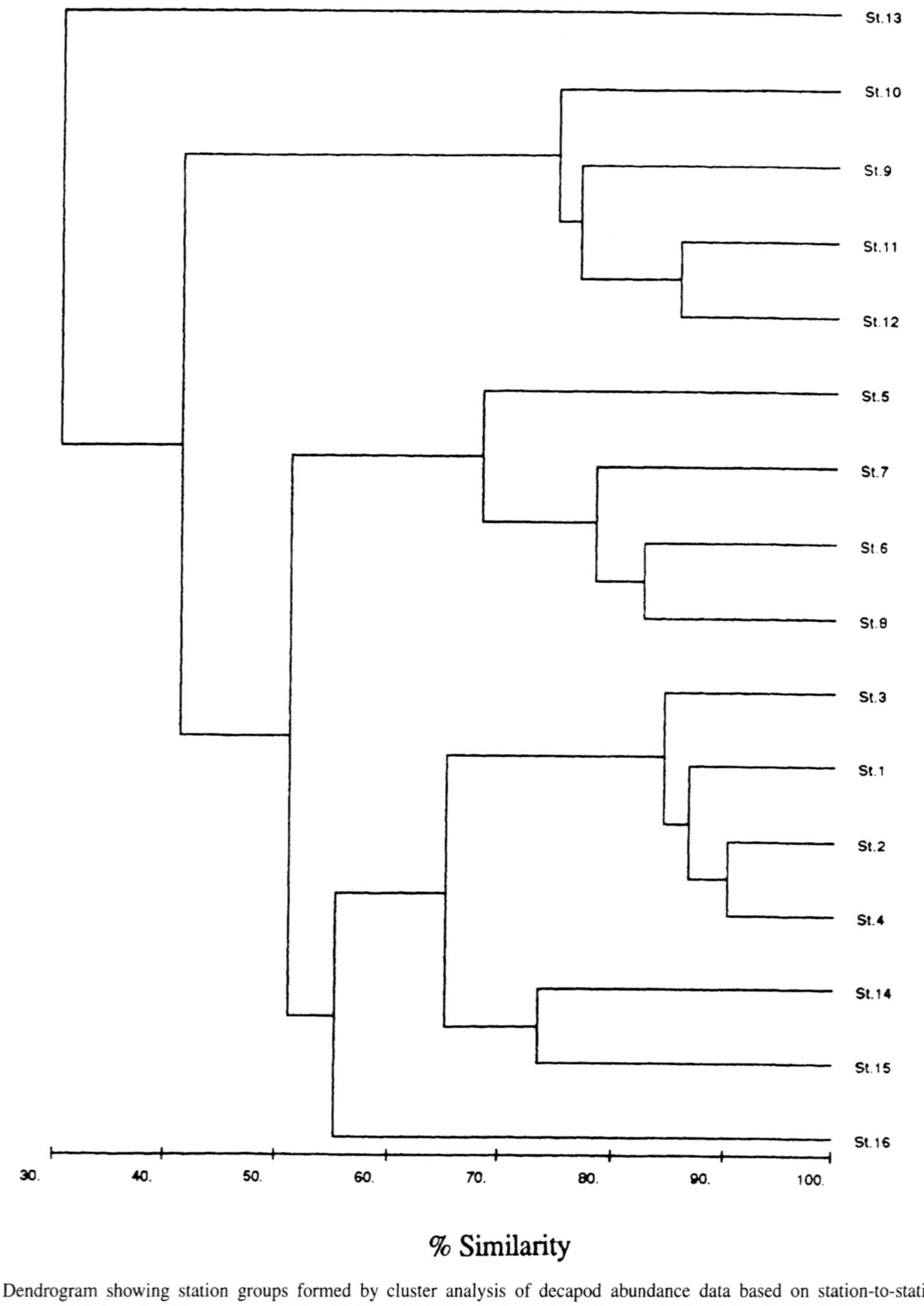

Figure 3. Dendrogram showing station groups formed by cluster analysis of decapod abundance data based on station-to-station faunal similarities as sampled from four European countries. Stations (St.) 1-4 = Ireland, 5-8 = Norway, 9-12 = Italy, 13-16 = U.K.

over the next four months of the trial. Given the many interstitial spaces available in 1 m^2 of cobble, this rate of survival is much lower than would be expected. In a similar experiment, Linnane et al., (2000) saturated enclosed cobble plots with *H. gammarus* postlarvae in an enclosed pond which was devoid of predators. Over a 9 month period, the density of lobsters remained constant at 16–18/m^2. Given the lack of success at

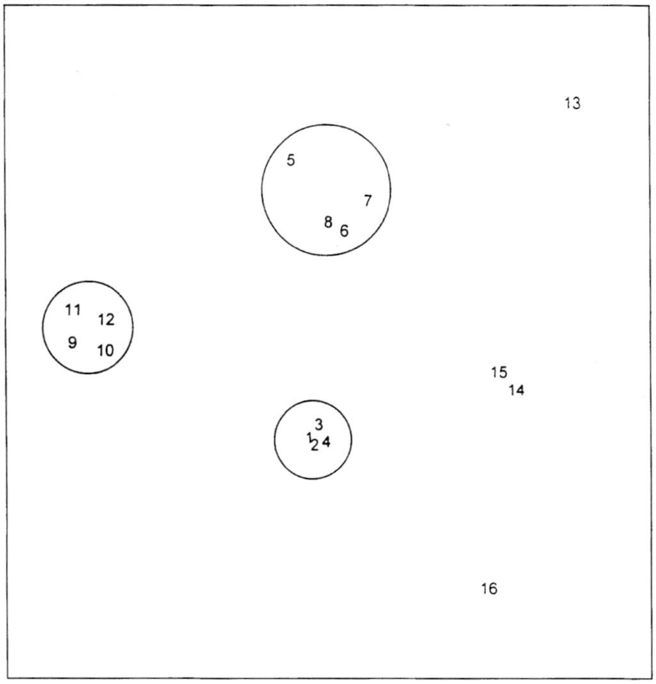

Figure 4. MDS configuration for standardised log-transformed decapod abundance data (stress = 0.10) from four European countries. Stations (St.) 1–4 = Ireland, 5–8 = Norway, 9–12 = Italy, 13–16 = U.K.

locating EBP European lobsters in the wild, both these studies suggest that there are factors other than space requirements which affect lobster recruitment.

Cobble-dwelling crustacean taxa appeared dominated by small sized individuals (<50 mm carapace length or width; CW or CL). In particular, the Irish and U.K sites were dominated by *P. longicornis* which recorded a CW range of 2–5.5 mm. This size range is within that of settling postlarval lobsters (Tully & Ó Céidigh, 1987) indicating that individuals may well be competing for similar sized interstitial openings. This adds further evidence to the suggestion that competitive exclusion by similar sized animals may occur.

The low density of decapod taxa in Italian waters is worth noting. Given the natural low density of lobsters within the Mediterranean sea, the inability to detect EBP animals is not surprising. However, predation pressure also seems to be particularly high in these waters with SCUBA divers noting the regular sightings of cephalopods (e.g. *Octopus* sp.) If areas without this predator were located, they might be very beneficial for future species interactions studies.

In conclusion, despite cobble substratum appearing to be highly suitable for European lobsters, the ability to locate EBP individuals within such habitat in the wild has proven unsuccessful. The possibility of extremely sparse distribution or indeed that cobble is not the preferred substratum of EBP lobsters cannot be ruled out. More extensive sampling of cobble sites can address this issue. Recently, intertidal crevices have been identified as a nursey area for juvenile lobsters (Linnane et al., 2000). This habitat is currently outside the range of suction sampling but warrents further investigation sub-tidally. However, if the competitive exclusion hypothesis is to be accepted, then it is clear that future research, in the form of small scale mesocosm experiments, must concentrate on both inter- and intra-specific species interactions. These studies will not only provide an insight into the efficacy of stock enhancement programmes but will benefit the understanding of fundamental biological processes that control the recruitment of *Homarus* populations.

Acknowledgements

Funding for this research project was received from the E.U. programme FAIR CT-96-1775.

References

Addison, J. T. & C. A. Bannister, 1994. Re-stocking and enhancement of clawed lobster stocks: a review. Crustaceana 67: 131–155.

Agnalt, A. L., G. I. van der Meeren, K. E. Jørstad, H. Næss, E. Farestveit, E. Nøstvold, T. Svåsand, E. Korsøen, & L. Ydstebø, 1999. Stock enhancement of European lobster (*Homarus gammarus*); a large scale experiment off south-western Norway (Kvitsøy). In Howell B., E. Moksness & T. Svåsand (eds), Stock Enhancement and Sea Ranching Fishing News Books. Blackwell Science Oxford, U.K.: 401–419.

Barshaw, D. E., 1989. Growth and survival of early juvenile American lobsters, *Homarus americanus*, on a diet of plankton. Fish. Bull. 87: 366–370.

Barshaw, D. E. & K. I. Lavalli, 1988. Predation upon postlarval lobsters, *Homarus americanus*, by cunners, *Tautogolabrus adspersus*, and mud crabs, *Neopanope sayi*, on three different substrates: eelgrass, mud and rocks. Mar. Ecol. Prog. Ser. 48: 119–123.

Childress, M. J. & W. F. Herrnkind, 1994. The behavior of juvenile Caribbean spiney lobster in Florida Bay: seasonality, ontogeny and sociality. Bull. mar. Sci. 54: 819–827.

Dybern, B. I., 1973. Lobster burrows in Swedish waters. Helgoländer weiss. Meeresunters. 24: 401–414.

Gendron, L. (ed.), 1998. Proceedings of a Workshop on Lobster Stock Enhancement Held in the Magdalen Islands (Québec) from October 29 to 31, 1997. Can. J. Fish. aquat. Sci. 244: xi + 135 pp.

Holthuis, L. B., 1974. The lobsters of the superfamily Nephropidea of the Atlantic Ocean (Crustacea: Decapoda). Bull. mar. Sci. 24: 723–884.

Hudon, C., 1987. Ecology and growth of post-larval and juvenile lobster, *Homarus americanus*, off Isle de la Madeleine (Quebec). Can. J. Fish. aquat. Sci. 44: 1855–1869.

Incze, L. S. & R. A. Wahle, 1991. Recruitment from pelagic to early benthic phase in lobsters, *Homarus americanus*. Mar. Ecol. Prog. Ser. 79: 77–87.

Lavalli, K. L. & D. E. Barshaw, 1989. Post-larval American lobsters (*Homarus gammarus*) living in burrows may be suspension feeding. Mar. Behav. Physiol. 15: 255–264.

Linnane, A., B. Ball, B. Munday & J. P. Mercer, 2000. On the occurrence of juvenile lobster *Homarus gammarus* in intertidal habitat. J. mar. biol. Ass. U.K. 80: 375–376.

Linnane, A., D. Mazzoni & J. P. Mercer, 2000. A long-term mesocosm study on the settlement and survival of juvenile European lobster *Homarus gammarus* L. in four natural substrata. J. exp. mar. Biol. Ecol. 249: 51–64.

Stein, R. & J. J. Magnuson, 1976. Behavioral response of crayfish to a fish predator. Ecology 57: 751–761.

Tully, O. & P. Ó Céidigh, 1987. The seasonal and diel distribution of lobster larvae *Homarus gammarus* in the neuston of Galway Bay. J. Cons. Ciem. 44: 5–9.

Wahle, R. A., 1992. Body-size dependant anti-predator mechanisms of the American lobster. Oikos 65: 52–60.

Wahle, R. A., 1998. A trans-atlantic perspective on *Homarus* recruitment and enhancement. In Van der Meeren, G. I. & O. Soldal (eds), Proceedings of the Seminar on the European lobster *Homarus gammarus* (L.), Kvitsøy, Norway, 1995. ÅPEN Press Bergen: 36–43.

Wahle, R. A. & L. S. Incze, 1997. Pre- and post-settlement processes in recruitment of the American lobster. J. exp. mar. Biol. Ecol. 217: 179–207.

Wahle, R. A. & R. S. Steneck, 1991. Recruitment habitats and nursery grounds of the American lobster (*Homarus americanus* Milne Edwards): a demographic bottleneck? Mar. Ecol. Prog. Ser. 69: 231–243.

Wahle, R.A. & R. S. Steneck, 1992. Habitat restrictions in early benthic life: experiments on habitat selection and in situ predation with the American lobster. J. exp. mar. Biol. Ecol. 157: 91–114.

Hydrobiologia **465:** 73–83, 2001.
G. Burnell (ed.), Coastal Shellfish – A Sustainable Resource.

Conflict between two inshore fisheries: for whelk (*Buccinum undatum*) and brown crab (*Cancer pagurus*), in the southwest Irish Sea *

Edward Fahy
Marine Institute, Fisheries Research Centre, Abbotstown, Castleknock, Dublin 15, Ireland

Key words: *Cancer pagurus*, *Buccinum undatum*, Irish Sea, Ireland, pot fishing, *Scyliorhinus*, whiting, cod, blue mussel, mackerel, scad, artificial bait, Korea, feeding behaviour

Abstract

Landings from the whelk fishery in the southwest Irish Sea rose from 63 t in 1990 to 6587 t in 1996 although landings declined to 2919 tonnes in 1998. Whelk are caught in pots baited with a combination of dogfish *Scyliorhinus* spp and brown crab *Cancer pagurus*. The input of bait crab is estimated at 7.2% of whelk landed. Most of the crab comes from southeast Ireland where some of it is a by-product of the claw fishery; some is targeted as bait crab, and this includes sub-size or recently moulted individuals. Fishing effort has been increasing on brown crab for human consumption also and there is concern about the sustainability of the crab fishery. Alternative bait trials examined a number of readily available fish species, some of them offals, whose attractiveness was evaluated by reference to crab and dogfish: whitefish (whiting and cod), pelagic species (mackerel and scad) and blue mussel. An artificial bait ration which included some crab meat, but 60% less than its natural equivalent, proved as effective as natural bait crab. A bait which consisted of a single species was less successful than two species fished together and some of the combinations (notably of whitefish) performed very effectively. Pelagic species performed poorly. The trade in brown crab for human consumption will always produce some form of discards which should be used; extending its use in an artificial bait is desirable.

Introduction

The whelk fishery in the south west Irish Sea has become one of the largest of Ireland's inshore fisheries in recent years, landings having risen from 63 tonnes in 1990 to 6537 tonnes in 1996 although they have fallen back slightly since (Table 1) as a result of trading difficulties, fluctuating prices and locally exhausted stocks. The expansion of the fishery arose from the depletion of stocks of Pacific species and the opening of markets in the Far East, particularly in Korea. An account of this fishery just before it peaked was provided by Fahy et al. (1995).

Whelk occur in patches and almost all of the fishery is located within 5 nautical miles of the shore. The animals are fished in pots fashioned from industrial plastic containers, baited with a combination of dogfish (*Scyliorhinus* spp, mainly *canicula*) and brown crab (*Cancer pagurus*). Pots are lifted every 24 h, weather permitting, when they are rebaited. The bait

Table 1. Landings and first sale prices of whelk from the fishery in the south west Irish Sea

Year	Tonnes landed	Total value (£ie)	Price (£ie)/ tonne
1990	63	12 689	203.02
1991	959	364 674	380.26
1992	2082	735 842	353.43
1993	2562	909 563	355.02
1994	4488	1 521 248	338.93
1995	5952	2 728 942	458.46
1996	6537	3 378 328	516.78
1997	3852	1 694 559	439.95
1998	2919	1 129 211	386.91

Source: Department of the Marine and Natural Resources.

* Paper delivered to the Third International Conference on Shellfish Restoration, Cork, September 1999

Table 2. Provenance of crab bought-in as whelk bait by Company B in 1996

Location of purchases	County	Crab Boxes	Crab kg	County totals kg	County totals as %
Malin Hd	Donegal	1	30	30	0.0
Howth	Dublin	5	150	150	0.1
Dunmore East	Waterford	31	930		
Helvic	Waterford	101	3030	3960	2.6
New Ross	Wexford	252	7560		
Cullenstown	Wexford	49	1470		
Duncormick	Wexford	484	14 520		
Fethard	Wexford	613	18 390		
Kilmore	Wexford	3073	92 190		
Rosslare	Wexford	337	10 110		
Slade	Wexford	37	1110	145 350	94.4
Arklow	Wicklow	152	4560	4560	3.0
	TOTALS	5135	154 050	154 050	100

Percentage of all records traced to port, 92.5.

ration in each pot consists of approximately 150 g of dogfish and 200 g of crab.

At its height, the whelk fishery occupied 80 vessels and while this number has declined by perhaps 50%, the remainder have continued to fish as intensively, some increasing the number of pots each year and occasionally the number of crew on a boat. The largest vessels set 1000 pots each day, using 500 crabs of a size acceptable for human consumption every 24 h as bait.

The work described here was undertaken to investigate the impact of the bait fishery on crab stocks and to seek alternative baits for whelk from readily available sources.

The use of brown crab as whelk bait

Shellfish purchasers who buy whelk also process other shell and fin-fish species and they supply bait to the fishermen when they collect landed whelk from the ports. Thus, it is possible to quantify the bait inputs to the fishery and to compare these with whelk output by reference to purchasing documentation.

Perusal of the records of a major purchaser of whelk in 1996 (Table 2) confirms that most of the crab distributed to fishermen as bait is caught locally (Fig. 1). The area in question – some 100 km of coastline extending west from the southeast corner of the island – is reckoned to produce approximately 7.5% of the total national landings of brown crab for human consumption.

An obvious source of crab for whelk bait are bodies which are discarded once the claws have been removed; crab claws are gathered for human consumption and are the main food product in the early months of the year, when the condition of the animals is poor (Fig. 2) (Fahy, unpublished). Whole crab is processed for human consumption mainly in the autumn. Both products have discards: whole crab discards include animals which have died before cooking etc but one would expect the crab claw trade to generate a larger volume of bait than whole crab processing.

The relationship between the purchase of bait crab and crab claws is explored in Table 3 which provides details of crab and whelk purchased by two processors (over a period of 5 and 1 years, respectively). Crab claws are raised to live weight by a factor of $*5$ (a conversion factor used by the Department of the Marine and Natural Resources, Ireland) and it is clear that bait crab might have been a by-product of the claw fishery in one year (1993) only. At that time total whelk landings were relatively low (Table 1). The data in Table 3 also enable a calculation of crab input as a percentage of whelk harvested. An average of 7.2% is estimated. This being the case, the consumption of bait crab by the whelk fishery in 1996, its year of peak landings, would have been 470 tonnes. There are however, factors which might increase the input of bait. A survey of whelk CPUE throughout the fishery shows it to be variable, ranging from <1 to occasionally 8.5 kg per pot (Fahy et al., in press). Whatever the yield, the bait ration remains the same. Hence, where the animals become less numerous as a result of overfishing, the amount of bait in the pots relatively increases and the author has witnessed an input of crab which amounted to 20% of the catch by weight on poor fishing ground.

Further clarification of the likely source of bait crab was undertaken by regressing the monthly purchased weights of bait crab on crab claws for the 6 years of records shown in Figure 3. The two correlate significantly (intercept=7155; x variable=3.08; $N=72$; R^2=0.303; $P<0.01$) but the wide scatter of data points suggests that bait crab does not originate from the crab claw fishery alone.

Results of an investigation of the size and whether bait crab bore claws is shown in Figure 4. As many as 50% of crabs of 12 cm maximum carapace width had not been declawed. A substantial number of animals

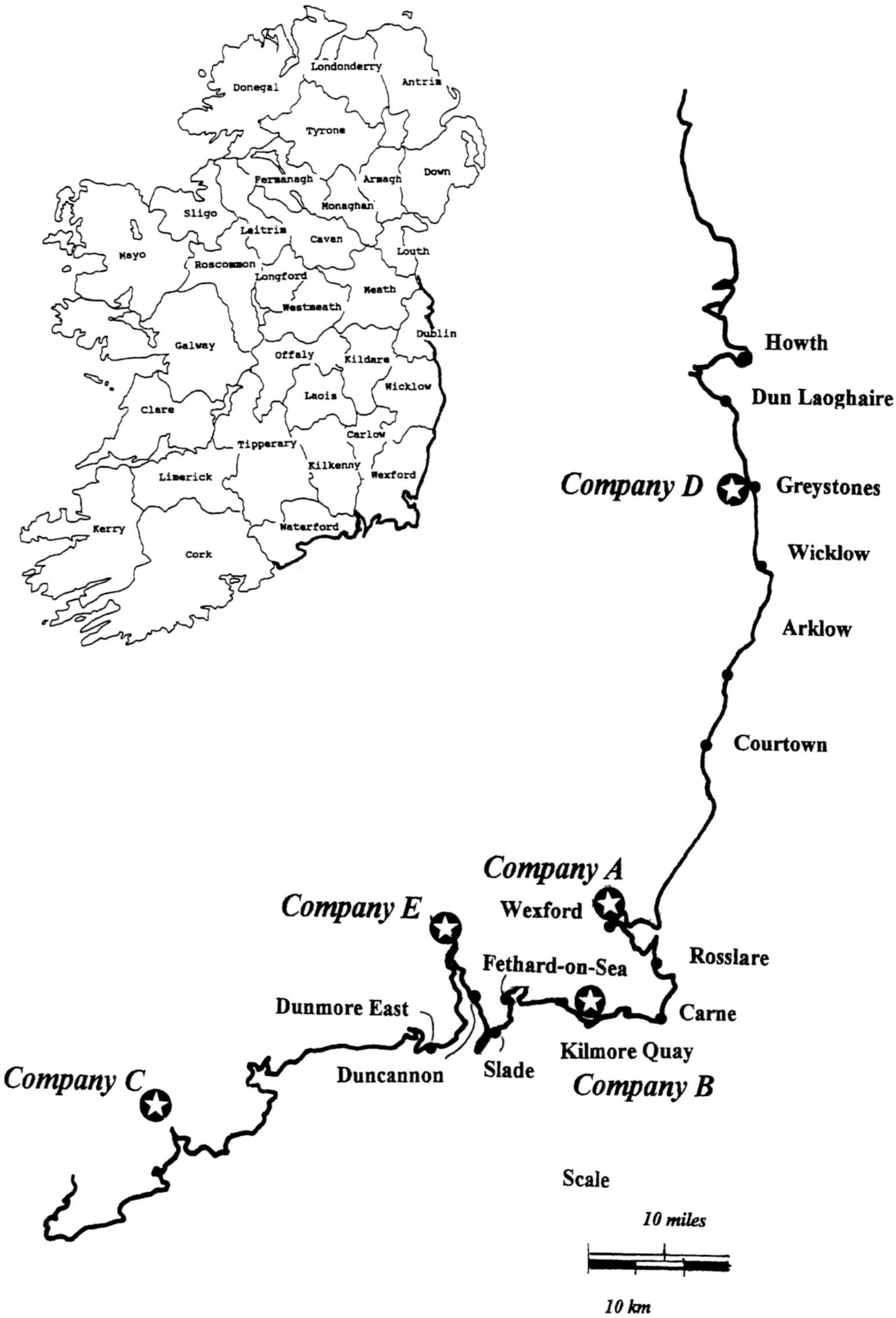

Figure 1. Map of the south eastern coastline showing the location of whelk purchasing companies ⋆ and landing places referred to in the text. Inset, a map of Ireland showing the counties from some of which crab were purchased.

Table 3. Quantities of whelk, bait crab and crab claws purchased by two shellfish processors; claws are raised to live weight for comparison with bait crab purchases

Year	Purchased whelk (kg)	Purchased bait crab (kg)	Bait crab as a percentage of whelk purchased	Purchased crab claws (kg)	Claws raised to live crab (kg)
From the records of Company B					
1992	1 220 857	102 930	8.4		
1993	1 997 673	104 430	5.2	22 025	110 125
1994	1 913 229	113 220	5.9	9570	47 850
1995	1 629 699	207 330	12.7	14 732	73 660
1996	2 444 683	166 560	6.8	22 083	110415
From the records of Company A					
1996	1 799 993	100 740	5.6		
Average,			7.2.		

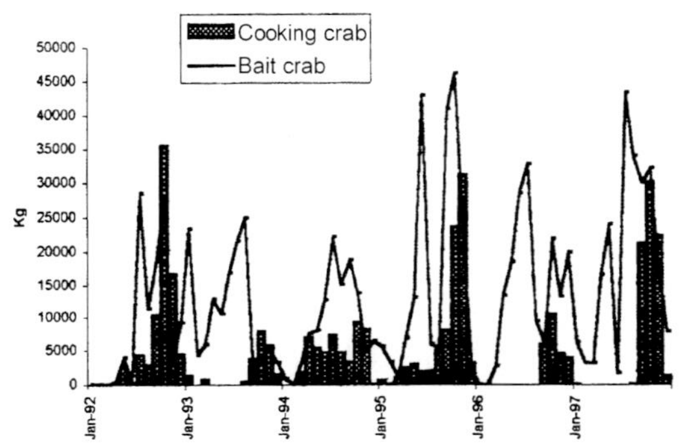

Figure 2. Purchase of cooking crab and bait crab by a processor during a 6-year period.

in the bait sample had one or both claws still attached to the body although the larger animals tended to be without both.

There is evidence that three additional sources of bait crab contribute to the whelk fishery. Soft crab which have not hardened up after moulting are the first of these. Large numbers of soft crab are captured in the early spring but they have little value. Some are retained for bait rather than being returned to the sea. Secondly, there have been reports of brown crabs being targeted for whelk bait rather than for human consumption and thirdly, there are reports of sub-size crabs (i.e. those having a maximum carapace width of less than 11.4 cm, 4.5 inches – cf. Ireland, Statutory Instrument No. 359 of 1951) being used for bait.

A substantial proportion of bait crab comes from southeast Ireland, centred on Kilmore Quay. In the peak year of the whelk fishery this amounted to landings of 470 tonnes. The total officially reported landings from south east Ireland usually approximate some 450 tonnes of brown crab annually.

Assessements of the south east brown crab stocks are ongoing. Fishing effort in the vicinity of Kilmore Quay (expressed in terms of number of fishing pots used) has increased from 4600 to 10 500 over a decade. The number of boats pursuing crab rose from 45 to 65 in the same period. Landings per unit effort (LPUE) expressed as kg brown crab/boat delivered to a processor or as kg per pot fished per day have both declined from 1990 but remained stable in the years between 1991 and 1996 (Fahy, unpublished).

There is concern among the fishing community for the long term sustainability of crab fishing in this area. Within that community efforts have been made to develop alternatives to the current bait combinations. The purpose of the work described here is to assess their prospects.

Materials and methods

Alternative bait trials were devised to seek an acceptable alternative to brown crab as a bait for whelk using commonly available fish of lower values and fish offals. An artificially manufactured bait was also tested. The following were the types, sources and conditions of baits tested:

(1) Brown crab *Cancer pagurus:* obtained from whelk processors, held frozen before use.
(2) Dogfish *Scyliorhinus* spp.: lesser spotted dogfish, *S. canicula* made up the vast majority of the spe-

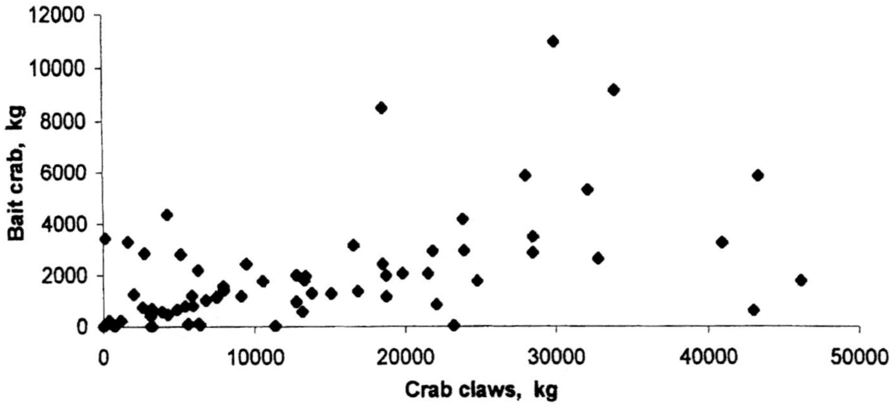

Figure 3. The relationship between monthly purchases of crab claws and bait crab by a processor who also purchases whelk.

cies used as bait, *S. stellaris* being occasionally included. Dogfish was frozen prior to use.

(3) Whitefish, whiting *Merlangus merlangius* supplied round and Cod *Gadus morhua* supplied entire with the fillets removed. This material had been frozen before use.

(4) Pelagic species, Scad *Trachurus trachurus* and Mackerel *Scomber scomber*. These animals were round and had been frozen prior to use.

(5) Mussel *Mytilus edulis*. Mussels were collected from along the lower harbour wall bordering the North Slob, Wexford Harbour. The mussels were small (mean length = 3.5 cm, standard deviation = 0.700, N=20). Flesh made up 22.3% of the total weight (wet) of the animals. The shells were broken open using hammers, some 400–500 g of shell and meat made up a bait ration.

(6) Artificial bait crab. This product has been developed by a company called 'Celtic Bait' which is based in Co Wicklow. The company is currently seeking a patent so the bait can be described in general terms only. The following are the terms used by the company: Celtic bait is a single piece, covered by a scent-soaking mesh containing measured ingredients which is totally biodegradable and contains no toxic products. It is easily stored being compact, a single ration weighing between 150 and 200 g. Its composition is intended to confer a longer working life because it dissolves more slowly than natural baits. It is intended to remain effective in a whelk pot for two days rather than one. The bait is composed of fish offals and up to 40% of its weight consists of brown crab mixed with gelling agents, binders and flavour enhancers. A single bait ration is said to contain 60% less crab meat than its natural equivalent.

Experimental design

Whelk pots were set in gangs of approximately 30 in an west–east direction across whelk patches; 360 pots were usually set together and their soak time was 24 h although bad weather extended this to 48 h on two occasions. Details of the settings are presented in Appendix 1.

A trial was designed to compare two baits or a combination of the two simultaneously. The repeated sequence of pots was: Bait A; Bait B; Baits A+B.

When a pot was hauled it was inspected for the presence of bait. When a bait had been washed out or the pot had been torn open by the current, its contents were not scored as a result. Otherwise, numbers of whelks in each pot were counted and the results were compared by *t*-test.

In the first trials in the series the total numbers of whelks per pot were compared. In response to a suggestion by fishermen that certain types of bait attracted small whelk, the yield of a pot was expressed as numbers of marketable whelk and numbers below the legal limit (this fishery is regulated by size limit, the critical dimension being 25 mm maximum width, which corresponds to a length from the top of the shell to the end of the siphonal canal of ca. 50 mm, c.f. Ireland, Statutory Instrument No. 278 of 1994).

Crab and dogfish are the staple bait components used in this fishery so fishing success is measured by comparison with these. When two baits were compared, the outcome was tested for significance and,

Table 4. Summary of trials of various baits against natural crab

Species tested against crab	*t*	*N*–2	*P*	Significant?	If significant percentage above or below crab alone
Dogfish	2.3004	246	0.01>0.025	Sign.	18
Dogfish and crab	6.8186	200	<0.0005	Sign.	92
Whiting	0.8361	238	>0.05	n.s.	
Whiting and crab	6.0999	239	<0.0005	Sign.	65
Mussel	3.7216	243	<0.0005	Sign.	−25
Mussel and crab	9.4563	236	<0.0005	Sign.	99
Scad	4.2142	236	<0.0005	Sign.	−51
Scad and crab	1.9548	243	<0.05	Sign.	30

where a significant difference in performance occurred, the success, or lack of success, of the second bait was expressed as a percentage of the first.

Results

The summarised results of trials comparing various baits against natural crab are set out in Table 4. Crab did not perform better than whiting but all other trials showed significantly different results. Single species comparisons provided the smallest differences. Crab alone was better than scad and, to a lesser extent, than mussel. Dogfish on its own was better than crab.

Combinations of two species greatly enhanced attractiveness. Dogfish and crab captured more whelk than crab alone as did the other three combinations including species which on their own performed poorly.

Trials using dogfish are summarised on Table 5. The test of crab against dogfish is the inversion of the result on Table 4. Dogfish performs in a similar way to crab against mussel and pelagic species and, as in Table 4, species combinations all perform better than single species even where one species on its own gave a very poor result. Pelagic species and dogfish did not perform significantly differently from dogfish alone. Cod performed well against dogfish and the combination of cod and dogfish provides the highest score on this Table.

Comparisons of natural crab with artificial bait crab are summarised on Table 6. Results from the first comparison of natural crab with artificial bait crab did not produce significantly different results but, in the course of hauling the pots it was observed that dogfish had been included as a filler component in the artificial bait crab rations. The trial was repeated as a straightforward comparison between natural crab and dogfish and artificial bait crab and dogfish. The results on that occasion were not significantly different either.

The summarised results for undersized whelk recorded in the bait trials are set out in Table 7. In only one, where mussel was compared against dogfish, was a higher proportion of small whelk associated with mussel.

Two series of potting results are set out in Figures 5 and 6. Figure 5 shows numbers of whelks captured in a trial which compared natural crab and dogfish with artificial bait crab and dogfish, both of which provided results which did not differ significantly. The numbers of whelks entering the pots are assumed to reflect the numbers in the whelk patch. Figure 6 is an example of two baits which are relatively unsuccessful on their own but which stimulate an enhanced response in combination. As in Figure 5, there would appear to be an increased density of animals at the eastern end of the patch but the abundance of animals on the ground is less confidently discerned.

Discussion

Hancock (1974) identified the ideal bait as one which attracts a target and repels others. Crabs (like some other crustacean species) will not move into the vicinity of a dead member of their own species so they avoid pots baited with dead crab. Whelk consume dead

Table 5. Summary of trials of various baits against dogfish

Species tested against dogfish	*t*	*N*–2	*P*	Significant?	If significant percentage above or below dogfish alone
Crab	2.3004	246	0.01>0.025	Sign.	−12
Crab and dogfish	5.4938	184	<0.0005	Sign.	63
Cod	5.7154	226	<0.0005	Sign.	96
Cod and dogfish	7.1633	216	<0.0005	Sign.	131
Mussel	4.4016	165	<0.0005	Sign.	−38
Mussel and dogfish	5.3535	154	<0.0005	Sign.	66
Pelagics	2.0557	245	0.01>0.025	Sign.	−25
Pelagics and dogfish	0.6055	237	>0.05	n.s.	

Table 6. Trials involving natural crab and artifical bait crab

	t	*N*–2	*P*	Significant?
Natural crab				
Artificial crab	1.30135	295	>0.05	n.s.
Natural crab and dogfish				
Artificial bait crab and dogfish	0.446481	231	>0.05	n.s.

crab although they are wary of approaching living ones. Lapointe & Sainte-Marie (1992) reported that whelk were slow to approach a bait in the presence of live crab.

These facts might have influenced the choice of crab as a bait for several reasons. When pots have to be left at sea for a prolonged period due to bad weather, the odour of the bait crab dissipates in time and live crab may enter a pot killing the whelk it contains. The contents of such pots must be discarded to avoid the possibility of contaminating the whelk meat in the processing factory with decaying animals. Problems of this kind might be reduced by fitting a narrower entrance to whelk pots. However these considerations are not believed to significantly influence the preference for crab as a bait in the Irish whelk fishery where, instead, it is valued for its greater attractiveness to the target species.

Fishermen report that some inshore areas, notably where benthic contours are close together and currents are strong, are highly productive of whelk and fish consistently well, but elsewhere there are instances of local depletion of whelk numbers. In the first years of the Irish Sea fishery, when whelk were abundant, fishermen report that the choice of whelk bait was unimportant. Fishermen in south east Ireland maintain that they discovered the strong attractive qualities of brown crab (*Cancer pagurus*) and dogfish (*Scyliorhinus* spp.) placed together in a whelk pot, as a result of experimentation although this combination of baits has been used in the Normandy whelk fishery since 1977 (Latrouite, pers commun.). Its high efficiency was in fact first demonstrated in the UK earlier (Hancock, 1974). Hancock noted that in Whitstable, Kent, fishermen used a combination of salted skate (*Raja* spp.) and shore crab (*Carcinus maenas*) to lure *Buccinum* into traps.

The bait trials described here demonstrate that various combinations of fish offals are attractive to whelk. Whelk are scavangers which find a wide range of taxa acceptable in the Irish Sea (Ramsay et al., 1997). Nickell & Moore (1992) described *Buccinum undatum* as a scavanger in the Clyde Sea area. Combining baits

Table 7. Percentage undersized whelks (<25 mm maximum diamater or 50 mm length) recorded in bait trails

Species combinations	t	N–2	P	Significant?	If significant percentage above or below species 1
Species tested against dogfish (species 1)					
Crab (species 2)	Not tested				
Crab and dogfish (species 2)	Not tested				
Cod (species 2)	1.1496	203	>0.05	n.s.	
Cod and dogfish (species 2)	0.2567	193	>0.05	n.s.	
Mussel (species 2)	2.0117	165	0.01>0.025	Sign.	40.5
Mussel and dogfish (species 2)	0.4244	154	>0.05	n.s.	
Pelagics (species 2)	0.5559	245	>0.05	n.s.	
Pelagics and dogfish (species 2)	1.3680	237	>0.05	n.s.	
Species tested against crab (species 1)					
Dogfish (species 2)	Not tested				
Dogfish and crab (species 2)	Not tested				
Whiting (species 2)	Not tested				
Whiting and crab (species 2)	Not tested				
Mussel (species 2)	0.7842	243	>0.05	n.s.	
Mussel and crab (species 2)	0.6935	236	>0.05	n.s.	
Scad (species 2)	0.3646	236	>0.05	n.s.	
Scad and crab (species 2)	0.1543	243	>0.05	n.s.	
Artificial bait crab (species 2)	0.8759	295	>0.05	n.s.	
Species tested against crab and dogfish (species 1)					
Aritificial bait crab and dogfish (species 2)	0.2663	231	>0.05	n.s.	

is also known to be effective for other target species (Franco et al., 1987).

The results obtained in these trials suggest that certain of the fish offals, notably whitefish, used in combination with dogfish, might be at least as successful as crab and dogfish. However the nature of the trials does not permit such precise comparison because the conditions under which each was conducted must be regarded as unique. McQuinn et al. (1988) demonstrated that current speed and direction define the size and area of attraction fished by a pot. Martel et al. (1986) showed that feeding activity varied annually in relation to reproductive behaviour; this can also be related to seasonal LPUE variations in the Irish Sea whelk fishery (Fahy et al., 1995). Capture rates by pots can be comparable with SCUBA estimates of whelk density (McQuinn et al., 1988), but presumably over the shorter term.

An immersion time of 24 h is probably the norm in pot fisheries in Irish inshore waters. Bennet (1974) found that baits for lobsters are not effective any longer. Daniel & Bayer (1987) stated that lobsters are attracted to amino acids of low molecular weight and that those released by herring bait reached a baseline level 12 h after leaching began. Fresh baits did not display high release rates whereas freezing and thawing causes cell rupture followed by a higher initial release rate.

Comparisons of artificial and natural baits frequently present the latter as more effective. Miller & van Heukelem (1988) tested various amino acids on crab species and found that while natural baits were more effective on the first day of immersion, the performance of artificials improved on the second. The artificial bait crab tested here was often in reasonably good condition after immersion for 24 h and it is thought to have been effective, though to a lesser extent, on the second day, according to fishermen.

There is little corroborative evidence available elsewhere on the efficacy of different kinds of bait for whelk. However, Kideys (1991) cites Duncan et al. (1990) who reported that dogfish and scallop was

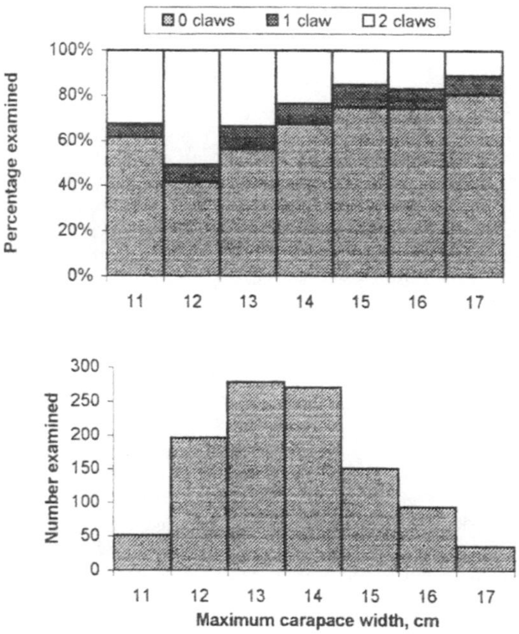

Figure 4. Examination of bait crab for the whelk fishery to ascertain the extent of claw removal within different size groups. The caption to the upper figure indicates the percentage of animals bearing 0, 1 or 2 claws.

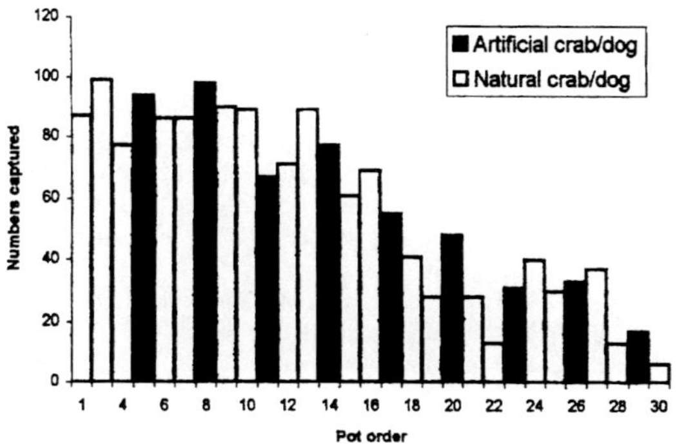

Figure 5. Transect through a whelk patch (west to east) illustrating the supposed density of animals on the ground as revealed by their response to two bait types which elicited a similar level of response.

a superior bait to herring and plaice, which is partly in keeping with findings reported here. Sainte-Marie (1991) on the other hand captured whelk for experimental purposes using herring as bait and McQuinn et al. (1988) also used herring and cod. Youm (1991) used anchovy, sardine, crab, shrimp and clam to capture *Neptunea arthritia*, a scavanging gastropod species.

As long as there is a human consumption fishery for brown crab there will be rejects from it, in the

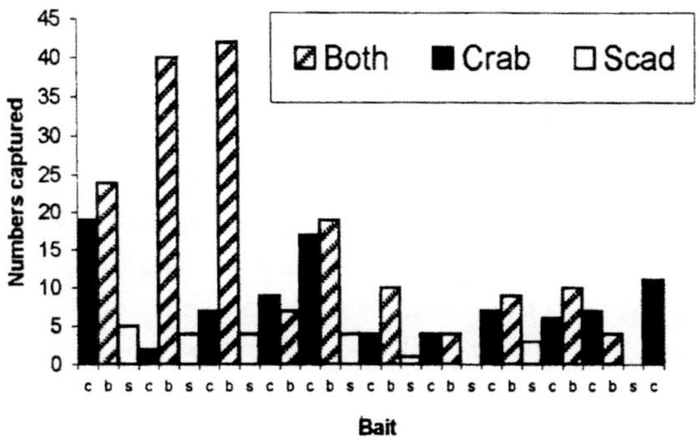

Figure 6. Transect through a whelk patch (west to east) illustrating the supposed density of animals on the ground as revealed by their response to two bait types and their combination, each element of which elicited a disimilar response.

form of discarded bodies from which the claws have been removed or as moribund or poorly conditioned animals intended for cooking. It is desirable that these products should not be wasted and a method of using them as bait is preferable to disposing of them. Artificial bait crab of the kind described here will stretch the product further although, in theory, identification of key amino acids and their substitution with artificial chemicals might, in time, supply the need. However, working on sea eels *Astroconger myriaster*, Youm et al. (1990) discovered that synthetic attractants derived from anchovy, although they were very good baits, were less successful than natural extracts from that species and Magalhães-Neto et al. (1996) made similar findings for lobster. Harada (1989) may have provided an explanation for the mechanism involved with the identification of animo acids as important components. Nucleic acid-related compounds were important contributors and in Molluscs glucides also had a role. Herada proposed that the combination of these compounds has a synergistic effect.

The only bait to attract significantly more juvenile (small) whelk was mussel in one trial only. Whelk of <20 mm are rarely captured in pots which would retain some, were they to enter. The behaviour of these small animals in early life is not known. Mussel reefs occur in the Irish Sea although they are periodically dredged and re-laid for ground culture in Wexford Harbour. Fishermen have reported certain fish species to be more abundant in the vicinity of these reefs and it is possible they provide habitat and feeding for juvenile whelk.

The co-existence of the two fisheries, for whelk and brown crab, offal and discards from the latter supplying bait for the former, is, in principle, a reasonable arrangement, provided the input of brown crab is confined to animals which are not specifically targeted for the purpose. Achieving a rational exploitation of crab in the circumstances is however difficult. The problem is not confined to south east Ireland, a similar dilemma being known in the Normandy whelk fishery (Latrouite, pers. commun.). The past decade has seen a considerable increase in fishing effort in Irish inshore waters. As species become over fished and landings decline, attention will have to be increasingly focussed on the enforcement of regulations and the provision of addition technical measures. The protection of undersized crab is an obvious starting point but consideration must be given to extending protection to recently moulted individuals and, eventually, to the advisability of harvesting crab claws only, thereby discarding 80% of the carcass. Prohibition of the use of crab as whelk bait would probably be unenforceable although crab is not indispensable to the capture of whelk. The promotion of artificial whelk bait, involving a lower consumption of brown crab, is a worthy development although it is likely to cost more than natural crab and thus to be a less popular option with fishermen.

Acknowledgements

I am grateful to the following who supplied information and data used in this work: Tom Roche of Kilmore Quay, Leslie Bates and Lorcan Barden of Sofrimar, Kilmore Quay, Richard Lett of Lett and Co., lately of Wexford, the Carr family of Carr and Sons, Tallow, Co Waterford and Michael and Mary Murphy of Saltee Shellfish, New Ross, Co. Wexford. The bait trials were carried out on vessels skippered by Derek Noble of Waterville, Co. Kerry, and by John O'Toole, Wexford. Alan Hegarty of Celtic Bait supplied the artificial bait crab. Fergal Nolan of Bord Iascaigh Mhara introduced me to Alan Hegarty and assisted the project in many ways. Finally, my colleagues Colm Duggan and Sorcha Wheatley read through a draft of the work and made valuable suggestions for its improvement, and Daniel Latrouite of IFREMER provided useful information on the whelk fishery in Normandy.

References

Bennet, D. B., 1974. The effects of pot immersion on catches of crabs *Cancer pagurus* L and lobsters *Homarus gammarus* (L) J. Cons. perm. int. Explor. Mer 35: 332–336

Daniel, P. C. & R. C. Bayer, 1987. Temporal changes in release rates and quality of *lobster (Homarus americanus)* feeding attractants from herring (*Clupea harengus*) baits. Mar. Behav. Physiol. 13: 13–27

Duncan, A., C. Gilling, G. Hughes, S. J. Hawkins & R. D. M. Nash, 1990. The viability of establishing a whelk fishery in Manx waters. A report submitted to the Isle of Man Government, Department of Agriculture, Fisheries and Forestry, by the University of Liverpool, 73 pp.

Fahy, E, E. Masterson, D. Swords & N. Forrest (in press), A second assessment of the whelk fishery in the south west Irish Sea with particular reference to its history of management by size limit.

Fahy, E., G. Yalloway & P. Gleeson, 1995. Appraisal of the whelk *Buccinum undatum* fishery of the southern Irish Sea with proposals for a management strategy Irish Fisheries Investigations B (42): 26 pp.

Franco, J. M. , A. Bjordal & S. Løkkeborg, 1987. Synergistic effects of different baits on the same hook in the longline fishery for hake (*Merluccius merluccius*) and torsk (*Brosme brosme*) ICES C.M. 1987/B:39.

Hancock, D. A., 1974. Attraction and avoidance in marine invertebrates – their possible role in developing an artificial bait J. Cons. perm. int. Explor. Mer 35 (3): 328–331

Harada, K., 1989. Feeding attractants for aquatic animals – VI. Attractant structure – activity and application to fishery. Seitai Kagaku/Ecol. Chem. 9(4): 55–69. Canadian translation of Fisheries and Aquatic Sciences No. 5518.

Kideys, A. E., 1991. The ecology of the common whelk, *Buccinum undatum* L off Douglas, Isle of Man, with particular reference to its ecological energetics. Ph.D. thesis, University of Liverpool.

Lapointe, V. & B. Sainte-Marie, 1992. Currents, predators and the aggregation of the gastropod *Buccinum undatum* around bait. Mar. Ecol. Prog. Ser. 85: 345–257

Magalhães-Neto, E. de O., R. H. de Queiroz. & M. Ogawa, 1996. Baits for lobster fishing. I – Tests of acceptance. Arq. Cin. Mar. Fortaleza 30 (1/2): 55–61.

Martel, A., D. H. Larrivée, K. R. Klein & J. H. Himmlman, 1986. Reproductive cycle and seasonal feeding activity of the neogastropod *Buccinum undatum*. Mar. Biol. 92: 211–221

McQuinn, I. H, L. Gendron & J. H. Himmelman, 1988 . Area of attraction and effective area fished by a whelk (*Buccinum undatum*) trap under variable conditions. Can. J. Fish. aquat. Sci. 45: 2054–2060.

Miller, R. E. & W. F. van Heukelem, 1988. Threshold levels of crab chemoreception for amino acids and results of field tests using three amino acids as attractants in artificial bait J. Shellfish Res. 7 (1): 127

Nickell, T. D. & P. G. Moore, 1992. The behavioural ecology of epibenthic scavenging invertebrates in the Clyde Sea area: laboratory experiments on attractions to bait in moving waters, underwater TV observations in situ and general conclusions Mar. Biol. Ecol. 159: 15–35

Ramsay, K., M. J. Kaiser, P. G. Moore & R. N. Hughes, 1997. Consumption of fisheries discards by benthic scavengers: utilization of energy subsidies in different marine habitats J. anim. Ecol.66: 884–896

Sainte-Marie, B. 1991. Whelk (*Buccinum undatum*) movement and its implications for the use of tag-recapture methods for the determination of baited trap fishing parameters Can. J. Fish. aquat. Sci. 48: 751–756

Youm, M.-G., Y.-J. Choi, S.-J. Kang, B.-D. Choi & C.-W. Cho, 1990. Attraction activities of fish to the baits – II Identification of chemical attractants for sea eels, *Astroconger myriaster*, in their natural habitats. Bull. Korean Fish. Tech. Soc. 26(4): 126–132

Youm, M.-G., 1991. Catching rate of trap due to the various baits Bull. Korean Fish. Tech. Soc. 27(4): 1–6

Appendix 1. Details of bait trials

Date	Bait trial	GPS co-ordinates				Depth (m)	Number of pots	Soak time (h)
		North	South	East	West			
15-Sep-97	Crab /dogfish	52 18 370	52 16 950	−6 10 740	−6 11 950	28–40	358	48
16-Oct-97	Crab/whitefish	52 18 350	52 16 830	−6 11 690	−6 12 690	26–38	391	24
25-May-98	Dog/whitefish	52 28 695	52 26 620	−6 11 332	−6 16 412	16–19	273	24
15-Sep-98	Crab/pelagics	52 26 485	52 24 613	−6 12 952	−6 14 642	23–28	359	24
16-Sep-98	Dogfish/pelagics	52 26 485	52 24 622	−6 13 048	−6 14 642	24–28	210	24
23-Nov-98	Crab/mussel	52 28 332	52 27 316	−6 14 319	−6 15 440	14–18	358	24
25-Nov-98	Dogfish/mussel	52 28 372	52 27 316	−6 14 319	−6 15 440	15–18	240	24
23-Mar-99	Crab/artificial bait crab	52 26 954	52 23 035	−6 15 050	−6 16 031	16–19	298	48
14-May-99	Crab & dog/art. crab & dog	52 28 335	52 26 798	−6 11 515	−6 13 489	13–19	239	24

Hydrobiologia **465:** 85–94, 2001.
G. Burnell (ed.), Coastal Shellfish – A Sustainable Resource.

The commercial potential of the common sea urchin *Echinus esculentus* from the west coast of Scotland

M.S. Kelly, P.V. Owen & P. Pantazis
Scottish Association for Marine Science, Oban, Argyll PA34 4AD, Scotland, U.K.
E-mail: mke@dml.ac.uk

Key words: Echinus esculentus, echinoid, gonad growth, diet, gonad colour

Abstract

Sea urchin roe is a valuable food product. The commercial potential of *Echinus esculentus* L. (Echinodermata: Echinoidea) was assessed both by surveying field populations and gonad growth in response to artificial feeds. Urchins were collected from four locations on the Scottish west coast between June and August of 1998. The roe content was found to be of variable quality. A relatively low percentage of urchins (34.9%) had roe which was considered marketable both in terms of yield and colour. Urchins with a marketable yield of roe were found where algal growth or encrusting organisms were most prolific. Good roe colours and high roe yields were not necessarily coincident. The data suggest an urchin fishery is likely to be impractical economically. In tank-based trials large *E. esculentus* were fed either commercially available salmon feed, a pelletised urchin diet or the macroalgae *Laminaria saccharina*. Urchins fed the artificial diets increased in test diameter during the 24-week experiment. At the end of the trial urchins fed salmon feed had a significantly higher gonad and alimentary index than those in the other treatments. *E. esculentus* is able to successfully exploit a diet rich in animal-origin oils. The diets also influenced gonad colour. The proportion of urchins with marketable gonad colours was highest (85%) in the treatment receiving salmon feed. As this species appears robust in culture and as artificial diets promote somatic and gonadal growth, an investigation into the practicalities of a fully farmed approach to gonad production may be worthwhile.

Introduction

The sea urchin *Echinus esculentus* L., often referred to as the common or edible urchin, is the largest echinoid found around the Scottish coast; specimens with test diameter approaching 10 cm are not uncommon (Nichols et al., 1985). It is found in a variety of habitat types from the littoral zone, where it may be exposed at low tides, to depths of ca. 40 m (Mortensen, 1927), although it has been recorded at depths exceeding 100 m (Comely & Ansel, 1989). There have been two previous investigations of *E. esculentus* populations on the west coast of Scotland (Comely & Ansell, 1988, 1989). Both were initiated in response to the anticipated exploitation of urchin stocks to supply the test for the curio trade. There is still a small fishery for urchin tests in Cornwall and the Scilly Isles. However this species is now attracting the attention of Scottish fishermen as a source of marketable urchin gonad.

The gonad of both male and female urchins, commonly referred to as 'roe', is a valuable food product. World urchin landings were in the region of 112 000 tonnes in 1996. The largest fisheries are in Japan, Chile and the USA, the latter two countries primarily exporting the product to markets in the far east. Smaller fisheries exist in Europe, mainly supplying domestic markets. The Chilean urchin fishery has increased dramatically in recent years to around 54 000 tonnes, largely due to the discovery of new fishing grounds. While it is unlikely that these catches can be maintained *Loxechinus albus* (Molina) from Chile currently exceeds *Strongylocentrotus* spp. from North America in terms of tonnage landed. However the fishery statistics clearly demonstrate that most of the world's urchin fisheries are fully or over exploited and it is generally accepted that further urchin fishing grounds are unlikely to be discovered (Keesing & Hall, 1998).

In the Japanese markets urchin roe is usually sold in skeins arranged on cedar wood trays. Each tray contains 280–300 g of roe worth up to £70 kg^{-1} (Hagan, 1996). The value of a tray is heavily dependent on the roe colour, the most desirable being a bright orange shade. To be in marketable condition the roe must also be well formed and firm.

Premium quality whole urchins can also command high prices (retailing at up to £12 kg^{-1}) in European markets (Grosjean et al., 1998). In Scotland, the native *Psammechinus miliaris* has already been identified as a potential aquaculture species and several investigations have been made into its growth rates, survivorship and roe production (Kelly et al., 1998a, 1998b; Cook et al., 1998). *E. esculentus* has not been exploited for its roe in Scotland in recent times, except in the Shetland Isles where for a brief period a fish processing company was exporting weekly shipments of up to 300 kg of live *E. esculentus* to European markets during the season when roe quality was acceptable (Penfold, 1995). Trading ceased following the 'Braer' oil spill, January 1993. *E. esculentus* gonad is generally not consumed in Europe and is considered to be of poor quality. However seasonal yields can be substantial and individual roe segments are of a similar size to that sought by the Japanese and Taiwanese markets. The increasing demand for urchin roe in these markets suggests it may be marketable there (E. Chu, Marketing Director, Seapro, Taiwan, pers commun.), but at present the low quality means that the cost of air freighting produce to the far east is not economically viable. Consequently, if quality control could be improved, it may become viable to market *E. esculentus* roe.

Many species of echinoids consume and thrive on prepared feeds; the topic has been extensively reviewed by Klinger et al. (1998). As the gonad also serves as a storage organ in some species (Kelly et al., 1998b) it may also be possible to achieve marketable gonad weights outside the normal maturing period by providing an excess of food. However, there is no information on the performance of *E. esculentus* fed artificial diets.

The aim of this study was to survey the range of gonad yields and colours in *E. esculentus* populations from a variety of habitat types on the west coast of Scotland. In addition, the gonad yield, condition and colour of urchins fed commercially prepared diets was also assessed. The commercial potential of roe from a wild stock fishery or produced under culture conditions was then assessed.

Methods

Field survey

The survey of gonad content and colour was conducted from June to August 1998. Twenty *E. esculentus* were collected from each of four different locations on the west coast of Scotland by SCUBA divers. The locations were Rhu Bay (Sound of Arisaig), Loch Melfort (south of Oban), Dunbeg Point (near Oban) and Portavadie (Loch Fyne) (Fig. 1). Site descriptions are given below. Two sites (10 urchins $site^{-1}$) were sampled at each location, to allow within location and between location comparisons. *E. esculentus* abundance was recorded at each site using Nature Conservation Council scales for semi-quantitative abundance of large solitary species: abundant = 1 or >1 0.1 m^{-2}; common = 1 or >1 m^{-2} and frequent = <1 m^{-2} but >20 individuals observed in that location (Davies, 1989). Similarly each site was scored in terms of the abundance of food items, including macroalgae and encrusting invertebrate communities as *E. esculentus* is known to feed on both plant and animal material (Lawrence, 1975; Emson & Moore, 1998). For algae and encrusting communities the following scoring system was used: abundant = plants / colonies mostly <50 cm apart; common = plants/colonies 50 cm – 1 m apart; frequent = plants /colonies 1–2 m apart; occasional = plants/colonies >2 m apart; rare = few plants/colonies present.

Site descriptions

Rhu Bay 1: Depth 10 m. *Laminaria digitata* forest growing on flat bedrock. *E. esculentus* collected from kelp stypes which had a dense covering of epiphytic red algae.

Rhu Bay 2: Depth 6 m. Steep slope comprised of large boulders, *E. esculentus* collected from rock faces below *L. digitata* zone.

Loch Melfort 1: Depth 14 m. Cliff, very little *L. digitata* present. *E. esculentus* collected from silted surfaces of steeply sloping sea bed and large boulders.

Loch Melfort 2: Depth 14 m. As above

Dunbeg Point 1: Depth 16 m. Sloping, muddy sand with widely separated (several metres apart) patches of drift and live kelps. *E. esculentus* collected from patches of kelp.

Dunbeg Point 2: Depth 14 m. Sloping rock face, 50°, covered in encrusting algae, bryozoans and barnacles also abundant. Adjacent to minor sewage outfall. *E. esculentus* collected from below the *Laminaria* zone.

Portavadie 1: Depth 8 m. Man-made lagoon, restricted exchange environment, boulder and scree slope giving way to concrete floor, silted up with muddy sand, *L. saccharina* present. *E. esculentus* collected from vertical wall descending to floor of lagoon.

Portavadie 2: Depth 12 m. Man made lagoon, restricted exchange environment, large boulders (1 m diameter) sitting on sand floor, near entrance to sea. *L. saccharina* abundant, *E. esculentus* collected from both boulder slope and lagoon floor.

Upon collection the urchins were placed in insulated boxes of sea water and transferred promptly to aquaria at the laboratory. The urchins were dissected after 48 h when the gut contents had been evacuated. Upon dissection the following parameters were recorded: test diameter (measured using calipers adapted by the addition of metal pins which fitted between the urchins' spines), total wet weight, eviscerated test weight and gonad weight. A gonad index (GI) was calculated using the formula: GI = 100 × (gonad weight/eviscerated test weight). The gonad colour was recorded using a Pantone colour card and then scored as being of marketable or unmarketable colour (Marketable colour codes 135, 1215, 1225, 134, 129 (yellow and orange); Unmarketable colour codes 452, 1255, 1245, 131, 4495 (brown and grey). A marketable gonad weight was taken to be a GI of >25% or a gonad weight of >25 g (Penfold, 1995).

The data generated were analysed using the multiple range Duncan test, the Student–Newman–Keuls test and the statistical package SPSS for Windows

Diet trials

E. esculentus were collected from the Dunbeg site in March, 1999, and maintained in sea water aquaria and observed for 3 weeks prior to the start of the experiment to allow them to acclimatise to the tanks. Any casualties resulting from handling or collection were removed. The urchins were then graded according to horizontal test diameter. Similarly sized urchins were then randomly assigned to one of nine 400 l aquaria each equipped with an independent sea water supply (flow rate 3.5l/minute). Each tank contained 30 individuals, mean test diameter over the nine aquaria was 94.00 mm, the maximum standard deviation around the mean in any tank was 8.02 mm. Ten of the 30 individuals in each tank were corralled away from the other urchins using a 2 × 2 cm mesh fence, which divided the tank. These urchins, of known test diameter, received the same treatment as their cohabitants but were maintained for estimation of somatic growth at the end of the trial. The remaining 20 urchins in each tank were destructively sub-sampled over the course of the experiment for estimation of gonad and alimentary indices.

Each tank was then randomly assigned one of three diets: (a) Sa ('Trouw' 8 mm extruded salmon feed containing 25 ppm of each of the pigments astaxanthin and canthaxanthin); (b) Ur (an experimental extruded urchin diet (Pantazis et al., 2000) manufactured by Redmills, Ireland with no added pigments); (c) Ls ('natural' kelp diet of *L. saccharina.*) There were three replicate tanks for each diet and the urchins were fed all three diets ad lib. Approximate diet formulations are given in Table 1.

The trials were conducted from March to August 1999. The tanks were cleaned weekly using a siphon to removed uneaten food and faeces. Four urchins were removed from each tank at approximately monthly intervals and the total wet weight, gonad weight, gut weight and gonad colour was recorded.

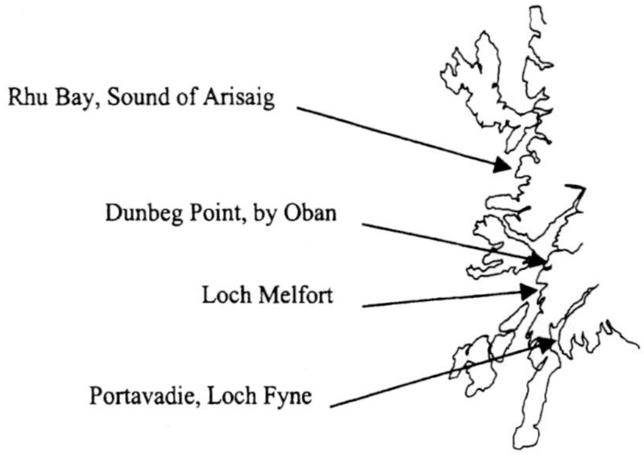

Figure 1. Collection sites on the West Coast of Scotland.

Table 1. Proximate analysis of the formulated diets (% on a dry matter basis)

	Urchin diet	Salmon diet
Crude protein	36.96	44.13
Crude lipid	5.1	31.37
Carbohydrates	47.84	14.063
Ash	7.93	9.40
Energy in kJ/g	16.24	23.72

GI, was calculated as described above and alimentary index (AI) was calculated as AI = 100 × (empty gut weight/eviscerated test weight). Differences in GI and AI between treatments on any given sample day were tested for significance using nested Analysis of Variance, Tukey test and the statistical package Minitab Version 12. To ensure compliance with the assumptions of ANOVA, percentage data were arcsine transformed and tested for normality (Anderson–Darling test) and homogeneity of variance (Bartletts test). The colour was assigned a number using Pantone colour card and a further colour category added (i.e., excellent, acceptable or unacceptable) to distinguish marketable roes from those which would be considered top quality. The colour codes were as follows; Excellent colour codes 135U, 1225U (bright orange); Acceptable colour codes 1215U, 124U, 129U, 134U (pale yellow and dull orange) and Unacceptable colour codes 452U, 1255U, 1245U, 131U, 4495U (brown or grey). The frequencies observed in each category on the last sample date were compared to those expected should the treatments not influence gonad colour (i.e. not differ from those observed in the initial sample) using the log-likelihood ratio (Zar, 1996).

One of the gonads from each of 10 urchins from each treatment was collected at the beginning, during (July) and at the end of the experiment. The gonad samples were preserved in Bouins fluid for histological examination of the reproductive state. A central portion of the preserved gonad was dehydrated in alcohol, embedded in paraffin wax and sectioned at 7 μm. The sections were stained with haematoxylin and eosin (H/E). The reproductive state of the urchins was then assessed following the descriptions of Fuji (1960) and Byrne (1990).

Results

Field survey

Comparison of the recorded parameters (Table 2) showed that there were significant differences in test diameter, total wet weight and GI both between locations and between sites at the same location. The largest urchins were collected from Rhu Bay, the GI of these urchins varied from a mean of 35.61–22.74, corresponding to a mean of 55.53 and 36.96 g of gonad per urchin respectively. Note that the largest urchins did not have the highest GI. The relatively small urchins from Dunbeg Point 2 (mean test diameter 90 mm) yielded a similar quantity of gonad (>36 g) as the urchins from Rhu Bay 2 (mean test diameter of 103.58 mm). The urchins from the two Loch Melfort sites and Portavadie 1 had the smallest test diameters and also the lowest gonad weights, which ranged between 11.28 and 22.74 g. The urchins from Rhu Bay 1 and Dunbeg Point 2 had significantly higher GI than those from the other sites. While GI was highly variable between individuals within a population, there was no significant difference in GI of the urchins from the six remaining sites. Generally the highest GI were found at the sites where food was abundant. However, at Portavaddie Site 2, the food (*L. saccharina)* was scored as abundant but the GI index of the urchins was relatively low.

Table 3 summarises the percentages of urchins found at each location which had marketable gonad in terms of yield and colour. Gonad colour varied from pale yellow to pale brown at all the sampled locations. The pale yellow coloured gonads would be acceptable for marketing in the far eastern (pers. commun. Eric Chu, Purchasing Director, Seapro Co. Ltd, Taiwan). The Portavadie population had the highest percentage of gonads with desirable colours but there was no obvious difference between this site and the others in terms of the possible food sources present. Overall a relatively small percentage (34.9%) of the urchins surveyed would have been considered marketable.

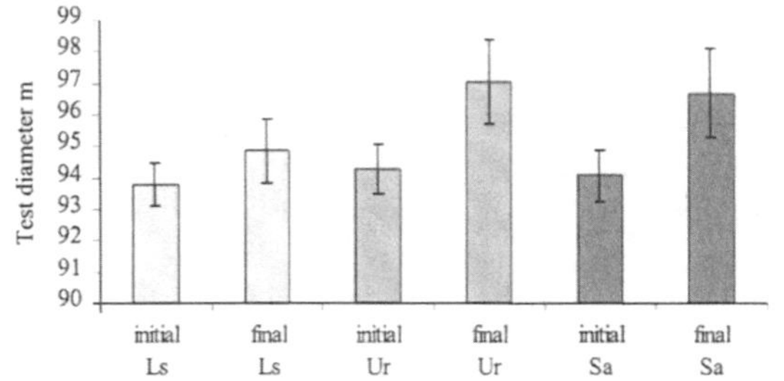

Figure 2. Initial and final mean test diameters of *E. esculentus* (n=30) fed either *Laminaria saccharina* (Ls), Urchin diet (Ur) or Salmon feed (Sa). Error bars represent 95% confidence limits.

Diet trials

Survivorship was high (97%) throughout the trial. The urchins adapted well to the pelletised diets, which they were able to trap and manipulate and appeared to feed on freely.

The data from the somatic growth trial showed that the test diameter of the urchins from treatments Sa and Ur both increased significantly over the course of the 24 week experiment (Fig. 2). The increase in test dia-

Table 2. Comparison of *E. esculentus* from different locations

Site/date	Echinus/ food abundance	Mean test diameter mm	Mean wet weight (g)	Mean gonad wet weight (g)	Gonad index (gonad weight / eviscerated test weight × 100)
Rhu Bay 1 17.06.98	C/A	99.73^b (9.34)	512.17^a (127.2)	55.53^a (17.68)	35.61^a (13.65)
Rhu Bay 2 17.06.98	F/C	103.58^a (3.79)	546.74^a (64.82)	36.96^b (12.12)	22.74^b (6.19)
Melfort 1 25.06.98	F/R	76.64^f (7.51)	235.76^c (63.63)	11.28^d (9.92)	16.17^b (12.41)
Melfort 2 25.06.98	F/R	77.35^f (7.69)	244.76^c (64.96)	11.72^d (5.45)	19.33^b (9.99)
Dunbeg Pt 1 31.07.98	C/O	93.82^c (8.2)	396.92^b (97.96)	28.68^c (9.84)	24.40^b (8.57)
Dunbeg Pt 2 31.07.98	C/A	90.03^d (7.02)	347.63^b (81.38)	38.81^b (9.28)	40.49^a (13.54)
Portavadie 1 14.08.98	F/C	81.51^f (11.57)	251.51^c (79.51)	22.74^d (19.26)	25.74^b (20.58)
Portavadie 2 14.08.98	F/A	86.40^e (7.73)	352.31^b (81.43)	29.42^c (11.03)	29.02^b (11.84)

Standard Deviation in parentheses. Numbers in the same column with the same superscript are not significantly statistically different. A, abundant; F, frequent; C, common; O, occasional; R, rare.

meter for the these treatments was Sa 2.58 mm, (SD 3.02, $n = 20$); Ur 2.75 mm, (SD 3.53, $n = 30$).

A high degree of variation in GI between individuals was recorded throughout the experiment, as evidenced by the large error bars (Fig. 3). The majority of urchins dissected over the duration of the experiment appeared ripe and gametes were extruded from the gonads on dissection. On the last sample day the remainder of the urchins in each tank were sacrificed ($n = 30$ and not 12 as on previous sample days) and significant differences were then found between treatments; urchins from treatment Sa had significantly higher GI than those from either Ur or Ls ($P<0.001$, $F = 0.5$, MS = 4873.0 df 2,5).

The pattern of AI was less variable, on the last two sample days urchins from treatment Sa had significantly higher AI than those in the other two treatments (Fig. 4) (27.09.99, $P = 0.001$, $F = 29.3$, MS = 19.38, df 2,5). The increased AI indicated that the urchins in treatment Sa were using the gut wall as an additional site for nutrient storage. Log-likelihood comparison of the colour distribution between the three categories showed that by the last sample date the colour distribution in all three treatments varied significantly from the original distribution suggesting that diet altered gonad colour. The colour distribution in the acceptable category had increased from 6 to 18 ($n = 30$) in all treatments (Table 4) and the percentage of unacceptable colours in all treatments had declined. The urchins fed the salmon food had the lowest number of unacceptable, the highest number in the excellent category and a total 85% of urchins with acceptable or excellent coloured gonads (Table 5). However, the presence of ripe gametes in the gonads also effects gonad colour; the sex of mature individuals could also

Table 3. Percentage of urchins ($n = 20$ per location) with marketable gonad yield (Gonad Index >25% or gonad weight >25 g), marketable gonad colour and with both marketable yield and colour

Site	Yield only (g) (%)	Colour only (%)	Yield and colour (%)	Total % marketable all sites
Rhu Bay	91	33	33	
Melfort	10.5	38	10.5	
Dunbeg	75	55	40	
Portavadie	50	85	50	
				34.9

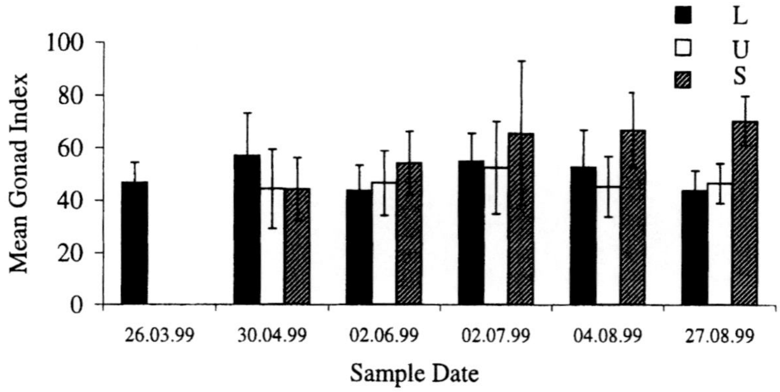

Figure 3. Mean Gonad Index (n=12, except 27.08.99 where n= 30) of *E. esculentus* fed either *Laminaria saccharina* (Ls), Urchin diet (Ur) or Salmon feed (Sa). Error bars represent 95% confidence limits.

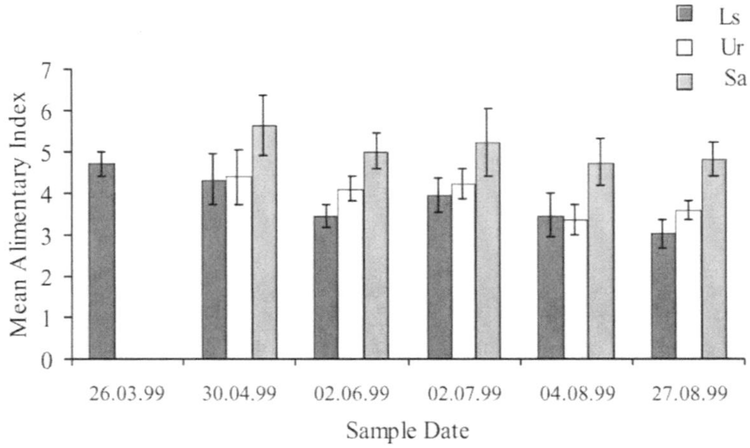

Figure 4. Mean Alimentary Index (n=12, except 27.08.99 where n= 30) of *E. esculentus* fed either *Laminaria saccharina* (Ls), Urchin diet (Ur) or Salmon feed (Sa). Error bars represent 95% confidence limits.

Table 4. Observed frequencies in gonad colour categories (U, unacceptable; A, acceptable; and E, excellent) initially and at the end of the experiment

Sample March 1999				Sample August 1999			Value of G
U	A	E		U	A	E	
15	6	9	*L. saccharina* (Ls)	8	18	4	G = 9.990**
			Urchin diet (Ur)	11	18	1	G = 14.304**
			Salmon food (Sa)	4.5	18	7.5	G = 11.282**

* Indicates a distribution having a significantly different Unacceptable:Acceptable:Excellent ratio from the initial sample. $G_{critical} = \chi^2_{0.01,2} = 9.21$ (Zar, 1996).

Table 5. Percentage of urchins (n = 30 per treatment) with marketable gonad yield (Gonad Index > 25% or gonad weight > 25 g), marketable gonad colour and with both marketable yield and colour

Treatment	Yield only g (%)	Colour only (%)	Yield and colour (%)	Total % marketable
L. saccharina (Ls)	93.3	73.0	70.0	
Urchin diet (Ur)	93.3	63.0	63.0	
Salmon food (Sa)	100.0	85.0	85.0	
Mean				72.8

be distinguished on the basis of colour, males tending to have paler coloured gonads.

Examination of the histological sections confirmed that the females were reproductively mature when collected in March. In females, mature ova (110–120 μm) were present in the lumen of the ascinal ducts and premature oocytes were also visible in a narrow

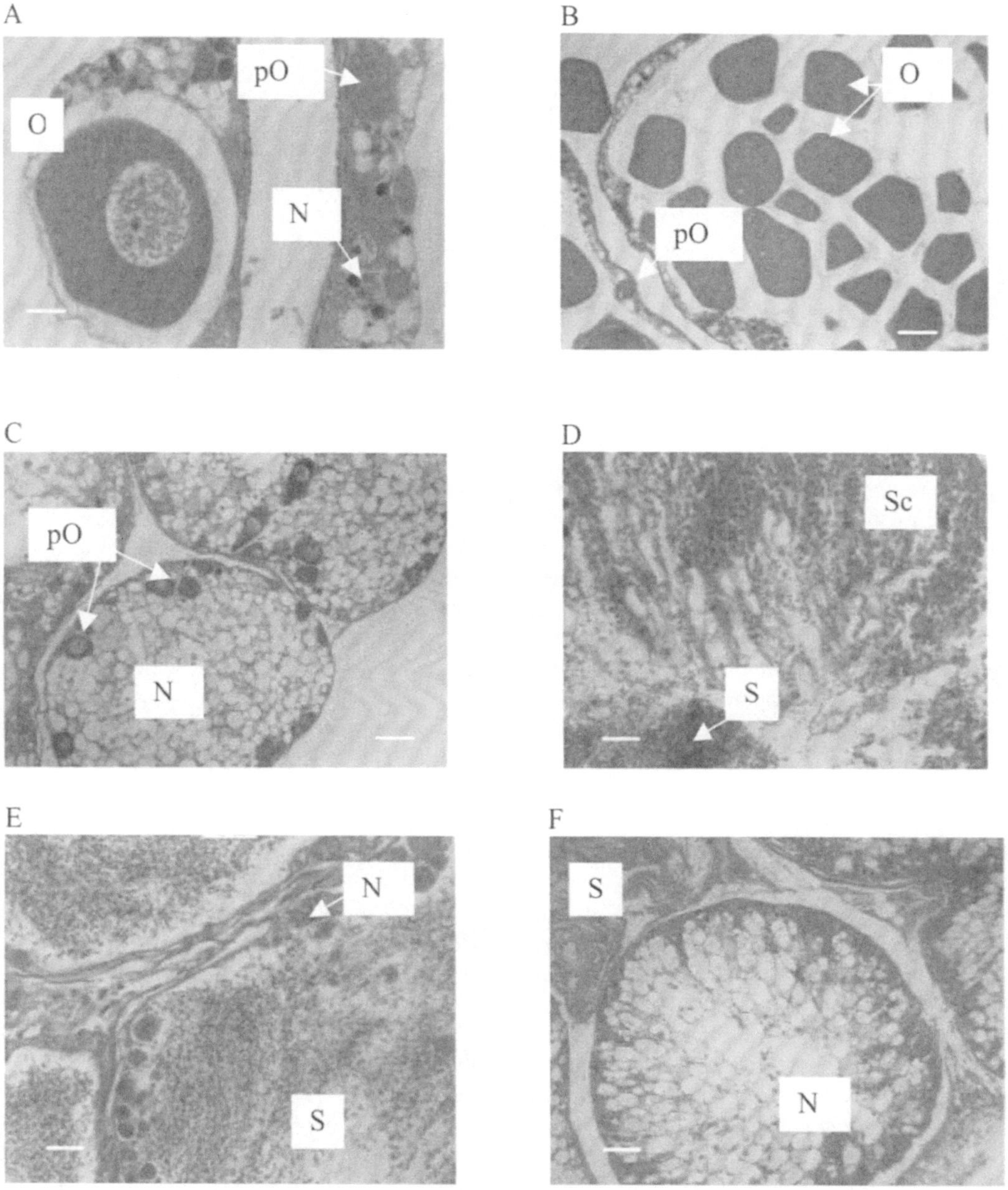

Figure 5. (A) Mature ovary, March 1999. Mature ova (O) project from the ascinal wall towards the lumen, premature oocytes (pO) are also present in the layer of nutritive phagocytes (N) surrounding the ascinal wall. Scale bar = 20 μm. (B) Mature ovary, July 1999. Mature ova (O) pack the lumen of the ascini. Some premature oocytes (pO) remain in the nutritive layer which is much reduced. Scale bar = 80 μm. (C) Recover phase ovary, August 1999. A loose network of nutritive cells (N) now fill the lumen, no mature ova remain. Part of the nutritive layer around the ascinal wall begins to stain more densely. Premature oocytes (pO) are present. Scale bar = 80 μm. (D) Mature testes, March 1999. Spermatozoa (S) are present in the ascinal lumen, but columns of spermatocytes (Sc) remain projecting from the ascinal wall. Scale bar = 20 μm. (E) Mature testes, July 1999. The columnar spermatocyte layer is reduced to a thin layer of nutritive cells (N), the lumen is packed with spermatozoa (S). Scale bar = 20 μm. (F) Partially spent testes, August 1999. A loose network of nutritive phagocytes (N) fill the lumen of the ascinus. Some spermatozoa (S) remain in the gonoduct. Scale bar = 80 μm.

layer of nutritive cells lining the ascinal wall (Fig. 5a). Examination of the histological sections from July showed females with mature ova present in the ascinal lumen and gonoduct. Some premature oocytes were still present (Fig. 5b). By the end of August 90% of the females examined appeared partially spent, there were no or few ripe ova remaining and the nutritive phagocytes lining the ascinal wall appeared more densely staining (Fig. 5c) indicating the ovaries were entering the recovery phase. The males examined were also reproductively mature from March; mature spermatozoa filled the lumen and columns of spermatocytes projected from the ascinal walls towards the lumen (Fig. 5d). In July the males were still mature, spermatozoa filled the ascinal lumen and were present in the gonoduct (Fig. 5e) and the nutritive layer had become very reduced. By August 50% of the males examined were spent, with no mature spermatozoa visible in the ascinal lumen or gonoducts (Fig. 5f). The remainder still had some spermatozoa present in the ascinal lumen or gonoduct. No distinction, histologically, in either reproductive (egg size, stage of maturity) or nutritive state (depth of nutritive phagocyte layer) could be made between urchins from the different treatments. No attempt to measure fecundity was made as mature females released eggs throughout the experimental period.

Discussion

Tank-based studies have shown that gonad growth in echinoids is related to food abundance as well as reproductive activity (Cook et al., 1998). It is also possible to relate gonad yield to food abundance in this field study. Populations with high GI were found where there was a prolific growth of either algae or encrusting organisms. It is, however, more difficult to attribute the higher percentage of marketable gonad colours found in a given population to the presence of any particular food type. As gonad maturity and spawning events vary between *E. esculentus* populations on the west coast (Comely & Ansell, 1989) differences in gonad yield and colour may also be caused by variation in reproductive status.

The data showed that a maximum of 50% of the urchins fished from any given location, during the season when the gonad was likely to be well formed, were actually marketable and that the occurrence of marketable populations is likely to be patchy. Ensuring a sufficiently high proportion of good quality individuals in any batch, without processing the product, would therefore be difficult. These observations correspond with those made on other urchin populations where gonad yield was found to vary in relation to food availability (Vadas, 1977) and illustrates the necessity for adopting roe enhancement or a fully farmed approach to obtaining consistent gonad quality.

The literature reviewed by Lawrence (1975) illustrates how sea urchins are often the determining factor with regard to the abundance and distribution of plants in shallow water marine environments. Jones & Kain (1967) concluded that the lower limit of the *Laminaria* zone was at least partly determined by the grazing pressure of *Echinus* species. The impact of over exploitation in echinoids can be compounded by the age structure of the target populations. The red sea urchin, *Strongylocentrotus franciscanus*, from Oregon and Washington, attains test diameters greater than 17 cm and recent estimates of their age from growth models places them in excess of 100 years old (Ebert, 1998). A study using the antibiotic tetracycline as a means of 'tagging' skeletal plates in the urchin test and monitoring the rate of growth (Gage, 1992) suggested that growth rates in adult *E. esculentus* are very low and that adults may survive to an age of 16 years or more. Neither the present study nor that of Penfold (1995) made any attempt to assess the impact of initiating a fishery for *E. esculentus*, often the dominant macroalgivore in inshore communities. The fact that this species is abundant in shallow water (<10 m) means it is particularly vulnerable to SCUBA diver collection. Without prior knowledge of the age structures of populations, and the factors influencing recruitment, collection of large numbers of *E. esculentus* could potentially lead to the populations being irrevocably damaged and to substantial changes in the flora and fauna of inshore ecosystems.

This is the first investigation of growth rates in *E. esculentus* fed formulated feeds. The recorded increase in test diameter over the duration of the trial in urchins fed formulated feeds was unexpected as these urchins were already large and at a stage where growth rates are reportedly slow in this species (Nichols et al., 1985; Gage, 1992). The data show that in this species an abundance of food permits both reproductive activity and somatic growth to continue over the reproductive season. All the diets sustained gonad mass throughout the trial.

Jorde & Klavestad (1963) investigated the gut contents of *E. esculentus* and found it primarily composed of algae, with very little animal material present. Later

Liemans & Dandrifosse (1972) showed that there were high levels of the digestive enzyme amylase in the stomach of this species, suggesting that it is best adapted to a carbohydrate diet. The observations of Comely & Ansell (1988) indicated that *E. esculentus* from the Scottish west coast fed predominantly on *L. saccharina*, its associated epiphytic growths and some encrusting organisms including barnacles. Consideration of these facts leads to the conclusion that this species is predominantly herbivorous. However, a more recent study from the Clyde sea (Emson & Moore, 1998) describes a population of *E. esculentus* which fed almost exclusively on barnacles and which had a higher GI than neighbouring populations feeding on either algae or detritus. The present study supports this observation confirming that *E. esculentus* can successfully exploit diets with high levels of animal-origin lipids.

Comely & Ansell (1989) conducted a detailed investigation into the reproductive cycle of *E. esculentus* on the Scottish west coast. Urchins were collected monthly from two locations and from within and below the *Laminaria* zone. All populations showed a clear annual cycle of gonad development and spawning, but the exact timing of gamete production and spawning events varied considerably between sites. In general, the gonad weights were highest in spring and early summer with urchins beginning to appear spent by September. In three of the four sites the females were predominantly ripe between April and June, while in the fourth they were ripe between July and August. The males seemed to ripen before the females, and there were not many months of the year when some ripe males could not be found. The histological examination of the gonads in the present study confirmed that ripe gametes could be found in males and females from March to August, although a proportion of the urchins appeared spent by late August. A prolonged spawning period has negative implications for marketing; gonads which containing mature gametes are soft and ooze on dissection, rendering the gonad unmarketable. However, as the *E. esculentus* adapted well to the pelletised feed it may be possible to use the artificial diets to enhance gonad growth outside the reproductive season and produce a gonad which is composed mainly of nutritive storage cells (Kelly et al., 1998) rather than gametes. If such gonad growth can be achieved the product would be available over the winter period when demand for urchin gonad is high.

Gonad colour in echinoids is the result of the assimilation and conversion of ingested carotenoid pigments. Tsushima & Matsuno (1990) found the most frequently occurring pigments in the gonads of 11 species of echinoids were echinone, derived from β-carotene and β-carotene. Galasko et al. (1969) found the main pigment in *Paracentrotus lividus* to be fucoxanthin, a derivative of fucoxanthinol. Both ß-carotene and fucoxanthinol are found in brown macroalgae such as *L. saccharina* (Davies, 1976). Goebel & Barker (1998) examined the pigment content of the New Zealand sea urchin, *Evechinus chloroticus*, with the aim of designing a diet to enhance gonad colour. The addition of carotenoid supplements to an artificial diet produced some alteration in gonad colour. Most notably, the addition of astaxanthin produced a significant improvement in gonad colour in experimental urchins when compared to urchins from field populations. However further research is required to fully understand the fate and expression of ingested pigments and to aid in the design of diets that enhance gonad colour. Feeding commercially prepared salmon feed containing astaxanthin resulted in dull orange rather than bright orange gonads in *Psammechinus miliaris* (Kelly et al., 1998b). The results of the present study indicate there is potential for designing diets to enhance gonad colour in *Echinus esculentus.* Feeding *E. esculentus* salmon feed containing astaxanthin and canthaxanthin produced a higher percentage of marketable gonad colours. However it is unclear to what extent these pigments influence gonad colour in this species.

In conclusion it appears the likelihood of an economically viable or sustainable fishery for this species is remote. The study shows that artificial diets can easily be used to enhance somatic and gonad growth in this species. Therefore, the practicalities of a full-farming approach, raising *E. esculentus* spat, and optimising the diet and grow-out facilities (either sea based or onshore) with the aim of marketing a 45 mm urchin with premium gonad quality, may be worthy of investigation.

Acknowledgements

This work was funded by Highlands and Islands Enterprise and the Highland Council. We thank the Director of the Scottish Association for Marine Science for the use of the facilities, Mr John Paterson, Shoreline International, Mr Gordon Anderson, Portavaddy Estates and the Dunstaffnage Marine Laboratory Dive

Team for facilitating urchin collection and Mr James Youlden for preparing the histological sections.

References

Comely, C. A. & A. D. Ansell, 1988. Population density and growth of *Echinus esculentus* L. on the Scottish West Coast. Est. coastal Shelf Sci. 27: 311–334.

Comely, C. A. & A. D. Ansell, 1989. The Reproductive cycle of *Echinus esculentus* L. on the Scottish West Coast. Est. coastal Shelf Sci. 29: 385–407.

Cook, E. J., M. S. Kelly & J. D. McKenzie 1999. Gonad development and somatic growth in the green sea urchin, *Psammechinus miliaris* fed artificial and natural diets. J. Shellfish Res 17 (5): 1549–1555.

Davies, B. H., 1976. Carotenoids. In Goodwin, T. W. (eds), Chemistry and Biochemistry of Plant Pigments. Academic Press, London, Vol. 2: 38–165.

Davies, M., 1989. Surveys of Scottish Sealochs, Lochs A'Chairn Bhain, Glendhu and Glencoul. Marine Nature Conservation Review report no. 983. University Marine Biological Station, Millport, Isle of Cumbrae, Scotland.

Ebert, T. A., 1998. An analysis of the importance of allee effects in management of the red sea urchin *Strongylocentrotus franciscanus*. In Mooi, R. & M. Telford (eds), Proceedings of the 9th International Echinoderm Conference, San Francisco. Balkema: Rotterdam: 619–627.

Emson R. H. & P. G. Moore, 1998. Diet and gonad size in three populations of *Echinus esculentus*. In Mooi, R. & M. Telford (eds), Echinoderms: San Francisco: Proceedings of the 9th International Echinoderm Conference, San Francisco, 1996: 641–644.

Gage, J. D., 1992. Natural growth bands and growth variability in the sea urchin *Echinus esculentus*: results from tetracycline tagging. Mar. Biol. 114 (4): 607–616.

Goebel, N. & M. F. Barker, 1998. Artificial diets supplemented with carotenoid pigments as feeds for sea urchins. In Mooi, R. & M. Telford (eds), Proceedings of the 9th International Echinoderm Conference, San Francisco. Balkema: Rotterdam: 667–672.

Galasko, G., J. Hora, T. P. Toube, B. Weedon, D. Andre, M. Barbier, E. Lederer & V. R. Villanueva, 1969. Carotenoids and related compounds. Part XXII Allenic carotenoids in sea urchins. J. Chem. Soc. C: 1264–1265.

Grosjean, P., C. Spirlet, P. Gosselin, D. Vaitilingon & M. Jangoux, 1998. Land-based, closed-cycle echiniculture of *Paracentrotus lividus* (Lamarck) (Echinoidea: Echinodermata): A long-term experiment at a pilot scale. J. Shell. Res. 17 (5): 1523–1531.

Hagen, N. T., 1996. Echinoculture: from fishery enhancement to closed cycle cultivation. World Aquacult. 27 (4): 6–19.

Jones, N. S. & J. M. Kain, 1967. From Lawrence, J. 1975. On the relationships between marine plants and sea urchins. Oceanogr. Mar. Biol. annu. Rev. 13: 213–286.

Jorde, I. & N. Klavestad, 1963. From Lawrence, J. 1975. On the relationships between marine plants and sea urchins. Oceanogr. Mar. Biol. Ann. Rev. 13: 213–286.

Kawakami, T., M. Tsushima, Y. Katabami, M. Mine, A. Ishida & T. Matsuno, 1998. Effect of b,b-carotene, b-echinenone, astaxanthin, fucoxanthin, vitamin A and vitamin E on the biological defense of the sea urchin *Pseudocentrotus depressus*. J. Exp. Mar. Biol. Ecol. 226, 165–174.

Kelly, M. S., C. C. Brodie, & J. D. McKenzie, 1998a. Sea urchins in polyculture: the way to enhanced gonad growth. Mooi, R. & Telford, M. Echinoderms: San Francisco: Proceedings of the 9th International Echinoderm Conference, San Francisco, 1996: 707–711.

Kelly, M. S., C. C Brodie, & J. D. McKenzie, 1998b. Somatic and gonadal growth of the sea urchin *Psammechinus miliaris* maintained in polyculture with the Atlantic salmon. J. Shellfish Res. 17 (5): 1557–1562

Keesing, J. K. and K. C. Hall (1998). Review of the status of world sea urchin fisheries points to opportunities for aquaculture. J. Shell. Res. 17 (5): 1597–1604.

Klinger, T. S., J. M. Lawrence, & A. L. Lawrence, 1998. Digestion, absorption and assimilation of prepared feeds by echinoids. Mooi, R. & Telford, M. Echinoderms: San Francisco: Proceedings of the 9th International Echinoderm Conference, San Francisco, 1996: 713–721.

Lawrence, J., 1975. On the relationships between marine plants and sea urchins. Oceanogr. mar. Biol. Annu. Rev. 13: 213–286.

Liemans, M & G. Dandrifosse, 1972. From Lawrence, J., 1975 On the relationships between marine plants and sea urchins. Oceanogr. mar. Biol. Annu. Rev. 13: 213–286.

Mortensen, T.H. 1927 Echinoderms of the British Isles. Oxford University Press

Nichols, D., A. A. T. Sime & G. M. Bishop, 1985. Growth in populations of the sea urchin *Echinus esculentus* L. (Echinodermata: Echinoidea) from the English Channel and the Firth of Clyde. J. Exp. Mar. Biol. Ecol. 86: 219–228.

Pantazis, P., M. S. Kelly, J. G. Connolly & K. D. Black, 2000. The effect of artificial diets on growth, lipid utilisation and gonad biochemistry in the adult sea urchin *Psammechinus miliaris*. J. Shellfish Res. 19(2): 995–1002.

Penfold, R., 1995. A report to evaluate the feasibility of a sea urchin processing industry in Shetland. North Atlantic Fisheries College, Port Arthur, Scalloway, Shetland, ZE1 OUN.

Tsushima, M. & T. Matsuno, 1990. Comparative biochemical studies of carotenoids in sea-urchins – I. Comp. Biochem. Physiol. 96B (4): 801–810.

Vadas, R. R,. 1977. Preferential feeding: An optimisation strategy in sea urchins. Ecol. Monogr. 47: 337–371.

Zar, J. H., 1996. Biostatistical Analysis. 3rd edn. Prentice Hall International, Princeton, NJ.

Hydrobiologia **465:** 95–102, 2001.
G. Burnell (ed.), *Coastal Shellfish – A Sustainable Resource.*

Implications of temporal and spatial variability in *Paracentrotus lividus* populations to the associated commercial coastal fishery

David K. A. Barnes* & Anne C. Crook
Department of Zoology and Animal Ecology, University College, Lee Maltings, Cork, Ireland
**Present address: Biological Sciences Division, British Antarctic Survey, High Cross, Madingley Rd, Cambridge CB3 0RT, U.K.*
E-mail: dkab@bas.ac.uk

Key words: echinoid fisheries, migration; population censi, Lough Hyne, *Paracentrotus lividus*

Abstract

Analysis of population structure and behaviour of a semi-enclosed and theoretically non-fished population of the echinoid *Paracentrotus lividus* at Lough Hyne marine nature reserve, Co. Cork, Ireland, showed distinct variability. Aspects of the behavioural variability in this species have significant consequences for the natural fishery harvest potential of this species. Whether or not, when and where individuals migrate to the upper surfaces of rocks varies on a local and regional scale as well as seasonally. The effect of this is apparent changes in (a) the population structure, (b) the population size and (c) the population centres of abundance. The population structure and size of the Lough Hyne population has changed substantially, as have those on a larger spatial scale, although the causative agents are debatable. Such changes are the result of large-scale mortality of adults and failure of cohort recruitment. Despite the undeniable pressures of overfishing at many localities, it seems likely there is a strong 'natural' and possibly predictable element to the variability in *P. lividus* population size and structure. For the restoration and sustainable harvest of Irish *P. lividus* aspects of its behavioural ecology need to be taken more closely into consideration.

Introduction

The productivity of coastal *Paracentrotus lividus* (Echinoidea) fisheries vary greatly between the localities in Europe (Ledireach, 1987). In addition, the scale and success of the fishery has varied on a national and local scale. This is particularly evident in the Mediterranean Sea since the early 1970s (Southward & Southward, 1975; Le Gall, 1987). Non-Mediterranean populations of *P. lividus*, in Britain, Ireland and Macronesian islands (e.g., the Azores), have more recently become subject to commercial exploitation (Byrne, 1990). The commercial landings of *P. lividus* on Irish coasts has, during the 1990s, undergone a similar decline to that observed in the Mediterranean Sea decades ago (Barnes et al., 1999).

Population censi have been carried out each summer at Lough Hyne, Co. Cork, Ireland, in a semi-continuous fashion for a number of decades (Barnes et al., 1999). In recent periods censi have been carried out using a variety of methods, including (a) observation and counting from a shallow draft boat using viewing boxes, (b) snorkelling and (c) detailed counts inspecting under rock surfaces. As *P. lividus* rarely occur below 2 m depth at this site (and elsewhere) these methods should provide realistic censi. The two principal modes of habitat use by *P. lividus*, subtidal migration from boulder under to upper surfaces and intertidal burrowing in rock, vary between locality and should have a strong influence on apparent population size and structure to visual censi. The populations burrowing in rock (some Mediterranean Sea, West Ireland and Azores) should be transparent to visual censi both in terms of population structure and size. Within a locality, the census method used has given population estimates more than an order of magnitude in difference. The 1994, 1995 and 1998 censi of the South Basin of Lough Hyne (Barnes et al., 1999), carried out using boat-board viewer, snorkel and snorkelling with rock turnover, gave typical relative estimates of 10–

13× and 33–56× respectively. Although clearly more accurate, the latter census has limited use as it cannot be directly compared to the historical censi at Lough Hyne (see Muntz et al., 1965; Kitching & Thain, 1983; Kitching, 1987).

The population of *P. lividus* resident in Lough Hyne, a marine nature reserve (and thus supposedly not fished), was noted to have changed in size in the late 1970s and early 1980s (Kitching & Thain, 1983). Over several decades the scale of this change was almost four orders of magnitude and appeared, prior to the 1980s, to fluctuate from 10 000s to 10s (Barnes et al., 1999). Superannual variability in echinoid population sizes and their ecological consequences are widely known from American coasts (see Elner & Vadas, 1990). Toxic dinoflagellate blooms have been found to be responsible for mass mortality of adult echinoids and *P. lividus* in particular (Cross & Southgate, 1980). Failure of mass spawning because of lowered sea temperatures (Fenaux, 1968), possibly linked to El Nino Southern Oscillation events (Barnes et al., 2001), may be important to annual variability in cohort success.

Censi, as have been carried out on the Lough Hyne population, are a practical method of assessing population size, structure and for potential fishery advice and population modelling. Such censi do, however, only provide a snapshot of a given population at that moment. We looked at various components (individual size, time of day, month) of diurnal migration, from under to upper boulder surfaces (Crook et al., 2000), in the Lough Hyne *P. lividus* population and investigated the influence on apparent population characteristics conferred to the snapshot observer. We also investigated variation in apparent population size with census methodology and population structure at three localities (Lough Hyne, South Ireland; Beara Peninsula, West Ireland; and São Miguel, the Azores).

Materials and methods

Observations of *P. lividus* migratory behaviour were made in situ (snorkelling and occasionally SCUBA) at Lough Hyne Marine Nature Reserve, Ireland, bimonthly from February to September 1999. The substratum in the 'North Basin' region of Lough Hyne (where the largest populations of *P. lividus* are now found) is characterised by rocks and boulders covered by a variety of algal species upon which they graze (Ebling et al., 1960; Kitching & Ebling, 1961). The ecology of Lough Hyne, including the role of *P. lividus* in the community, has been well described in a series of papers summarised by Kitching (1987). Three essentially similar sites were selected in the North Basin of Lough Hyne (Fig. 1). Observations of *P. lividus* were also made using similar methods (snorkelling) at other localities in Ireland (Beara Peninsula) and the Azores (São Miguel Island).

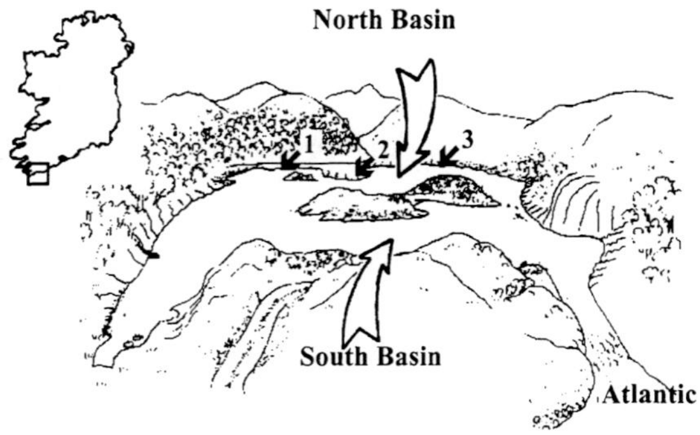

Figure 1. Lough Hyne. The position of Lough Hyne in South west Ireland and sites 1, 2 and 3 in the North Basin. Lough Hyne is approximately 0.85 km long by 0.6 km wide.

At Lough Hyne the proportion and size of *P. lividus* exposed were counted in nine replicate randomly placed 0.5-m^2 quadrats at each of three sites at the North Basin of Lough Hyne. The sizes (test diameters) of all individuals were measured using Vernier callipers. Similar methods were used to assess size frequency data for Lough Hyne South Basin, Beara Peninsula and Azores *P. lividus* populations. At Lough Hyne this was carried out across a 24-h period over 8 months at four distinct time periods: 'midday' (12:00–14:00 h), 'dusk' and 'midnight' (12:00–02:00 h) and 'dawn'. The times of the crepuscular periods varied with season.

An assessment of daily and seasonal activity of potential predators of *P. lividus* (e.g., *Marthasterias glacialis, Asteria rubens, Necora puber, Cancer pagurus* and *Carcinus maenus*) was also estimated at the three sites using randomised 1-m^2 quadrats. These predator observations were made using nine replicate 1-m^2 quadrats per site for each of the four time periods. Actual predation was recorded when sea stars were feeding on intact tests (with spines still attached) or (if crabs were feeding on *P. lividus*) there was signs of cracking by crab chelae. All instances of predation and freshly killed *P. lividus* individuals were recorded for a preliminary assessment of importance at this site and for comparison with elsewhere.

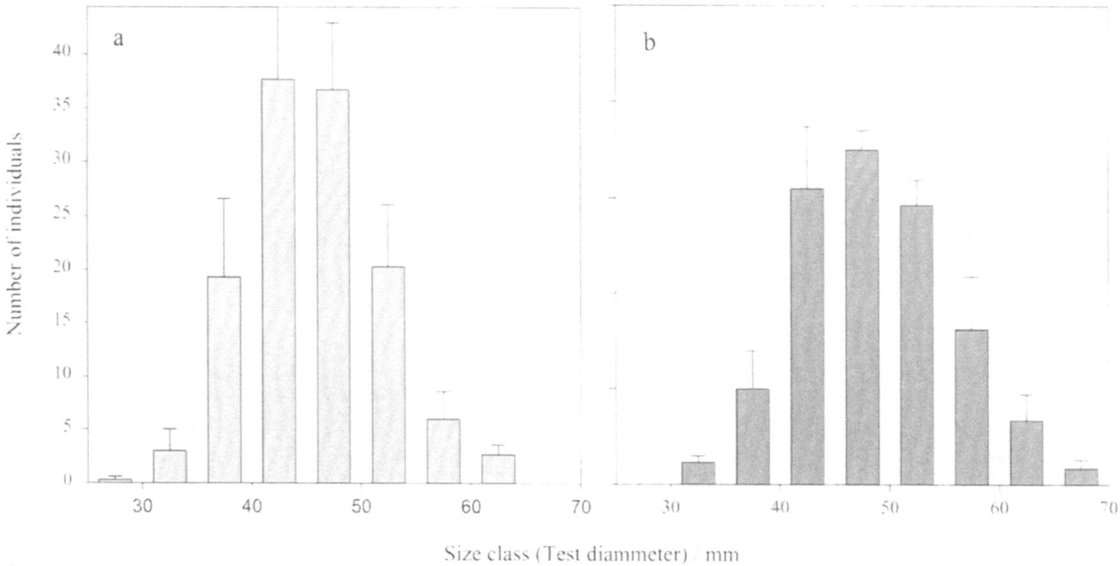

Figure 2. Paracentrotus lividus. Size frequency distribution at Lough Hyne during periods of (a) daylight (n=378) and (b) darkness (n=397). Data is presented as mean number of individuals sampled per site with standard deviation. Data did not significantly differ with site so is shown pooled between sites.

Results

The apparent population structure (of exposed or visible individuals of *P. lividus*) differed between the day and night periods at Lough Hyne (Fig. 2a,b). The median size (and thus age) of individuals was greater at night than by day (Kruskal–Wallis, H = 103.8, P < 0.001). The apparent population size, when estimated by observing exposed individuals (i.e., those individuals on the top surfaces of boulders), also changed between dark and light periods, increasing from the former to the latter (6.7/m^{2}and 15.8/m^2, respectively). Both (apparent) population structure and size also changed along a seasonal time frame (Fig. 3a, b). Censi recorded in autumn or winter months would suggest a larger population than similar censi in spring or summer months. In addition to apparent population changes, real population changes have occurred on an annual and superannual basis (Fig. 4a, data from Muntz et al., 1965; Kitching & Thain, 1983; Kitching, 1987; Barnes et al., 1999). There has also been a real shift in the population size structure between 1977 and 1999; recent populations being dominated by larger (older) individuals (Fig. 4b).

The apparent (and real) structure of *P. lividus* populations varied on both local and wider spatial scales. The size of individuals in the South Basin, Lough Hyne were typically larger than those of the North Basin, Lough Hyne and those of the surrounding coast of west Cork were larger still (Fig. 5a). The population at São Miguel Island in the Azores, in contrast, was dominated by very young individuals (Fig. 5b). The 1999 census estimated the Lough Hyne population at similar levels to the estimates of 1994 and 1995, 1998 (Barnes et al., 1999).

In the present study potential predators were much more active during dark periods than light (Kruskal–Wallis, H= 23.2, P<0.001), but instances of *P. lividus* predation were ≈1% (by choice of prey item) for *Marthasterias glacialis* and less for all other potential predators monitored. No significant variation was found in predator density at Lough Hyne between sites or months during either daytime (midday) or dark (night) periods (Fig. 6); (Kruskal–Wallis, all P > 0.05).

Discussion

Most populations of the echinoid *P. lividus*, for which data is available, have shown declines from medium to catastrophic levels (Southward & Southward, 1975; Le Gall, 1987). Typically this is considered to be through fisheries demand outstripping supply, ease of

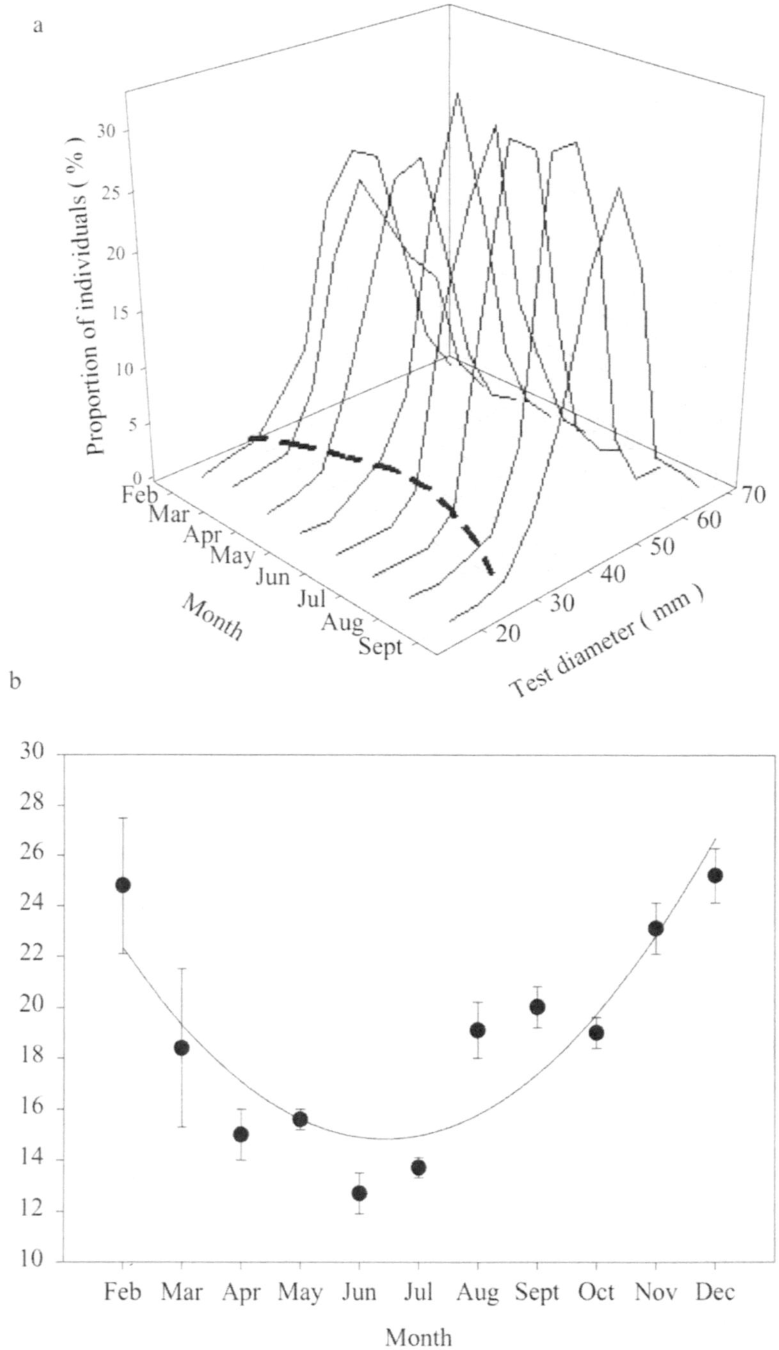

Figure 3. Paracentrotus lividus. (a) Size frequency distribution at Lough Hyne (in daylight) with month (n>120 per month). Data did not significantly differ with site so is shown pooled between sites. The dashed line illustrates the decrease in small individuals visable during summer months. (b) Density of individuals at Lough Hyne (daylight) with month. Data are shown as mean with standard error (n=54 per month).

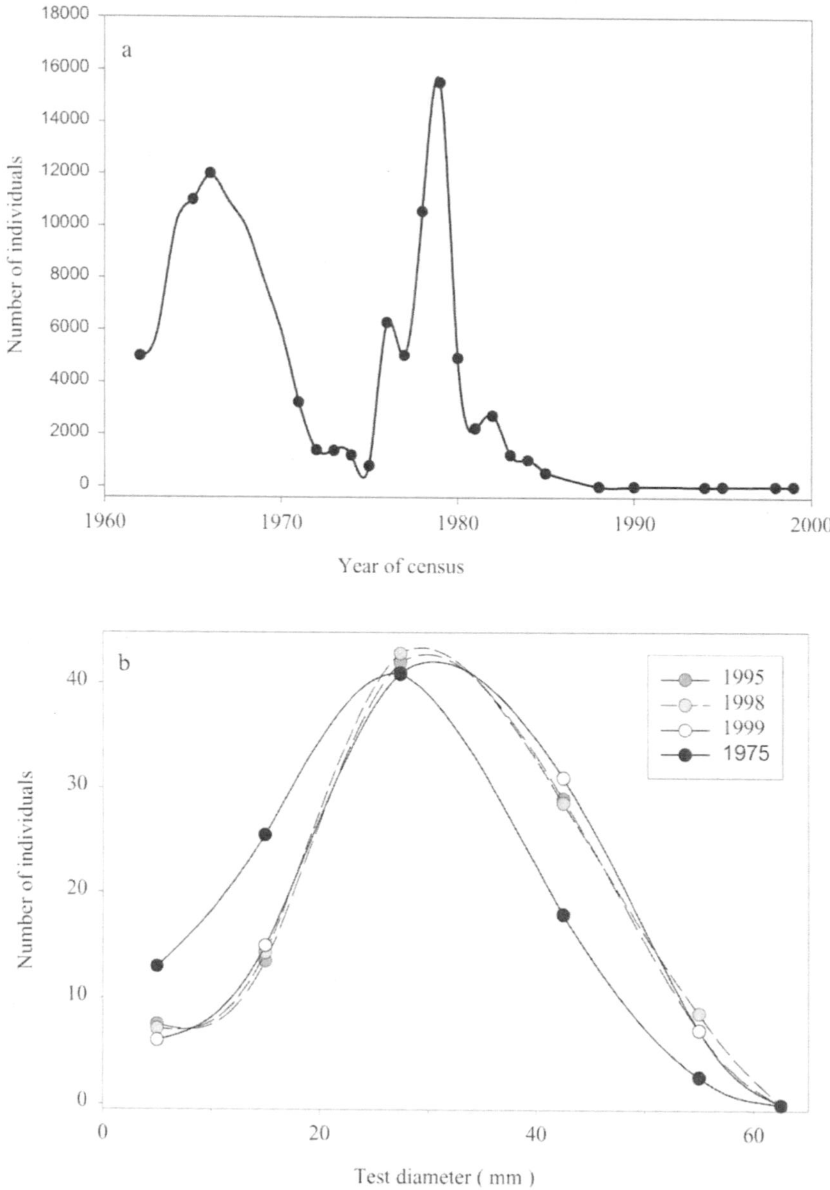

Figure 4. Paracentrotus lividus. Annual variability in populations at Lough Hyne (in daylight) in (a) numbers of individuals and (b) size frequency. All data from Barnes et al. (1999) and references therein except 1999 data (from present study).

access to their shore habitat and subsequent overfishing. Both *P. lividus* and other echinoid species are, however, known to have natural fluctuating (usually cyclical) patterns of abundance (Elner & Vadas, 1990; Barnes et al., 1999). The establishment and continuation of visual censi is thus an important first step to estimation and modelling of standing stock, recruitment and sustainable fishery levels. Interpretation of censi and population structure measurements must, as demonstrated in the present study, be carried out with care and appreciation of the behavioural ecology of the species at each locality.

Whilst censi should reveal accurate numbers in burrowing (intertidal) *P. lividus* populations, boulder dwelling populations (such as at Lough Hyne) would be underestimated because of diurnally migrating in-

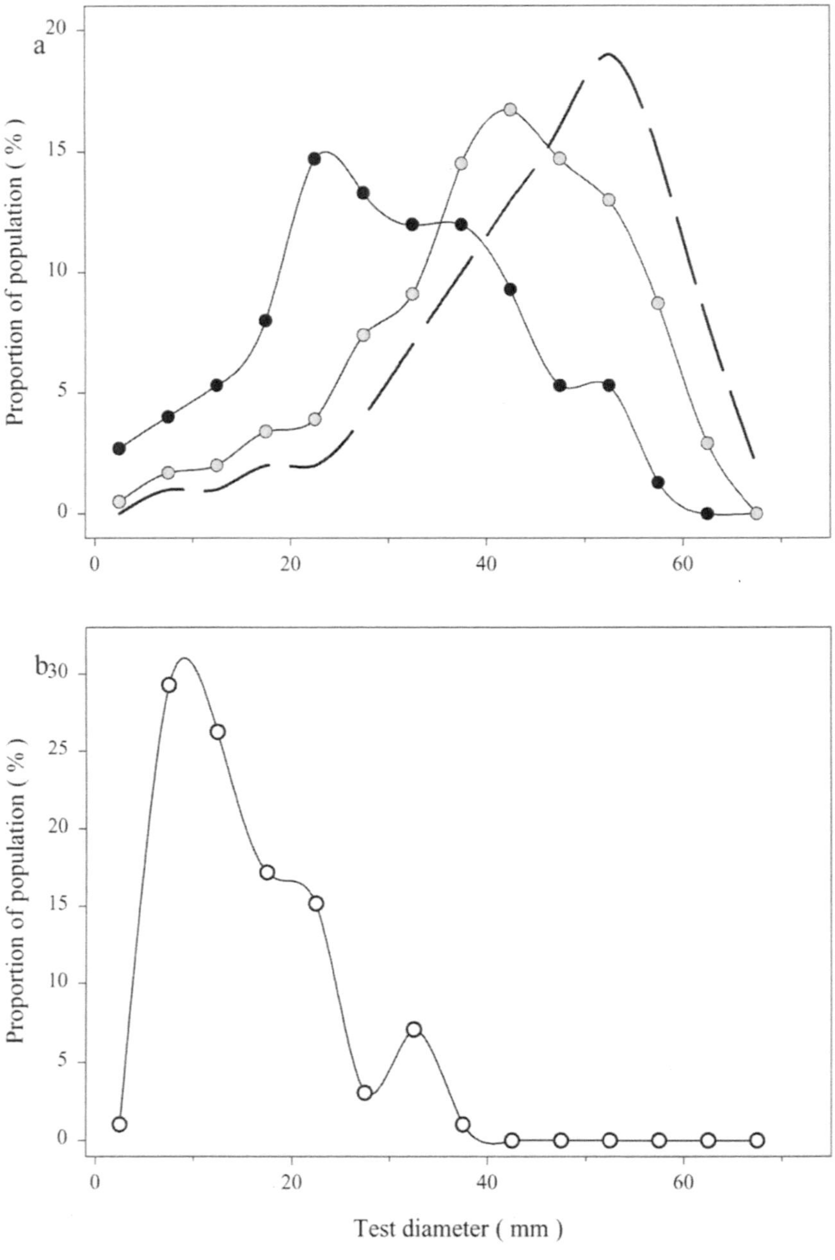

Figure 5. Paracentrotus lividus. Size frequency variability in populations with locality (a) Ireland: black circles, Lough Hyne North Shore; grey circles, Lough Hyne South Shore and dashed line, Beara Peninsula; and (b) São Miguel Island, Azores.

dividuals. Circadian patterns of activity have not only been documented in *P. lividus* elsewhere (Dance, 1987) but also in many other tropical echinoid species (Thorton, 1956; Nelson & Vance, 1979). This error can be reduced as the proportion of each size class involved in migration (which is different for most size classes) can be calculated (see Crook et al., 2000). Mediterranean censi have the opposite problem in that the migration of *P. lividus* from the under to upper boulder surfaces is nocturnal (Dance, 1987). That a

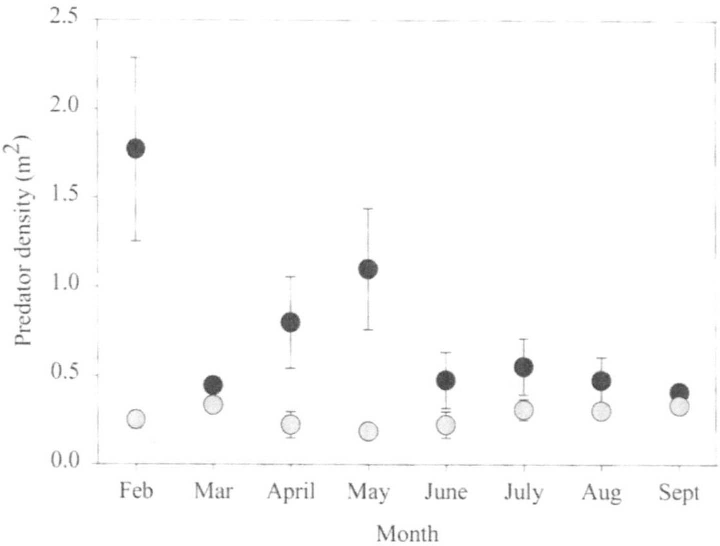

Figure 6. Marasterias glacialis, Asterias rubens, Necora puber, Cancer pagurus and *Carcinus maenus*. Predator density with month at Lough Hyne. Data is presented as mean and standard error (pooled between sites). Symbols are diurnal (grey circles) and nocturnal (black circles).

seasonal pattern to *P. lividus* migration also occurs at Lough Hyne (and possibly elsewhere) complicates the process but again this variation seems to be predictable and therefore can be included in a potential fisheries model. Studies of longer-term patterns in echinoid activity have centred on the population increases of *Strongylocentrotus* species on the Atlantic and Pacific coasts of America (see Elner & Vadas, 1990, for review). Although there are clearly substantial annual, and possibly decadal, patterns to *P. lividus* recruitment (at least at Lough Hyne) these do not seem to be predictable other than a possible link to El Nino Southern Oscillation events (Barneset al., 2001).

It has been well established that a number of echinoids are sensitive to UV radiation (Häder et al., 1998) and experiments at Lough Hyne have linked various behaviours of *P. lividus* to degree of light exposure (Crook et al., 1999). As *P. lividus* typically occurs in less than 2 m depth this species may be more vulnerable than many echinoids to potentially harmful UV light penetration (McFarland, 1986). Irradiation maybe a source of mortality and, along with the geology of rocks and exposure of coast, be a determinant of various aspects of the species ecology, not least as to whether the habit is burrowing or migrating. We suggest that this might be a valuable area to pursue for *P. lividus* as well as for echinoids in general despite the timescale over which work in this area has already been carried out (e.g., von Uexküll 1897; Mortensen, 1943; Millott, 1956; 1975; Sharp & Gray, 1962; Crook et al., 1999).

Apart from potential algal (Cross & Southgate, 1980) and sea temperature influences (Barnes et al., 2001) other factors influencing *P. lividus* population size and structure would include fishing and predation pressures. Censi of predators at Lough Hyne is restricted to surveys in 1955 and in 1990 (Little et al., 1992). As with censi of *P. lividus*, such surveys have been of a visual, non-intrusive nature but the censused species may have daily, seasonal and spatial confounding factors. In addition, in years of decreased echinoid abundance (as at present) increased algal growth potentially decreases the effectiveness of traditional non invasive censi. Observed natural mortality and predation of *P. lividus* is low at Lough Hyne, suggesting that this, locally, is not a major contributory factor to population decreases. This may not, however, be the case for the Mediterranean Sea where fish predators may have more influence on *P. lividus* mortality (Dance, 1987). Construction and calibration of censi, preferably inclusive of test size measurements, in the vicinity of fishery localities should ultimately be a valuable predictive tool in suggesting annual effort or catch levels. To increase the effectiveness and success of potential models the link between maximum sea temperatures, mass spawning and cohort success in *P. lividus* needs to be established (Fenaux, 1968; Barnes et al., 2001). Whether closed areas, such as marine reserves like Lough Hyne, are effective in supply of viable recruits to surrounding fished areas, as has been effective with finfish (see Kennedy, 1990; Roberts & Polunin, 1993; Attwood et al., 1997), is currently unknown.

Acknowledgements

We thank Declan O'Donnell for Lough Hyne permits and are grateful to James Bell, John Bohane, Nora Buttimer, Amy Greenwood, Maria Long, Ben Maughan and Emma Verling for aiding work at Lough Hyne. We would also like to thank Áina Healy and Julianna O'Callaghan for help with data collation and Bob McNamara for manufacturing some of the field equipment used.

References

Attwood, C.G., J.M. Harris & A.J. Williams, 1997. International experience of marine protected areas and their relevance to South Africa. S. Afr. J. mar. Sci. 18: 311–332

Barnes, D.K.A., S. Steele, D. Maguire & J. Turner, 1999. Population dynamics of the urchin *Paracentrotus lividus* at Lough Hyne, Ireland. Proceedings of the 5th European Echinodermata conference, Milano, Italy: 427–431.

Byrne, M., 1990. Annual reproductive cycles of the commercial sea urchin *Paracentrotus lividus* from an exposed intertidal and a sheltered subtidal habitat on the west coast of Ireland. Mar. Biol. 104: 275–289.

Crook, A. C., E. Verling & D. K. A. Barnes, 1999. A comparative study of the covering reaction of the purple sea urchin, *Paracentrotus lividus* (Lamarck), under laboratory and field conditions. J. mar. biol. Ass. U.K. 79: 1117–1121.

Crook, A. C., M. Long & D. K. A. Barnes, 2000. Quantifying daily migration in the sea urchin *Paracentrotus lividus*. J. mar. biol. Ass. U.K. 80: 177–178.

Cross, T.F. & T. Southgate, 1980. Mortalities of fauna of rocky substrates in south west Ireland associated with the occurrence of *Gyrodinium aureolum* blooms during autumn 1979. J. mar. biol. Ass. U.K. 60: 1071–1073

Dance C., 1987. Patterns of activity of the sea urchin *Paracentrotus lividus* in the Bay of Port-Cros (Var, France, Mediterranean). Mar. Ecol. 8: 131–142.

Ebling, F.J., M. A. Sleigh, J.F. Sloane & J. A. Kitching, 1960. The ecology of Lough Hyne VII. Distribution of some common plants and animals of the littoral and shallow sublittoral regions. J. Ecol. 48: 29–53

Ebling, F.J., A. D. Hawkins, J. A. Kitching, L. Muntz, & V. M. Pratt, 1966. The ecology of Lough Hyne XVI. Predation and diurnal migration in the *Paracentrotus* community. J. anim. Ecol. 35: 559–566.

Elner, R. W. & R. L. Vadas, 1990. Inference in ecology: The sea urchin phenomenon in the Northwestern Atlantic. Am. Nat. 1990: 108–125.

Fenaux, L., 1968. Maturation des gonades et cycle saisonnier des larves chez *A. lixula, P. lividus* et *P. mirrotuberculatus* (echineds) a villfranche-sur-mer. Vie Milieu. 19: 1–52.

Häder, D.P., H. D. Kumar, R. C. Smith & R.C. Worrest, 1998. Effects on aquatic ecosystems. J. Photochem. Photobiol. B. 46: 53–68.

Henderson, S. T., 1970. Daylight and its spectrum. American Elsevier, New York.

Kennedy, A.D., 1990. Marine reserve management in developing nations: Mida Creek – a study from East Africa. Ocean Shoreline Mgmt. 14: 105–132

Kitching, J.A., 1987. Ecological studies at Lough Hyne. Adv. Ecol. Res. 7: 115–186.

Kitching, J.A. & F. J. Ebling, 1961. The ecology of Lough Hyne X: the control of algae by *Paracentrotus lividus* (Echinoidea). J. anim. Ecol. 3: 373–383.

Kitching, J.A & V.M. Thain, 1983. The ecological impact of the sea-urchin *Paracentrotus lividus* (Lamark) in Lough Ine, Ireland. Phil. Trans. r. Soc. London B 300: 513–552

Ledireach, J.-P., 1987. Le Peche des oursins en Mediterranee: Historique, Techniques, Legislation, Production. In Boudouresque, CF (ed.), Colloque International sur *Paracentrotus lividus* et les oursins comestibles. GIS Posidonie publ. Marseille: 335–362

Le Gall, P.,1987. Le peche des oursins en Bretagne. In Boudouresque, C. F. (ed.), Colloque International sur *Paracentrotus lividus* et les oursins comestibles. GIS Poisidonie publ. Marseille: 311–324.

Little, C. D. Morritt & P. Stirling, 1992. Changes in the fauna and flora of Lough Hyne. Irish Naturalists' Journal 24: 87–95.

McFarland, W. N., 1986. Light in the sea – correlations with behaviours of fishes and invertebrates. Am. Zool. 26: 389–401.

Millott, N., 1956. The covering reaction of sea urchins I. A preliminary account of covering in the tropical echinoid *Lytechinus variegatus* (Lamarck), and its relation to light. J. exp. Biol. 3: 508–523.

Millott, N., 1975. The photosensitivity of echinoids. Adv. Mar. Biol. 13: 1–52.

Mortensen, T., 1943. A monograph of the Echinoidea. Vol. III. 2. Camarodonta I. Reitzel, Copenhagen.

Muntz, L., F.J. Ebling & J.A. Kitching, 1965. The Ecology of Lough Ine xiv. Predatory activity of large crabs. J. Anim. Ecol. 34: 315–329

Nelson, B. V. & R. R. Vance, 1979. Diel foraging patterns of the sea urchin *Centrostephanus coronatus* as a predator avoidance strategy. Mar. Biol. 51: 251–258.

Roberts, C.M. & N.V.C. Polunin, 1993. Marine Reserves: simple solutions to managing complex fisheries? Ambio 22: 363–368

Sharp, D.T. & I. E. Gray, 1962. Studies on factors affecting the local distribution of two sea urchins, *Arbacia punctulata* and *Lytechinus variegatus*. Ecology 43: 309–313.

Southward, A. & E. Southward, 1975. Endangered urchins. New Sci. 66: 70–72.

Thorton, I. W. B., 1956. Diurnal migrations of the echinoid *Diadema setosum* (Leske). Br. J. Anim. Behav. 4: 143–146.

von Uexküll, J., 1897. Ueber reflexe bei den Seeigeln. Z. Biol. 34: 298–318.

Hydrobiologia **465:** 103–114, 2001.
G. Burnell (ed.), Coastal Shellfish – A Sustainable Resource.

The impact of scallop drags on sea urchin populations and benthos in the Bay of Fundy, Canada

S.M.C. Robinson, S. Bernier & A. MacIntyre
Biological Station, Dept. Fisheries and Oceans, 531 Brandy Cove Rd., St. Andrews, New Brunswick, E5B 2L9, Canada

Key words: sea urchins, *Strongylocentrotus droebachiensis*, dragging, dredging, impacts, Bay of Fundy

Abstract

The fishery for the green sea urchin (*Strongylocentrotus droebachiensis*) began in New Brunswick in the late 1980s in response to increasing demand from Asian markets, primarily Japan. Fishing was initiated by members of the scallop industry and the harvesting practices were based on familiar fishing gear (scallop drags or dredges) and similar operations occurring simultaneously in Maine. Because of the potential impact of scallop gear on the shallow water populations in areas not traditionally fished, a study was initiated in 1993 to document (1) the proportion of sea urchins damaged during the harvesting operation, (2) the impact on and subsequent recovery time of the associated benthic flora and epifauna, and (3) the impacts on the bottom substrate. Two sites were chosen for the study based on a split-plot design with a control and experimental plot at each site. In each experimental plot, a towing lane was created parallel to shore. Divers used a fixed line transect method to survey the control and treatment plots prior to and after the harvesting operation done with commercial scallop Digby drags. Further diver-based surveys were done 3 and 6 months later. Results showed a significant decrease in sea urchin densities and an increase in the number of broken sea urchin tests after the harvesting operation at the Passamaquoddy Bay and Grand Manan sites (3–15%). On both sites, there were significant changes in the density of mobile predators such as lobsters, crabs, whelks and sculpins. Lobster density declined to zero after dragging, but the lack of body parts suggested they moved out of the area. Whelks were observed to move into each of the experimental areas, likely in response to probable increase in available food. The dragging operation increased the breakage rate of the kelp, *Laminaria longicruris*. From the study we concluded that there were short-term impacts from dragging, but the observable effects on the bottom from the single dragging event were gone in less than 3 months. The longer-term effects of dragging were unknown.

Introduction

As fisheries for new non-traditional species develop, older, familiar technology and techniques are commonly employed to harvest the new organisms. This shifting of gear from one species to another for commercial harvesting may be more cost-effective for initial start-up, but it may sometimes have some undesirable effects on the target species and habitat, and therefore caution is warranted. Some types of scallop fishing gear are more intrusive than others (Cover, 1994). Bottom fishing in experimental situations and commercial fishing zones has been shown to decrease the species number, richness and diversity in certain areas (Thrush et al., 1995; Collie et al., 1997).

The development of the green sea urchin (*Strongylocentrotus droebachiensis,* O.F. Müller) fishery in southwestern New Brunswick was typical of many newly exploited species. The fishery initially started off Campobello Island as an extension to the US market through Lubec, Maine, when local fishers could not supply the demand. The Campobello fishing boats were generally multi-purpose in nature (i.e., they fished for several species such as lobster, scallops, herring, etc.) and used the existing harvesting technology. Therefore, fishers were familiar with the scallop drag, it worked well, the gear was used in Maine to harvest urchins, and they could easily sell their product.

Standard inshore scallop gear is based on the original Digby design where a bag made of steel rings

Figure 1. Photo of scallop gear and towing bridle used for dragging sea urchins.

and rubber washers is attached to a square steel frame which is then attached by chains or rods to a steel towing bar (Fig. 1). There can be several of these bags attached to the bar and the frames can be modified to include teeth on the tops and bottoms to improve the digging efficiency. A full set of gear can be 5 m wide and weigh over 300 kg. When fishing scallops, the gear is towed with a steel cable slowly in deeper water (>25 m) over a stretch of bottom for 10–15 min. For sea urchins, the operation is a little different. The boat will manoeuvre into shallow water, drop the drags to the bottom and then tow off on a track roughly parallel to shore into the deeper water. Tow times last from 5 to 7 min. The drags are then hauled aboard and the catch is emptied onto a dumping table where the large sea urchins are removed and put into plastic fish boxes. The smaller urchins, rocks and other by-catch are dumped back into the water. The boats often fish in groups of two to five and will tow right after one another over the same track. The same piece of bottom will be fished many times until the catch rates drop off.

Divers have often noticed the tracks left on the bottom by the drags, and several groups including fishers, biologists and managers, have expressed concern about the potential effects that the drags may be having on the sea urchin populations, the substrate and the other associated species. Studies on the effects of scallop drags on lobster populations (Jamieson & Campbell, 1985; Robichaud et al., 1987; Roddick & Miller, 1992) have concluded that there is relatively little fishery-related impact as the overlap between scallop habitat and lobster habitat is minimal. However, this situation may change as scallop gear begins to fish sea urchin habitat, which is more similar to that of the lobster. Therefore, a study was initiated to examine the effects of the scallop drags on the green sea urchin populations and its associated habitat and species.

Materials and methods

A diver-based survey design was chosen for this study because the water depth was shallow and the bottom was composed of boulder fields and ledge. These traits made working from a vessel with traditional grabs or corers impractical. While using divers increased the logistic effort substantially, it allowed us to collect high-quality information on the bottom flora and epifauna. We used a split-plot sampling design (Winer, 1991) to quantify the effects of the drags (two plots with five transects per plot and 30 quadrats per transect). This was done at two study sites to also compare trends in different habitats.

The two study sites were chosen based on preliminary dive surveys, one around Grand Manan, in Cheney's Passage (44° 39.7′ N, 66° 43.5′ W) and the other in Passamaquoddy Bay off Ministers Island (45° 05.9′ N, 67° 01.8′ W) (Figure 2). Each study site was divided into two rectangular plots; one as a control and the other as a treatment where dragging took place. The plots measured 100 m long by 50 m wide and were located in a depth of 4–12 m at mean low water. All plots were parallel to the shore and were marked at each corner with a buoy anchored with a bag of cement. This created a towing lane in the experimental plot.

For logistic reasons, two different commercial sea urchin boats with scallop gear were chartered during this study (Figure 1). The following are the dimensions of the scallop fishing gear used.

Ministers Island; Total bar width: 2.5 m
four bags: 0.46 × 0.30 × 0.76 m deep.
four teeth per bag

Grand Manan; Total bar width: 5.1 m
seven bags; 0.61 × 0.25 × 0.69 m deep.
five teeth per bag

At the Ministers Island site a total of 36 tows was conducted within the 100 × 50-m experimental plot. This was estimated to be equivalent to a moderate towing disturbance from the commercial fishery. The plot was resurveyed after the first 18 tows and again after the 36. Surveys were conducted in the same manner at all times. The boats worked parallel to shore, inside the plot, always in the same direction, dragging from south to north and covering the maximum width possible. The vessel used on Grand Manan was fished in the same manner as described above except that having a total gear width twice that used at the Ministers Island site, it only did 18 tows. After eight tows the plot was surveyed and after 10 more resurveyed again.

The diver-based survey was done in the experimental plot prior to, midway and following the dragging operation, and in the control plot prior to and following the dragging of the experimental plot. These surveys were performed again after periods of 3 and 6 months. To survey the plot, two 100-m nylon main lines were laid on the bottom by diver between the two corners along the longest sides of the rectangular plot (100 × 50 m). Two divers would attach each end of a 50-m transect line (marked at every meter) to the two main lines. Each diver had a quadrat measuring 1 × 0.5 m and would survey one side of the transect by laying a quadrat every two meters along the transect starting at 10 m from where it was attached to the main line. Surveying was initiated between the 10- and 40-m marks of the transect in order to ensure the area had been impacted by the drags. All epifaunal macro-invertebrates were counted by diver and recorded underwater. The percentage of bottom covered by macrophytes was also recorded. Each diver assessed 15 quadrats along the transect for a total of 15 m^2 of sea floor surveyed along one 30-m transect line by the team of two divers. Additionally, bottom characteristics, such as numbers of rocks dislodged, were recorded. After the first transect was completed, it was moved 20 m to the next mark on the main lines. Five transects per plot were made in the same manner. To make handling the 50-m transect easier when moving it to the next position, two small buoys were attached near the middle thus preventing it from snagging on the bottom. These were removed prior to surveying.

During the course of the study, sea urchins were sampled at different stages of the experiment to de-

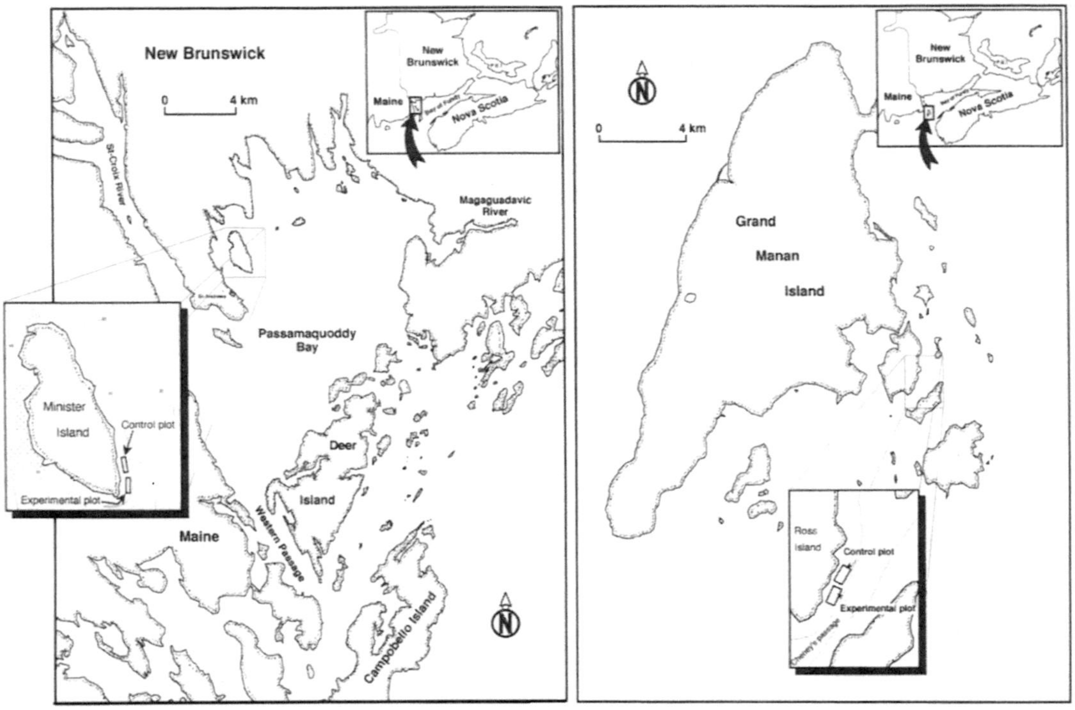

Figure 2. Location of the Ministers Island and Grand Manan study sites in the Bay of Fundy showing the control and experimental plots.

termine the population size frequency. Inside each plot, 400 urchins were collected at random in four different areas (100 per area) and included all sizes present. The urchins were measured on board the fishing vessel for test diameter to the nearest millimetre and then returned to the plots. These samples were taken in both the experimental and control plot prior to and following the dragging operation, and after a periods of 3 and 6 months.

Results

Over the course of the five survey periods at the two study sites, a total of 26 species were surveyed during the dive operation. Most of these species were relatively low in number and therefore, our analyses dealt with only seven of the most abundant species. A total list of species can be seen in Tables 1 and 2.

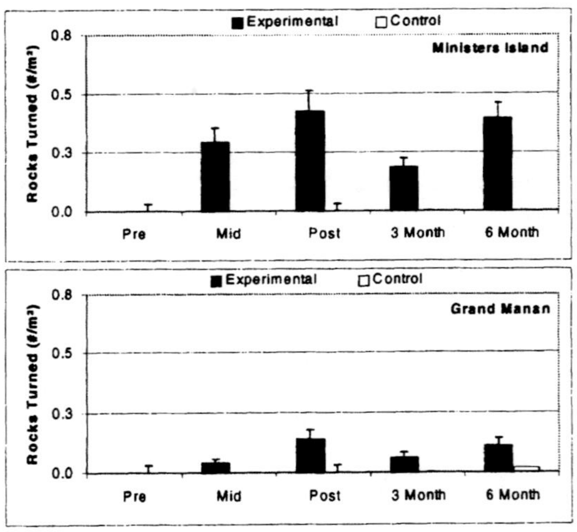

Figure 3. Diver survey results on the mean number of rocks turned over during the dragging operation from Ministers Island and Grand Manan.

Bottom impacts

There was a noticeable impact of the scallop drags on the bottom, based on the number of dislodged rocks observed during the dive survey (Fig. 3). At Ministers Island, the number of rocks dislodged per square meter significantly increased from 0.0 m^{-2} during the

Table 1. List of mean densities by plot from Ministers Island of various species surveyed by divers over the duration of the study

		Pre				Mid				Post				3-month				6-month			
		Exp.		Cont.		Exp.		Cont.		Exp.		Cont.		Exp.		Cont.		Exp.		Cont.	
Scientific name	Common name	*x*	SD	*x*	SD	*x*	SD	*x*	SD	*x*	SD	*x*	SD	*x*	SD	*x*	SD	*x*	SD	*x*	SD
Aeolidia papillosa	Nudibranch	0	0	0	0	0.01	0.03			0	0	0	0	0	0	0	0	0	0	0	0
Asterias vulgaris	Common starfish	0.30	0.19	0.64	0.85	0.33	0.14			0.40	0.18	0.52	0.20	0.39	0.20	0.35	0.14	0.12	0.11	0.37	0.13
Balanus sp.	Barnacle	0.06	0.11	0.11	0.20	0.00	0.00			0.01	0.03	0.05	0.12	0.15	0.21	0	0	0.09	0.21	0	0
Buccinum undatum	Whelk	0.41	0.27	1.13	1.10	0.83	0.50			1.37	0.19	0.65	0.36	0.24	0.14	0.31	0.23	0.53	0.34	0.39	0.28
Cancer irroratus	Rock crab	0.04	0.05	0.07	0.12	0.04	0.04			0	0	0.01	0.03	0	0	0.01	0.03	0	0	0	0
Carcinus maenas	Green crab	0.01	0.02	0.03	0.03	0	0			0.01	0.03	0.03	0.06	0	0	0	0	0	0	0	0
Cerianthus borealis	Burrowing anenome	0.08	0.07	0.15	0.24	0.08	0.14			0.09	0.08	0.24	0.23	0.01	0.03	0	0	0.08	0.11	0.08	0.11
Cucumaria frondosa	Sea cucumber	13.96	9.06	17.96	42.77	12.03	7.32			11.44	6.53	5.96	1.90	9.08	5.26	4.56	4.15	10.49	6.75	7.11	1.98
Echinarachnius parma	Sand dollar	0.01	0.02	0.01	0.02	0.00	0.00			0	0	0	0	0	0	0	0	0	0	0	0
Euspira heros	Moonsnail	0	0	0	0	0	0			0	0	0	0	0	0	0	0	0	0	0	0
Halichondria panicea	Breadcrumb sponge	0.07	0.09	0.14	0.21	0.01	0.03			0.03	0.04	0	0	0.01	0.03	0.12	0.14	0	0	0.01	0.03
Halocynthia pyriformis	Sea peach	11.34	7.71	29.71	29.66	7.84	4.85			7.03	4.20	17.01	3.43	6.07	2.88	12.61	1.89	9.23	5.41	13.52	2.52
Hemitripterus americanus	Sea raven	0.01	0.02	0.01	0.03	0	0			0	0	0	0	0	0	0	0	0	0	0	0
Henricia sanguinolenta	Blood star	0.05	0.07	0.08	0.15	0.03	0.04			0.01	0.03	0.01	0.03	0	0	0	0	0	0	0	0
Homarus americanus	Lobster	0.08	0.13	0.05	0.05	0.05	0.07			0	0	0.05	0.07	0.03	0.04	0.05	0.06	0	0	0.01	0.03
Lycodes sp.	Eelpout	0	0	0	0	0.01	0.03			0.01	0.03	0	0	0	0	0	0	0	0	0	0
Metridium senile	Anemone	0.02	0.03	0.06	0.07	0.01	0.03			0.03	0.04	0.07	0.05	0.01	0.03	0.03	0.04	0	0	0.01	0.03
Mogula sp.	Sea squirt	0.03	0.06	0.05	0.10	0	0			0	0	0	0	0.01	0.03	0	0	0	0	0	0
Myoxocephalus octodecemspinosus	Longhorn sculpin	0.27	0.11	0.71	0.78	0.33	0.14			0.48	0.21	0.60	0.25	0.04	0.04	0.03	0.04	0.03	0.04	0.03	0.04
Mytilus edulis	Mussel	0.01	0.02	0.01	0.03	0	0			0	0	0	0	0	0	0.01	0.03	0.01	0.03	0	0
Ophiopholis sp.	Brittle star	0.01	0.03	0.03	0.05	0.01	0.03			0.01	0.03	0	0	0	0	0	0	0.04	10.06	0.11	0.17
Pagurus acadianus	Hermit crab	0.13	0.09	0.37	0.40	0.15	0.12			0.16	0.10	0.13	0.08	0.05	0.06	0.13	0.12	0	0	0.03	0.04
Placopecten magellanicus	Scallop	0.05	0.05	0.12	0.14	0.04	0.04			0.03	0.04	0.03	0.04	0.01	0.03	0.01	0.03	0.01	0.03	0.01	0.03
Pseudopleuronectes americanus	Winter flounder	0.00	0.00	0.04	0.07	0.01	0.03			0.03	0.06	0.05	0.07	0	0	0.01	0.03	0	0	0	0
Psolus fabricii	Scarlet Psolus	0.04	0.06	0.08	0.12	0.03	0.06			0.01	0.03	0	0	0.03	0.04	0	0	0.07	0.05	0	0
Sabella spp.	Fan worm	0.13	0.08	0.30	0.38	0.40	0.67			0.20	0.09	0.15	0.06	0.05	0.09	0.05	0.06	0.05	0.03	0.01	0.03
Tubularia sp.	Hydroids	0.03	0.05	0.05	0.11	0.00	0.00			0	0	0.01	0.03	0	0	0	0	0	0	0	0
Ulvaria subbifurcata	Radiated shanny	0.07	0.06	0.18	0.20	0.07	0.05			0.07	0.05	0.11	0.06	0	0	0.01	0.03	0.01	0.03	0.01	0.03

Table 2. List of mean densities by plot from Grand Manan of various species surveyed by divers over the duration of the study

		Pre				Mid				Post				3-month				6-month			
		Exp.		Cont.		Exp.		Cont.		Exp.		Cont.		Exp.		Cont.		Exp.		Cont.	
Scientific name	Common name	*x*	SD	*x*	SD	*x*	SD	*x*	SD	*x*	SD	*x*	SD	*x*	SD	*x*	SD	*x*	SD	*x*	SD
Aeolidia papillosa	Nudibranch	0.03	0.04	0.15	0.11	0.04	0.06			0.05	0.06	0.03	0.0 4	0	0	0.03	0.06	0	0	0	0
Asterias vulgaris	Starfish	0.40	0.37	0.92	0.44	0.37	0.10			0.25	0.29	0.23	0.20	0.27	0.16	0.27	0.23	0.29	0.12	0.13	0.26
Buccinum undatum	Whelk	0.59	0.28	0.27	0.12	0.40	0.16			1.00	0.25	0.40	0.21	0.76	0.45	0.25	0.31	0.45	0.25	0.40	0.29
Cancer irroratus	Rock crab	0.05	0.06	0.05	0.07	0.07	0.05			0.16	0.10	0.03	0.06	0	0	0	0	0	0	0.03	0.04
Cerianthus borealis	Burrowing anenome	0.05	0.09	0.04	0.04	0.04	0.06			0.07	0.07	0	0	0.05	0.06	0.03	0.04	0.07	0.05	0.08	0.06
Chiridota laevis	Silky cucumber	0.03	0.04	0.03	0.06	0	0			0.03	0.04	0.05	0.12	0.25	0.37	0.04	0.06	0.01	0.03	0.15	0.14
Cucumaria frondosa	Sea cucumber	0.08	0.09	0.00	0.00	0.01	0.03			0.01	0.03	0.03	0.04	0	0	0	0	0.01	0.0	30.03	0.06
Echinarachnius parma	Sand dollar	0	0	0	0	0	0			0	0	0	0	0	0	0	0	0	0	0.01	0.03
Euspira heros	Moonsnail	0.01	0.03	0	0	0	0			0.03	0.04	0	0	0	0	0	0	0	0	0	0
Halichondria panicea	Breadcrumb sponge	0.01	0.03	0.03	0.06	0.01	0.03			0.03	0.04	0.01	0.03	0	0	0	0	0	0	0	0
Haliclystus auricula	Stalked jellyfish	0.01	0.03	0	0	0.01	0.03			0	0	0	0	0	0	0	0	0.01	0.03	0	0
Halocynthia pyriformis	Sea peach	0.07	0.05	0.03	0.04	0.03	0.04			0.07	0.15	0.01	0.03	0.03	0.04	0.01	0.03	0.09	0.11	0.03	0.04
Henricia sanguinolenta	Blood star	0.01	0.03	0	0	0.01	0.03			0.03	0.04	0.03	0.04	0	0	0.11	0.20	0	0	0	0
Lycodes sp.	Eelpout	0.01	0.03	0	0	0	0			0	0	0	0	0	0	0	0	0	0	0	0
Metridium senile	Anemone	0	0	0	0	0	0			0	0	0	0	0	0	0	0	0.01	0.03	0	0
Mogula sp.	Sea squirt	0	0	0	0	0	0			0	0	0	0	0.01	0.03	0	0	0	0	0	0
Myoxocephalus octodecemspinosus	Longhorn sculpin	0.15	0.11	0.15	0.11	0.15	0.07			0.07	0.09	0.09	0.15	0	0	0.03	0.06	0.09	0.06	0.03	0.04
Mytilus edulis	Mussel	0.80	0.51	0.44	0.46	0.39	0.36			0.41	0.20	0.32	0.26	0.63	0.67	0.21	0.31	0.52	0.59	1.12	1.52
Ophiopholis sp.	Brittle star	0	0	0	0	0	0			0	0	0	0	0	0	0	0	0.01	0.03	0.01	0.03
Pagurus acadianus	Hermit crab	0.23	0.15	0.21	0.13	0.09	0.17			0.28	0.13	0.23	0.15	0.07	0.07	0.1 1	0.09	0.09	0.08	0.49	0.28
Placopecten magellanicus	Scallop	0.04	0.04	0.15	0.12	0.03	0.04			0.00	0.00	0.17	0.13	0.05	0.06	0.09	0.04	0.01	0.03	0.12	0.09
Pseudopleuronectes americanus	Winter flounder	0	0	0	0	0	0			0.01	0.03	0	0	0	0	0	0	0	0	0	0
Sabella spp.	Fan worm	0.01	0.03	0.03	0.04	0.01	0.03			0.07	0.12	0.03	0.06	0.03	0.04	0	0	0.05	0.06	0.05	0.03
Terebratulina septentrionalis	Lamp shell	0.05	0.06	0	0	0.03	0.04			0.03	0.04	0.01	0.03	0	0	0	0	0	0	0	0
Tubularia sp.	Hydroids	0	0	0.01	0.03	0	0			0.09	0.08	0	0	0.01	0.03	0	0	0	0	0	0
Ulvaria subbifurcata	Radiated shanny	0.04	0.04	0.01	0.03	0.01	0.03			0.03	0.04	0.04	0.06	0	0	0	0	0.03	0.06	0.09	0.14

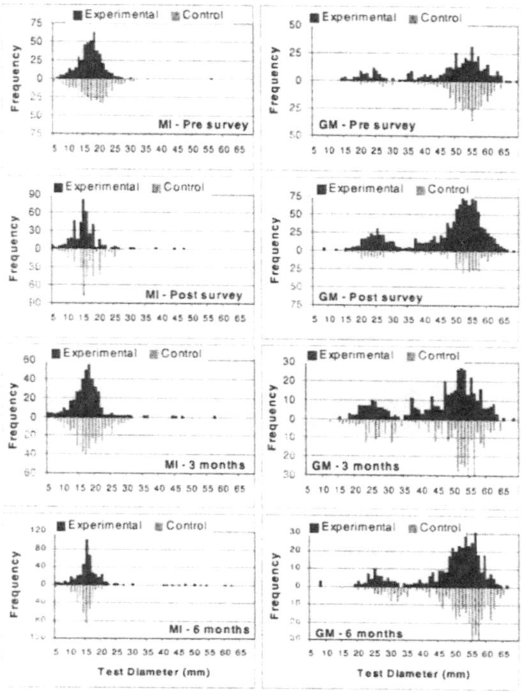

Figure 4. Size frequencies of sea urchins collected by divers before (pre), after (post), 3 months and 6 months after dragging in the Experimental and Control plots at Ministers Island and Grand Manan.

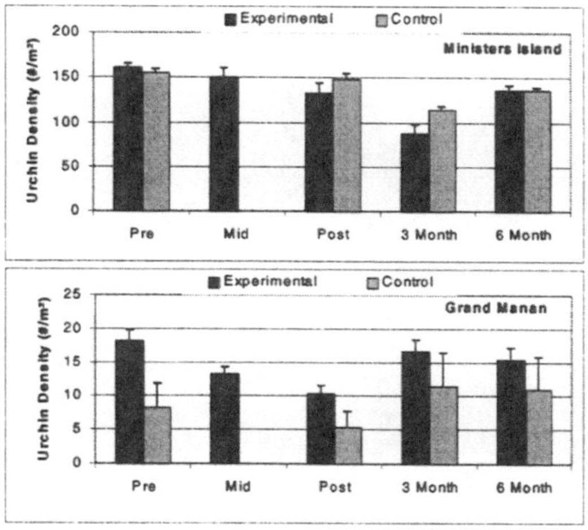

Figure 5. Mean sea urchin density (#m^{-2}) from the dive survey at the pre- mid- and post periods during the dragging operation and 3 and 6 months after for Ministers Island and Grand Manan.

pre-survey to a maximum of about 0.4 m^{-2} at the post-survey (P<0.05, Student's t-test). Evidence of rocks that had been dislodged were still obvious 6 months later as the line on the rocks showing where they had been embedded in the sediment matrix was easily observable by the diver. The number of rocks per square meter turned over at the Grand Manan site was less than that at the Ministers Island site but there was still a significant increase compared with the pre- (0 m^{-2}) and the post-survey (0.17 m^{-2}). Again, as at Ministers Island, signs of dislodged rocks were evident after 6 months.

Sea urchins

The populations of sea urchins at both study sites were very different with respect to size frequency distribution (Fig. 4). The Ministers Island site was comprised almost exclusively of small animals with a mode around 15 mm. There were some animals found up to 60 mm but they were relatively rare (minimum commercial size is 50 mm). There were no significant differences in size between the control and experimental site nor over the four sampling periods (P>0.05), except for the pre-survey control plot where there was a significant trend to slightly larger sizes (1.5 mm difference). There was a bimodal distribution of sea urchins at the Grand Manan site with the first mode at 25 mm and another at approximately 53 mm. The majority of the animals sampled were in the upper size class category. Again, no significant differences in mean size were observed between the control and experimental plots except for the 6 month survey where animals from the control plot were larger (P<0.05, Students t-test) (2 mm difference). However, there was a large difference between the average density of sea urchins within the plots. At the Ministers Island site, pre-survey urchin densities were slightly over 160 animals m^{-2} (Fig. 5). At the Grand Manan site, pre-survey urchins were an order of magnitude lower at about 15 animals m^{-2}.

There was a significant reduction in sea urchin density as a result of the dragging operation at both sites (Fig. 5) between the experimental and control plots and over time. At the Ministers Island site, the sea urchin densities dropped significantly in the experimental plot from 160 m^{-2} during the pre-survey to about 140 m^{-2} at the post-survey (P<0.05, Students t-test). There was no significant change in the control population during this period. The density of urchins were significantly different between the control and experimental plots at the post and 3-month mark (P<0.05, ANOVA). There was a significant drop in the urchin density from the pre- or post-surveys

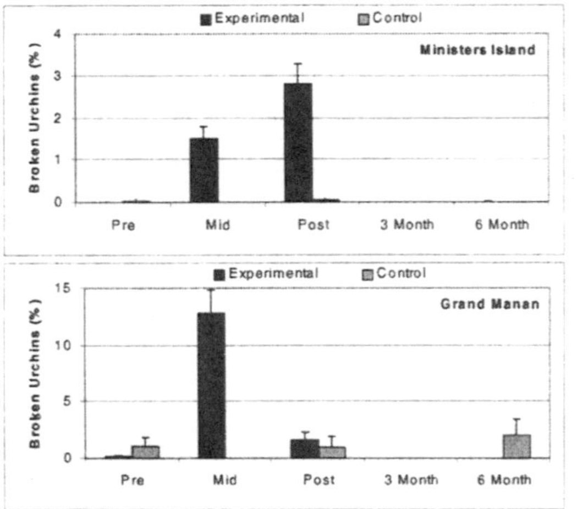

Figure 6. Mean percentage of broken sea urchin tests from the dive survey at the pre- mid- and post periods during the dragging operation and 3 and 6 months after for Ministers Island and Grand Manan.

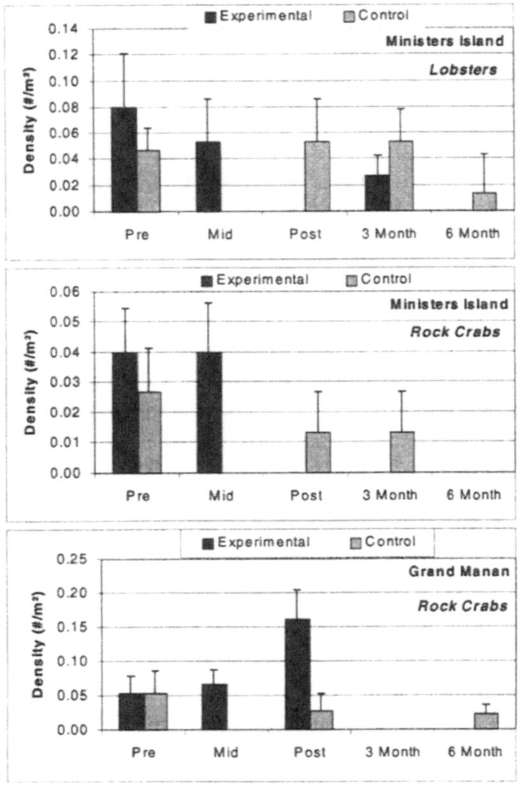

Figure 7. Mean lobster (*Homarus americanus*) and rock crab (*Cancer irroratus*) density (#m^{-2}) from the dive survey at the pre- mid- and post periods during the dragging and 3 and 6 months after for Ministers Island and Grand Manan.

in both plots at 3 months and a subsequent increase at 6 months ($P<0.05$, Students t-test). At the Grand Manan site, the sea urchin densities significantly decreased from about 18 animals m^{-2} down to 10 animals m^{-2} during the post-survey in the experiment plots ($P<0.05$, Students t-test). The experimental plots had significantly higher densities than the control ($P<0.05$, Students t-test). Again, there was no significant change in the control population over time. The sea urchin densities seemed to recover from the initial drag impact as densities increased in the experimental plots by the 3- and 6-month survey. There was no significant difference between the control and experimental plots from the post, 3-month or 6-month survey.

The percentage of broken sea urchin tests significantly increased over the course of the dragging operation at both sites ($P<0.05$, ANOVA) (Fig. 6). At the Ministers Island site, the percentage of broken urchins found during the survey increased from zero to almost three percent during the post-survey. At the Grand Manan site, there was an increase from 0 to 12% in broken tests found from the pre-survey to the midpoint survey, but the percentage of broken test dropped back to about two percent in the post-survey. Broken tests were also observed in both plots in the pre-survey. The percentage of broken tests declined back to zero in the experimental plots at both sites for the 3- and 6-month sampling periods. The Grand Manan site had significantly higher percentages of broken urchin test than the Ministers Island site in both the experimental and control plots.

Other species

Grand Manan was the only study site with any appreciable macrophyte coverage. This was expected based on previous observations in the region. During the dragging operation, the density of broken kelp stipes (*Laminaria longicruris*) increased significantly in the experimental plot from 0.01 m^{-2} during the pre-survey up to 0.27 m^{-2} at the post-survey ($P<0.05$, Students t-test). There were no significant changes in the control plot during this period. No differences were seen between the experimental and control plots after three months.

There was a drop in the lobster (*Homarus americanus*) population in the experimental plot at Ministers Island from 0.8 m^{-2} during the pre-survey to 0 m^{-2} at the post-survey ($P<0.05$, Students t-test). There was no significant change in the lobster population in the control plot (Fig. 7). No lobsters were found in either the experimental or control plots on Grand Manan.

Rock crabs (*Cancer irroratus*) also showed a response to the dragging operation, albeit different at each site. At the Ministers Island site, there was a significant drop in the density of rock crabs at the experimental plot in the pre-survey to the post-survey ($P<0.05$, Students t-test) while there was no significant change for the control plot (Fig. 7). At the Grand Manan site, there was a significant increase in the density of rock crabs in the experimental plot from the pre- to post-survey ($P<0.05$, Students t-test) while again there was no significant change in the control plot. The Grand Manan site showed no significant difference between the experimental and control densities at either the three-month or six-month sampling period. At the Ministers Island site, the pattern was not as clear.

The sea cucumber (*Cucumaria frondosa*) densities at the pre-survey were approximately 180-fold higher at the Ministers Island site (15 m^{-2}) than at the Grand Manan site (0.08 m^{-2}) (Tables 1 and 2). While, all the trends in the mean densities were down over the course of the dragging operation, the differences in sea cucumber densities from the pre-to post-survey were not significant at either site. There were also no observable differences at the 3- and 6-month sampling period at either site. However, the experimental plot at Ministers Island had significantly higher densities of sea cucumbers than the control plot.

There was also a 250-fold difference in the densities of the sea peach (*Halocynthia pyriformis*) between the Grand Manan (0.06 animals m^{-2}) and the Ministers Island (15 animals m^{-2}) sites (Tables 1 and 2). There were no significant differences in the densities between pre- and post-drag surveys in the control or experimental plots at either site. Diver observations indicated seeing detached sea peaches in the control plot, presumably having drifted from the experimental plot.

The densities of the northern whelk (*Buccinum undatum*) during the pre-survey were more similar at both study sites, ranging from 0.5 to 1.0 m^{-2} (Fig. 8). At both sites, densities of the whelk significantly increased in the experimental plot over the duration of the dragging operation ($P<0.05$, Students t-test). In the control plot at the Ministers Island site, there was a concurrent significant drop in the density of whelks ($P<0.05$, Students t-test). There was no significant

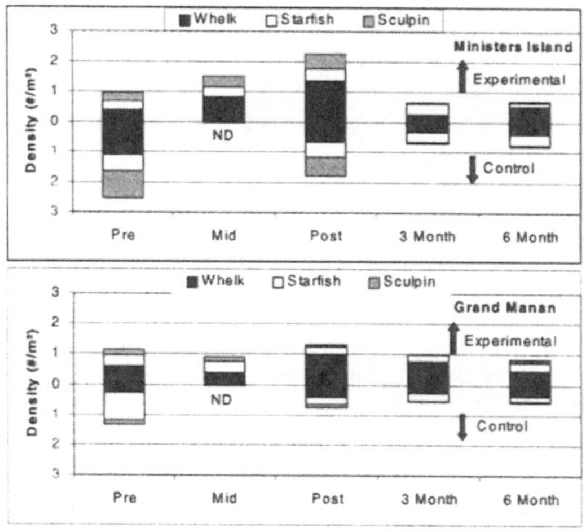

Figure 8. Mean northern whelk (*Buccinum undatum*), common starfish (*Asterias vulgaris*) and sculpin (*Myoxocephalus octodecemspinosus*) density (#m^{-2}) from the dive survey at the pre- mid- and post periods during the dragging and 3 and 6 months after for Ministers Island and Grand Manan.

change in the density of whelks in the control plot at the Grand Manan site.

Densities of the common starfish (*Asterias vulgaris*) were also similar at both sites during the pre-survey ranging from 0.25 to 0.8 m^{-2} (Fig. 8). There were no significant changes in density in the experimental plot at either site from the pre- to post-survey.

Sculpin (*Myoxocephalus octodecemspinosus*) densities were slightly higher at the Ministers Island site (0.5 m^{-2}) than at the Grand Manan site (0.15 m^{-2}) (Fig. 8). Sculpin densities significantly increased in the experimental plot at the Ministers Island site from the pre- to the post-survey ($P<0.05$, Students t-test), while there were no significant changes in the control plot. The sculpin densities in both plots were greatly reduced in the 3- and 6-month samples. There were no significant changes in either plot at the Grand Manan site. Sculpin densities also tended to be lower at the 3- and 6-month samples.

Discussion

The immediate impact of dragging on the bottom was the physical interaction of the drags on the substrate. As the drag moved over the bottom, it dislodged the rocks and boulders imbedded in the sediment matrix and rearranged existing interstitial spaces. The degree of impact was basically dependent on the size of the boulders on the bottom as very large boulders shifted substantially less than smaller ones. This may explain the observation that the Ministers Island site with its smaller boulders showed more bottom impact than the Grand Manan site. The impact of scallop dredges or drags on the bottom in Australia was also observed by Currie & Parry (1999) who reported that dredges tended to modify the bottom substrate by flattening the plots. The persistence of the disturbance marks on the dislodged rocks may be partly due to the fact that this study was conducted after the recruitment period of most species. Therefore much of the newly exposed surface available for settlement by planktonic larvae and plants was not yet re-colonised.

The two sites chosen for the study were substantially different in the sea urchin populations. It should be noted that the two sites were chosen not to act as statistical replicates for the effects of dragging on the populations (this is handled by the split-plot design at each site), but to compare trends in the responses of the animals to the impact. The Ministers Island site was comprised mainly of large numbers of small individuals below legal size (50 mm) and was typical of many sites found off the mainland of New Brunswick. The Grand Manan site, in contrast, had much lower densities of larger animals typical for that area. Unfortunately, in Grand Manan, it turned out that the experimental plot had higher densities of sea urchins than the controls (Fig. 5), but since the control densities did not change during dragging, we believe the comparisons are still valid. At both study sites, the dragging operation resulted in the immediate reduction of urchin densities from the pre-to post-survey compared to the controls, but populations seemed to recover within three months. Because of the short-term duration of the study, the recovery of population numbers was undoubtedly due to the movement in of existing animals from outside the plots.

The densities of sea urchins over time at the Ministers Island site suggested possible seasonal movement patterns in the population as the densities significantly decreased in both experimental and control plots from the post to the 3-month survey and then increased again by the 6-month survey (Fig. 5).

There was a certain amount of incidental mortality to sea urchins caused by the dragging operations. At the Grand Manan site, dragging killed over 12% of the animals based on results at the mid-survey (Fig. 6). The reduction in broken tests observed at the post-survey (2%) probably was current related. Bad weather forced us to delay our post-survey 10 days and it is likely that the normally high currents and storm surge washed many of the broken shells from the site. At the Ministers Island site, the percentage of broken urchins increased up to about 3% by the post-survey. The differences in mortality at the two sites may be related to the substrate size and also the size of the sea urchins. It is possible that larger animals sitting on larger rocks that are less likely to move when hit with the drag would suffer more crushing forces than smaller animals that might have more tumbling action. However, it may also be possible that many of smaller animals that were killed by the drags were smashed beyond recognition and thus were not included in the mortality count by the diver. At both sites, the percentage of broken tests had dropped to 0 m^{-2} within three months indicating the need to conduct studies on incidental fishing mortality early in the monitoring period at locations with high currents. Grand Manan had higher percentages of broken tests in the pre-survey as well as the control plots suggesting that there is either more predation happening in the area or the tests from the larger animals persist longer.

The other organisms analysed in this study could be basically divided into two groups: sedentary and more active animals. The sedentary group was represented by the kelp, sea peaches and sea cucumbers. For the sea peaches and sea cucumbers there was no demonstrable effect of the dragging based on changes in density in the experimental plot from the control. However, observations by divers of detached or crushed animals in the experimental plots were common. Observations were also noted in the control plot of detached sea peaches that had drifted there with the tide (not counted during the post-surveys). The fate of these animals is largely unknown although some predators, such as whelks, were observed feeding on damaged sea cucumbers. The dragging affected the kelp populations as the number of broken kelp stipes increased significantly from the pre-survey to the post-survey. The number of broken kelp stipes dropped at the 3- and the 6-month survey in the experimental plot, but this was expected as the winter season approached with a die-back of kelp and the loss of plants due to strong storms and surge.

The more mobile species were predators (lobsters, crabs, whelks, starfish and sculpins). Lobster populations were only found at the Ministers Island site and probably escaped the dragging operation. In the experimental plot, lobster densities declined from 0.08 m^{-2} in the pre-survey down to 0 m^{-2} in the post-survey and no broken lobster carapaces or claws were observed. There was no change in the control plot. It is possible that the drags may have modified lobster den structures by shifting rocks. At the Ministers Island site, the rock crabs and sculpins also probably escaped the drags as the densities again dropped from pre-survey to the post-survey in the experimental plot with no change in the control. On Grand Manan however, the rock crabs increased in density, while sculpins showed no change in the experimental plot after the dragging operation. This may have been in response to increased food levels and perhaps relatively less habitat disruption. These observations corroborate findings for fish and invertebrates in other trawling and dredging studies (i.e. Caddy, 1973; Kaiser & Spencer, 1994) that showed an increased attraction by predators to trawled and dredged (disturbed) sites. Sculpins showed a decreasing density over time. This is likely related to seasonal distribution patterns of the animals in shallow water, as they tend to move into deeper water in the winter months (Scott & Scott, 1988).

The northern whelk showed a classic opportunistic predatory response to the dragging operation. At Ministers Island, densities of the whelk increased in the experimental plot while densities decreased in the control plot following dragging suggesting that there was a general movement toward the impacted area to exploit the increased food resource. At Grand Manan, the same response was observed at the experimental plot after the dragging operation, but there was no significant change in the control plot. These data support observations by Ramsay et al. (1998) for northern whelk movements after fishing operations had occurred. The gear in this study, however, appeared have little or no direct impact on the whelks in comparison to Ramsay & Kaiser (1998) who reported that whelks that were rolled or tumbled during the fishing operation could have higher mortality rates through predation.

In summary, we found our experimental dragging operation caused a short-term impact to substrate and the associated species. However, the changes in the densities were generally short-lived and the differences between the control in experimental plots were generally gone after three months. This conclusion was similar to that of Currie & Parry (1996) who also found that changes to the community from scallop dredging was generally short-lived and within one recruitment cycle. Since this study, the use of scallop drags in the sea urchin fishery have been discontinued and replaced with a lighter drag (Green drag) specifically designed for sea urchins.

The short-term impacts created by this study (where a single perturbation was applied) are probably similar to those created by the natural environment through storms. Storms (hurricanes, typhoons, etc.) can have dramatic effects on shallow water habitats by decimating large sections of the bottom habitat such as reefs (Walsh, 1983). These storms are often part of a normal climatic cycle and the marine populations have adapted their life history strategies to them (Meeder & Meeder, 1989). In this study, we noticed that lobsters moved out of the impacted area. Walsh (1983) saw similar observations for fish after a storm impacted a reef. Damage to the bottom is generally patchy in nature and this spatial variability will allow for recolonisation from nearby relatively unaffected areas (Blythell et al., 1993). A study by Kroencke et al. (1998) concluded that storms had no impact on long-term changes in macrofaunal communities off Germany.

However, long-term fishing patterns may not fit this model of regular disturbances with minor impacts. Continual fishing in a specific area can modify the bottom so that certain features are destroyed (i.e., den structures for refuging animals, three dimensionality of the substrate, etc.). Jones (1992) suggests that "the greater the frequency of gear impact on an area, the greater the likelihood of permanent change". Therefore, the results of this study should be viewed from the perspective of short-term impacts only. Longer term changes to habitat and ecosystems will require more work and a great deal more investment of time and fiscal resources. Dealing with these changes will likely result in pooling expertise from fishing gear manufacturers, harvesters, educators and groups dealing with coastal remediation, ecosystems and animal production (i.e., aquaculture).

Acknowledgements

The authors thank the Grand Manan Fisherman's Association and the Canada-New Brunswick Cooperative Agreement for Fisheries and Aquaculture for sponsoring this study. We gratefully acknowledge the help of Mr. Kenny Stuart from Deer Island and Mr. David Green from Grand Manan with the dragging portion of the study and their contribution of practical knowledge on the project. Earlier drafts of the manuscript were read and improved by G. Thayer, D. Wildish and D. Robichaud. Thanks to two anonymous reviewers for their comments. Ms. Brenda Best assisted in the preparation of some of the figures.

References

Blythell, J. C., M. Blythell & E. H. Gladfelter, 1993. Initial results of a long-term coral reef monitoring program: Impact of Hurricane ugo at Buck Island Reef National Monument, St. Croix, U.S. Virgin Islands. J. exp. mar. Biol. Ecol. 172(1–2): 171–183.

Caddy, J.F., 1973. Underwater observations of tracks of dredges and trawls and some effects of dredging on a scallop ground. J. Fish. Res. Bd Can. 30: 173–180.

Collie, J. S., G. A. Escanero & P.C. Valentine, 1997. Effects of bottom fishing on the benthic megafauna of Georges Bank. Mar. Ecol. Prog. Ser. 155: 159–172.

Cover, P. & D. Sterling, 1994. Scallop dredging: an engineering approach. Mem. Queensl. Mus. 36: 343–349.

Currie, D.R. & G.D. Parry, 1996. Effects of scallop dredging on a soft sediment community: a large-scale experimental study. Mar. Ecol. Prog. Ser. 134: 131–150.

Currie, D.R. & G.D. Parry, 1999. Impacts and efficiency of scallop dredging on different soft substrates. Can. J. Fish. aquat. Sci. 56: 539–550.

Jamieson, G.S. & A. Campbell, 1985. Sea scallop fishing impact on American lobsters in the Gulf of St. Lawrence. Fish. Bull. USA 83: 575–586.

Jones, J. B., 1992. Environmental impact of trawling on the seabed: a review. N.Z. J. Mar. Freshwat. Res. 26(1): 59–67.

Kroencke, I., J. W. Dippner, H. Heyen & B. Zeiss, 1998. Long-term changes in macrofaunal communities off Norderney (East Frisia, Germany) in relation to climate variability. Mar. Ecol. (Prog. Ser.). 167: 25–36.

Kaiser, M.J. & B.E. Spencer, 1994. Fish scavenging behaviour in recently trawled areas. Mar. Ecol. Prog. Ser. 112: 41–49.

Meeder, J. F. & L. B. Meeder, 1989. Hurricanes in Florida Bay: a dominant physical process. Bull. mar. Sci. 44(1): 518.

Ramsay, K. & M.J. Kaiser, 1998. Demersal fishing disturbance increases predation risk for whelks (*Buccinum undatum* L.). J. Sea Res. 39: 299–304.

Ramsay, K., M.J. Kaiser & R.N. Hughes, 1998. Responses of benthic scavengers to fishing disturbance by towed gears in different habitats. J. exp. mar. Biol. Ecol. 224: 73–89.

Robichaud, D.A., A.M. Williams & D.E. Graham, 1987. Characteristics of the St. Marys Bay lobster stock in relation to scallop gear impact. Can. MS Rep. Fish. aquat. Sci. 1955: iv + 17 pp.

Roddick, D.L. & R.J. Miller, 1992. Spatial and temporal overlap of the American lobster (*Homarus americanus*) and sea scallop (*Placopecten magellanicus*) as related to the impact of inshore scallop dragging. Can. J. Fish. aquat. Sci. 49: 1486–1492.

Scott, W. B. and M. G. Scott, 1988. Atlantic Fishes of Canada. Can. Bull. Fish. aquat. Sci. 219: 731 p.

Thrush SF, J. E. Hewitt, V.J. Cummings & P. K. Dayton, 1995. The impact of habitat disturbance by scallop dredging on marine benthic communities: what can be predicted from the results of experiments? Mar. Ecol. Prog. Ser. 129: 141–150.

Walsh, W. J., 1983. Stability of a coral reef fish community following a catastrophic storm. Coral Reefs 2(1): 49–63.

Winer, B. J., D. R. Brown & K. M. Michels, 1991. Statistical Principles in Experimental Design. McGraw-Hill Inc., New York. 1057 pp.

Hydrobiologia **465:** 115–127, 2001.
G. Burnell (ed.), Coastal Shellfish – A Sustainable Resource.

Short-term impact of blue mussel dredging (*Mytilus edulis* L.) on a benthic community

P. Dolmer*, T. Kristensen, M. L. Christiansen, M. F. Petersen, P. S. Kristensen & E. Hoffmann
Danish Institute for Fisheries Research, Charlottenlund Castle, DK-2920 Charlottenlund, Denmark
**E-mail: pdo@dfu.min.dk*

Key words: mussel dredging, *Mytilus edulis*, impact study, benthic fauna, sediment texture, mussel growth
Additional keywords: multivariate analysis, *Crangon crangon*, anthropogenic effects, trophic structure, experiment design, seabed topography, commercial and experimental dredging, infauna, mussel condition, sediment organic content, short-term impact, dependent treated and control areas

Abstract

The short-term effect of mussel dredging in a brackish Danish sound was studied. A commercial dredging track was identified and an analysis of the species composition inside the track and at an adjacent control area showed that dredging changed the community structure by reducing the density of polychaetes. In order to investigate the extent and the duration of the dredging impact experimental dredging was conducted. The experimental dredging removed 50% of the mussels in two dredged areas. Immediately after dredging, a significantly lower number of species was measured inside the mussel beds in dredged areas compared to control and boundary areas. This effect lasted for at least 40 days. The analysis of the species composition showed that the dredged area had a significantly lower density, particularly of polychaetes compared to the boundary area. An increased number of species was recorded outside the mussel beds just after dredging, but this effect lasted for less than 7 days. After dredging, brown shrimps, *C. crangon* invaded the dredged areas. This species is an important predator of smaller invertebrates, and it is suspected that it was feeding on small vulnerable polychaetes exposed at the sediment surface after dredging. The dredging process was observed to form 2–5-cm deep furrows in the seabed, but the sediment texture and the organic content of the sediment was not affected. The biomass accumulation of individual blue mussels was significantly lower in the dredged area compared to the boundary area. This indicates that the disturbance of the mussel bed structure reduced growth and that the lowering of intraspecific food competition caused by a reduced density of mussels did not increase the accumulation of biomass in the mussels which remained in the dredged area.

Introduction

The increasing fishing effort and technical developments in marine fisheries such as the use of heavy gear and powerful vessels have raised concern about their impact on marine ecosystems: do we exploit the populations in a sustainable way and what is the impact on the ecosystem? The use of dredges, beam- and otter-trawls is responsible for the major part of fishery induced impact on the seabed and benthic community (see review by Jennings & Kaiser; 1998, Hall 1999).

The extended dredging effort for bivalves in nearshore waters has in the last ten years resulted in a number of impact studies including studies on changes in seabed topography and sediment structure (Hall et al., 1990; Pravoni & Giovanardi, 1994; Kaiser & Spencer 1996), the effect of resuspension of sediment (Riemann & Hoffmann, 1991; Dyekjær et al., 1995) and the effect of dredging on macrofauna (Hall et al., 1990; Eleftheriou & Robertson, 1992; Pravoni & Giovanardi, 1994; Trush et al., 1995; Currie & Parry, 1996; Hoffmann & Dolmer, 2000). The impact of fishing gear on seabed structures and the benthic community depends on both sediment type and current regimes. The impact is low or insignificant in mobile sandy sediments, whereas significant effects have been demonstrated in stable sediments (Shephard, 1983; Kaiser & Spencer, 1996).

The impact of dermersal fisheries are difficult to analyse because of the complexity of benthic communities and variability due to natural and anthropogenic effects (Gislason, 1994). Currie & Parry (1996) showed that the reduction in density of benthic invertebrates after dredging was small compared to annual changes. Commercial fisheries may induce long-term changes at established fishery grounds, changing the benthic community species composition towards a higher density of disturbance tolerant species. Impact studies in such areas may then fail to demonstrate or underestimate significant short-term effects because the invertebrate population is already changed.

Apart from the direct effect of dredging, the mussel fishery causes a number of indirect effects on the ecosystem. The target species in many bivalve fisheries constitute important or even dominant elements in suspension-feeding benthic communities which control the phytoplankton biomass in some areas (Møhlenberg, 1995; Meeuwig et al., 1998; Dolmer, 2000a,b). Reductions in bivalve densities as effect of the fishery may reduce the ecosystem's filtration capacity and the potential removal by grazing of phytoplankton. Fishing activity may therefore influence the structure of the ecosystem and energy pathways from the primary producers to the next trophic level as also reported from higher trophic levels (see Jennings & Kaiser, 1998).

The aim of this study is to analyse the impact of mussel dredging and the mechanical disturbance of the seabed on the benthic community. The hypothesis is that dredging may reduce the species richness and change the species composition in dredged areas. Blue mussels not removed by the dredge might be expected to increase their growth as the density and the intraspecific food competition is reduced in dredged areas (Frechete & Bourget, 1985b; Smaal et al., 1986; Petersen et al., 1997). In addition, dredging might be expected to change the sediment composition by the resuspension and export of fine particles and organic matter away from dredged areas.

Study-site

The effect of mussel dredging on the benthic community was investigated in a Danish sound Limfjorden (Fig. 1) on dredge tracks from a commercial fishery and from controlled experimental dredging. In Limfjorden (1575 km^2) blue mussels have been commercial fished since the beginning of this century (Kristensen, 1997). Around 750 $km^2$of the sound is today exposed to fishery. During the first decades the landings were small (<2000 tonnes $year^{-1}$, wet-weight). During the 1970s to 1980s the landings started to increase, and in the 1990s exceeded more than 100 000 tonnes $year^{-1}$ corresponding to 15% of the total stock in the area (Dolmer et al., 1999). The mussels are dredged by 51 licensed vessels restricted to waters deeper than 2 or 3 m and to mussels with shell lengths above 4.5 cm. The mussels are fished by the Dutch type of mussel dredge which is 1.8 m wide and weighs approximately 250 kg (see Kristensen, 1997). The fishery works in rotation between different areas within the fished area (see Dolmer et al., 1999) and the area used for experimental dredging in autumn 1996 was not dredged since the winter 1993–1994.

Methods

Observations from a commercial dredge-track

During spring 1996, blue mussels were fished by commercial vessels in the area shown in Figure 1. In June the area was examined by use of SCUBA-diving and a recent dredge track was identified at 7 m of water depth, from which the macro infauna was quantitatively sampled by a 0.1-m^2 van Veen grab. Benthos was sampled inside the dredge track and from an unaffected area 5–10 m apart the dredge track (n=5). The sampling process was inspected by use of diving in order to ensure that the samples were obtained from the correct areas. The sample fraction >1 mm was preserved in 2% buffered formalin, and the invertebrates were identified in the laboratory.

Experimental dredge-tracks

In June 1996 four study sites (20×20 m) were established at 7.4 m of water depth (Fig. 1). Each study site was organised in a coordinate system marked by plastic sticks (10×10 m). In order to test the impact of dredging on the dredged areas and on the adjacent boundary areas half of the area of two of the study sites were experimentally dredged in September separating the study sites into two dredged and two undredged boundary areas. The two other study sites served as control areas. This design then included three categories of areas: two areas affected by dredging (D1 and D2), two undredged boundary areas adjacent to the

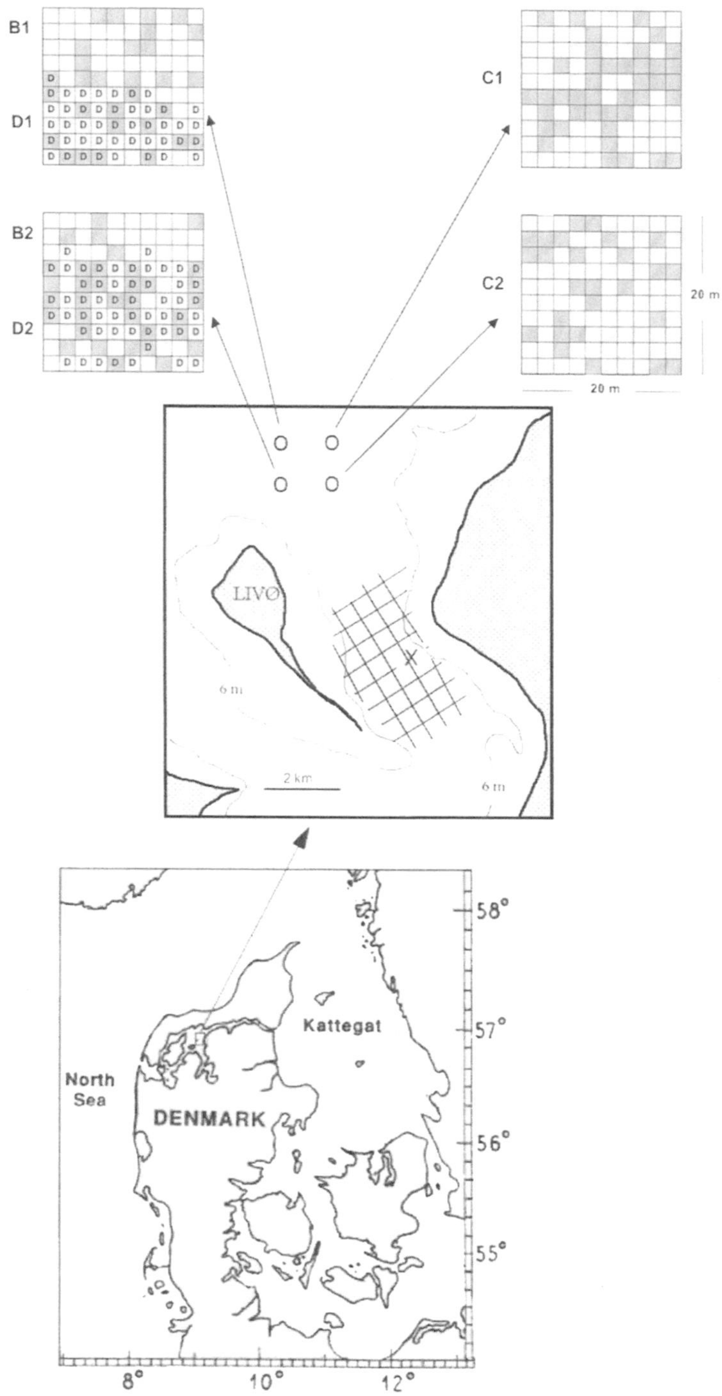

Figure 1. Map of Limfjorden showing the position for the study on commercial dredge tracks (X) and the positions of the study sites for the dredging experiment (◯). The crossed area mark the exploited fishing grounds in spring 1996. The upper part of the figure shows the distribution of mussel beds (grey areas) and dredging activity at the two dredged areas (D1 and D2). Area B1 and B2 serve as boundary area adjacent to the dredged areas and C1 and C2 serve as controls. The dimensions of each study area is 20 × 20 m.

dredged areas (B1 and B2) and two undredged control areas (C1 and C2) (Fig. 1). At all four study sites populations of blue mussels formed a mosaic of 1–15 m^2 large mussel beds interrupted by bare mud flat of approximately same size. The distribution of mussel beds was recorded at all coordinates marked by plastic sticks. The mean density (±2 S.E.) of the mussels at the four study sites ranged from 1.5±1.6 kg m^{-2} (wet weight) at site B1–D1 to 3.3 ±2.0 kg m^{-2} at site B2–D2. The two dredged areas were dredged in September by a 1:2 scaled model of the commercial mussel dredges (1 m wide, 80 kg). Each of the two areas was dredged 12 times and the mussels were dumped outside the areas. Just after dredging, the sites were inspected by use of diving and the dredged area was mapped according to the plastic stick coordinate system.

The infauna in the dredged and boundary areas was sampled 20 days before dredging and 0, 7 and 40 days after dredging, and the control areas were sampled 20 days before, on day 0 and day 40 after the dredged sites were dredged. In each area (D1, D2, B1, B2, C1 and C2) 6 samples of the infauna were obtained by a 0.1-m^2 van Veen grab with three samples from inside a mussel bed and three samples from outside a mussel bed. Before the experimental dredging took place six positions – three inside mussel beds and three outside mussel beds – were identified in the two dredged area D1 and D2 and subsequently used for sampling of the dredged areas. The coordinates of the six positions were mapped out as distances to two fixed rods close to the edge of the study sites. Sampling of the positions at the dredged sites was inspected by use of diving to ensure that the samples were actually obtained from the fixed positions. At boundary and control areas, three samples from inside mussel beds and outside mussel bed, respectively, were obtained by random sampling. Samples were preserved and treated as previously described.

Data analysis of bottom samples

In the experimental dredging study, the number of species in each sample was analysed by a three-way nested ANOVA as a function of time, study site and area nested to study site. Samples obtained inside and outside the mussel beds was analysed separately. Before the tests, data was ln-transformed in order to ensure variance homogeneity. The variance homogeneity was tested with F_{MAX}-tests. The ANOVA test was followed by a Tukey multiple comparison procedure. Since there was no data from the control areas at day 7, data from dredged and boundary areas for this data was tested by a *t*-test.

The species compositions, in samples from the commercial dredge track study and the dredge experiment, were analysed by use of the PRIMER package (Plymouth Routines in Multivariate Ecological Research, Clark and Warwick 1994). A Bray–Curtis similarity coefficient matrix was calculated on fourth root transformed data and separate clusters were identified on multidimensional scaling plots (MDS). Differences between species compositions were tested by the analysis of similarities permutation test (ANOSIM). When the ANOSIM test indicated significant differences between areas the SIMPER analysis was performed on the data in order to determine which taxa contributed to the dissimilarity, and the significance of density differences of the taxa was pairwise tested by nonparametric Mann–Whitney tests, in order to handle variance heterogeneity. The samples from the commercial dredge tracks were analysed in one-way ANOSIM tests where the species composition in samples from dredged bottom was compared to control samples.

The design of the dredge-experiments include sampling from four geographical separated study sites. Due to the heterogenity in species composition among the sites, they could not be compared directly. The species compositions at the two control areas were analysed in two-way crossed ANOSIM tests, with bottom type (inside/outside mussel beds) as one factor and time as the other factor. Where significant effects of bottom type was observed, one-way ANOSIM tests were used to analyse the effect of time on samples from mussel beds or sandy bottom separately. The species composition in samples from dredged and boundary areas was analysed in two-way crossed ANOSIM tests with dredging (dredged or boundary samples) as one factor and time as the other factor. Only samples obtained after dredging were included in the tests. Separate tests were conducted on samples obtained inside and outside mussel beds. When the analyses indicated a significant difference between dredged and boundary areas the effects of time were analysed in one-way ANOSIM tests including all sampling periods. The tests where followed by pairwise comparisons, but since the total number of possible permutations in these tests was low the lowest detectable significance level was 10%.

Density of Brown shrimps

The brown shrimp, *Crangon crangon*, was a quantitatively important predator of smaller animals such as polychaetes at the study sites. Shrimp density was quantified from pictures taken by a Nikonos V camera mounted with a 35-mm lens. The photos were taken from a fixed height (1m) above the sea bed (sampling area, 0.24 m^2; n=10) and the shrimp density was counted from slides in a stereo microscope. At dredged and boundary areas the sediment surface was photographed 2 days before dredging and 0, 7 and 40 days after dredging, whereas the control areas were photographed 0 and 40 days after dredging. Samples, where mussel beds covered more than 25% of the photographed area were excluded from analysis. In order to handle variance heterogeneity the data was $\ln(x+1)$ transformed and the densities of the shrimp in boundary areas and dredged areas were pairwise tested by t-tests.

Mussel growth and condition index

The density and size distribution of mussels were recorded by sampling of mussels using a frame (sampling area, 0.28 m^2; n=10). The size distributions were established from recordings of shell lengths (n=126–401) to the nearest 0.1 mm. Furthermore, the ash-free dry weight of the soft part (AFDW) of 25 individuals sampled from each of the six areas at day 0 and 40 days after dredging was measured. First, the soft parts were dried at 100°C during 24 h, and secondly ashed at 550°C during 4 h and the AFDW estimated. A condition index was estimated as $CI=W/L^3$, where W is the AFDW and L is the shell length. The condition index was tested as a function of time and treatment at the dredged and boundary area by use of a two-way ANOVA. Before the analysis the data was balanced and tested for homeoscedasticity by the F_{max}-test. The ANOVA were followed by a Tukey multiple comparison procedure. The specific growth rate (μ) was estimated as $\mu=\ln(W_t/W_0)t^{-1}$, where W_0 and W_t are mean dry weight at day 0 and day t.

Sediment texture

The effect of dredging on the sediment structure was analysed from sediment cores (diameter, 5 cm) sampled from the three different categories of areas just after dredging and after 40 days. The sediment was wet sieved according to the Wentworth scale (Φ-notation) and the fractions dried at 100°C during 24 h. The fraction smaller than the smallest sieve (63 μm) was allowed to sedimentate and the sediment was dried. The weight of the sediment fractions was plotted as cumulative frequencies and the median diameter was calculated. Furthermore, the organic content of the sediment was measured by ashing the sediment at 650°C for 4 h.

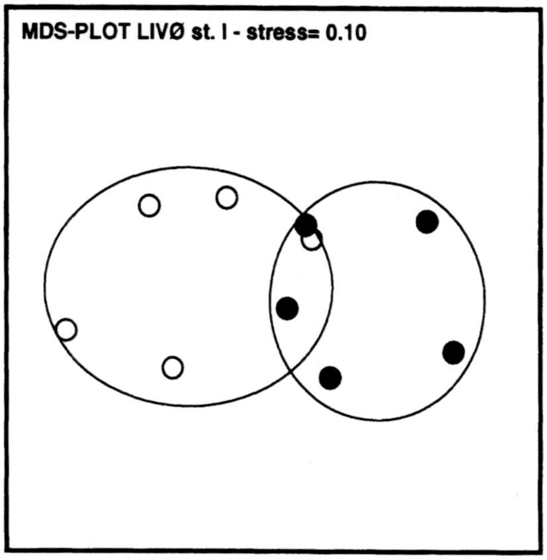

Figure 2. MDS plots of the samples inside a commercial dredge track (●) and from an unaffected adjacent area (○).

Table 1. Density of the taxa, which contribute most to the species composition dissimilarity in control and dredged bottom.

Taxa	Density±2 S.E. (indiv. m^{-2})		
	Control	Dredged	P
Commercial dredgetrack			
Capitellidae	134.0±190.5	6.0±8.9	0.028*
Philine aperta	8.0±13.0	48.0±29.5	0.066
Harmothoe imbricata	56.0±62.3	4.0±5.5	0.264
Corbula gibba	28.0±25.9	152.0±92.0	0.012*
Neanthes succinea	10.0±7.1	0.0±0.0	–
Scoloplos armiger	48.0±60.6	10.4±20.7	0.380

Differences were tested by pairwise Mann–Whitney tests and the significances are indicated as P-values. *P<0.05; **P<0.01.

Results

Observations from commercial dredge-track

The effects of mussel dredging on the benthic fauna were investigated by comparing the species composition in dredge-tracks created by commercial mussel

Table 2. Three-way nested analysis of variance of number of species inside and outside the mussel beds as a function of time (20 days before and 0 and 40 days after dredging), study sites and area (control – boundary – dredged) nested to study site.

Source	DF	SS	MS	F	P
Inside mussel beds					
Time	2	0.7045	0.3522	2.27	0.115
Site	1	0.1112	0.1112	0.72	0.402
Area (site)	4	3.1223	0.7806	5.04	0.002
Time×site	2	0.5068	0.2534	1.63	0.207
Error	44	6.8204	0.1550		
Total	53	11.2651			
Outside mussel beds					
Time	2	0.1109	0.0554	0.40	0.674
Site	1	0.5814	0.5814	4.18	0.047
Area (site)	4	1.7452	0.4363	3.14	0.024
time×site	2	0.5982	0.2991	2.15	0.128
Error	44	6.1190	0.1391		
Total	53	9.1547			

vessels versus the unaffected control area. A MDS–plot (Fig. 2) and a ANOSIM test indicated that the species composition in samples from the dredged seabed was significantly different from the composition in the control samples (n=5, P=0.032). The SIMPER routine determined which taxa that contributed to the observed dissimilarity between dredged and control areas. When a Mann–Whitney test indicated significant differences in density (P<0.05) the species or higher taxa are underlined in the following text. The polychaetes Capitellidae, *Harmothoe imbricata, Neanthes succinea* and *Scoloplos armiger* had a reduced density in the dredged seabed, whereas the ophistobranch, *Philine aperta*, and the bivalve, *Corbula gibba*, showed an increased density (Table 1).

Experimental dredge-tracks

Visual inspections by diver of the experimental dredged areas showed that the topography of the sea bottom was changed after mussel dredging by the formation of 2–5-cm deep furrows. The mussel beds were destroyed, and mussels not removed by the dredge showed a homogeneous distribution over the dredged area. At B1–D1, 45% of the study site was dredged and at B2–D2, 54% was dredged (Fig. 1).

A total of 39 species and in addition 13 higher taxa of benthic invertebrates were observed in the

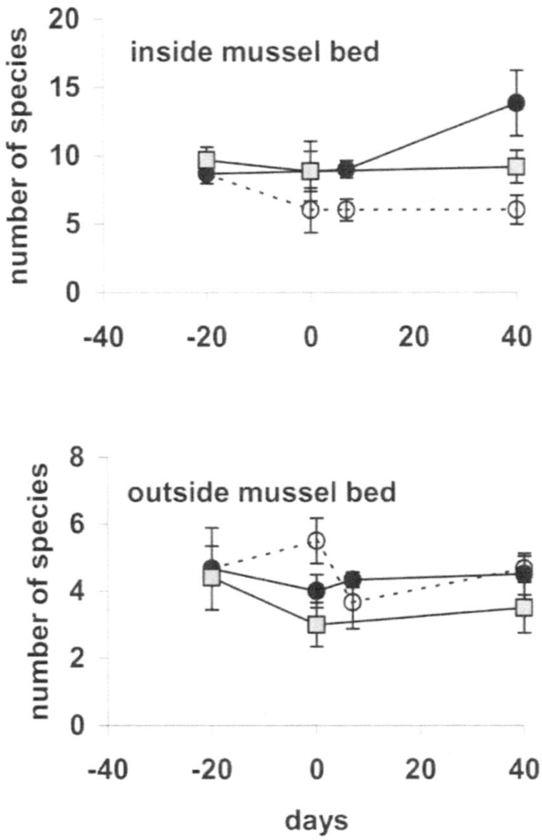

Figure 3. Total mean number of species (± S.E) in samples inside the mussel beds (upper) and outside the mussel beds (lower). The samples were collected in control areas (■), boundary areas (●) or dredged areas (○).

experiment. The ANOVA analysis showed that the number of species were significantly different between dredged, boundary and control areas in samples from inside mussel beds and outside mussel beds whereas no effects of time could be demonstrated (Table 2). No effect of site was observed in samples from inside mussel beds, whereas a significant effect of study site was demonstrated in the samples from outside the mussel beds. At sites in the boundary and dredged areas the mean number of species per sample (± 2 S.E.) was 8.7±1.4 species inside the mussel beds 20 days before the experimental dredging (Fig. 3). Tukeys procedure showed, that the number of species in the control areas (9.7±1.0 species) was significant larger (P<0.05). Outside the mussel bed only 4.7±2.4 species per sample was observed at time −20, and no difference was measured among the three area types (B–D–C) (P>0.05). A significantly lower number of

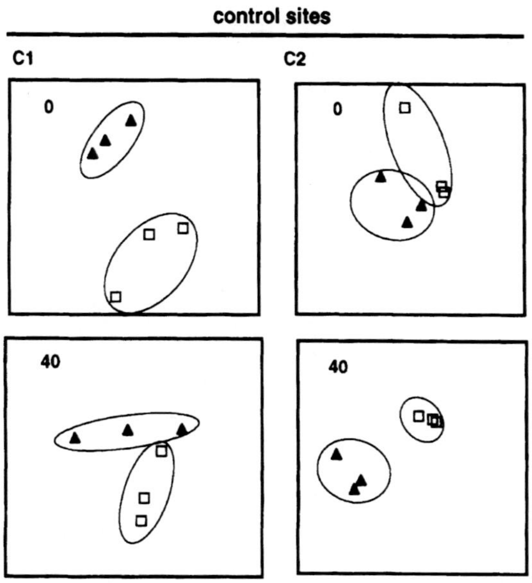

Figure 4. MDS plots of the samples inside the mussel beds (▲) or outside the mussel beds (□) in the two control areas.

Table 3. Density of the taxa inside mussel beds, which contribute most to the species composition dissimilarity in control bottom and dredged bottom.

Taxa	Density±2 S.E. (indiv. m^{-2})		
	Boundary	Dredged	*P*
Station B1–D1, inside mussel beds			
Harmothoe imbricata	81.1±47.2	7.8±7.7	0.001**
Polydora cornuta	20.0±17.3	1.1±2.4	0.017*
Rissoa albella	21.1±17.5	4.4±5.1	0.097
Scoloplos armiger	8.9±9.0	1.1±2.4	0.055
Philine aperta	113.3±42.6	100.0±52.7	0.678
Nereimyra punctata	6.7±6.1	1.1±2.4	0.116
Neanthes succinea	1.1±2.4	6.7 ±6.1	0.116
Actica islandica	1.1±2.4	4.4±3.7	0.140
Nepthys hombergi	24.4±8.7	21.1±10.3	0.554
Corbula gibba	90.0±63.4	77.8±39.0	1.000
Polydora ciliata	4.4±5.1	0.0±0.0	–
Mysella bidentata	6.7±9.4	0.0 ±0.0	–

Differences were tested by pairwise Mann–Whitney tests and the significances are indicated as *P* values: $^*P<0.05$; $^{**}P<0.01$.

species was observed inside the mussel beds in the dredged areas compared to the boundary areas and the control area ($P<0.001$) at time 0 after dredging. Outside the mussel bed an increased number of species was measured in the dredged areas compared to the boundary areas ($P<0.001$) and the controls ($P<0.05$). Seven days after the experimental dredging the number of species inside the mussel beds was still lower in the dredged area compared to the boundary area (*t*-test, $P<0.05$), whereas no difference was observed outside the mussel beds (*t*-test, $P>0.05$). Forty days after dredging the species richness inside the mussel beds was significant larger in the boundary area than in the control area ($P<0.001$), whereas the density in the dredged areas were significant lower ($P<0.001$). Outside the mussel beds the number of species in the control area was lower than that measured in the other two areas ($P<0.05$).

Control study sites

A MDS analysis indicated that the samples from inside and outside the mussel beds at the two control sites formed separate clusters (Fig. 4). ANOSIM tests showed that the benthic fauna inside the mussel beds at the two control areas was significantly different from that outside (P=0.003 and 0.005). At site C1 no effects of time on species compositions could be demonstrated (P=0.61), whereas a significant effect of time was observed at site C2 (P=0.01). At both stations the mussel beds had higher densities of the polychaetes *H. imbricata*, *S. armiger*, *Polydora cornuta*, capitellids, *nereids* and nepthyids, and the prosobranch, *Rissoa albella*, whereas higher densities of the ophistobranch, *P. aperta*, and the bivalves, *Arctica islandica, Corbula gibba* and *Cerastoderma edule* was observed outside the mussel beds.

Dredged study site: inside mussel beds

At the two study sites exposed to experimental dredging, samples of the infauna in dredged and boundary areas formed separate clusters (Fig. 5). The ANOSIM tests of samples from sites B1–D1 and B2–D2 indicated a significant difference between dredged and boundary areas (P=0.002 and 0.001, respectively). The analysis of the effect of time at site B1–D1 indicated a significant effect at both dredged and boundary areas (P=0.028 and P=0.046, respectively). Pairwise tests of data from the dredged area indicated that the samples obtained 20 days before dredging were different from the other samples (P=0.1 lowest detectable significant level), whereas samples obtained after dredging at day 0, 7 and 20 were not significantly different from each other (P=0.3–0.5). Pairwise tests of data from the boundary area indicated that samples obtained 7 days after dredging were different (P=0.1) from samples obtained 20 days before and 40 days after dredging. At site B2–D2 the pairwise tests of effect of time indicated that the samples ob-

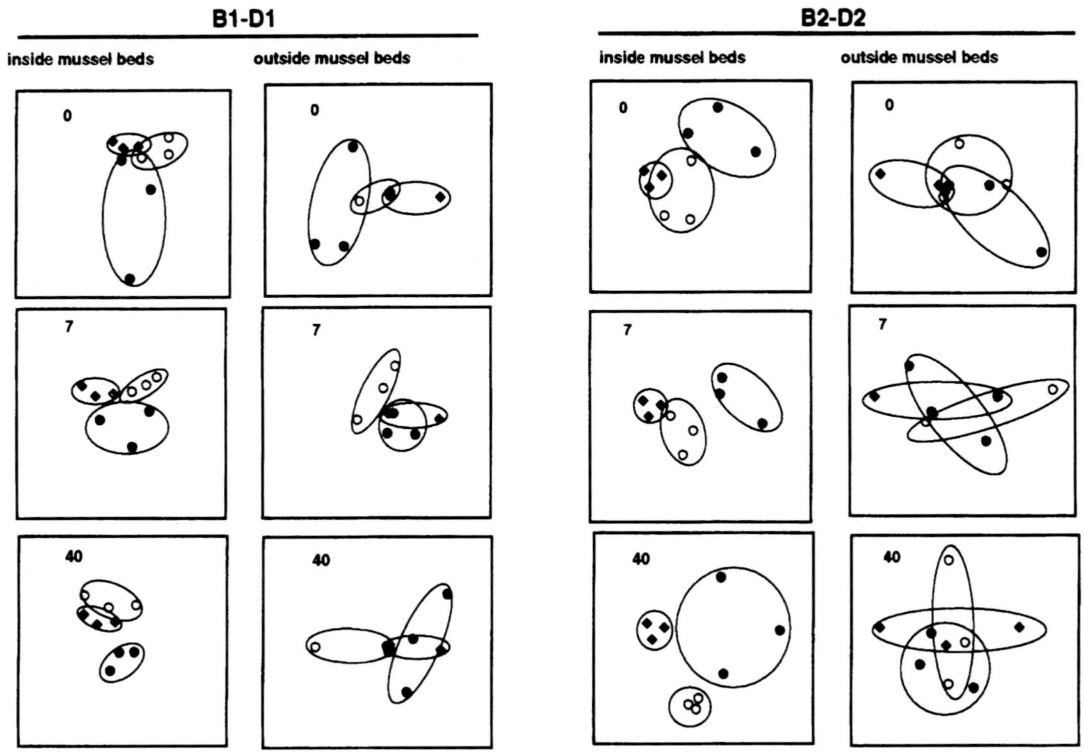

Figure 5. MDS plots of the samples from the two dredged study sites B1–D1 and B2–D2. The samples were collected inside or outside the mussel beds just after dredging (time 0), 7 and 40 days after dredging. The two study areas were separated in dredged areas (●) and boundary areas (○). In addition, the samples taken after the experimental dredging are compared to samples collected 20 days before the dredging (♦).

tained before dredging was different from the samples obtained after dredging (P=0.1) in both dredged and boundary areas indicating that the dredging process also affected the boundary area. The SIMPER analyse of data from study site B1–D1 showed reduced densities of the prosobranch *R. albella*, the bivalves *Mysella bidentata* and *C. gibba*, the polychaetes *H. imbricata*, *P. cornuta*, *Polydora ciliata*, *S. armiger*, *Nepthys hombergi*, and *N. punctata* in dredged seabed compared to the boundary area. A few species had an increased density in the dredged area including the ophistobranch, *P. aperta*, the bivalve, *Arctica islandica* and the polychaete *N. succinea* (Table 3).

Dredged study sited: outside mussel beds

Samples from boundary and dredged areas outside the mussel beds formed overlapping clusters in the MDS analysis, indicating little or no effect of dredging on the species composition (Fig. 5). At site B2–D2 no difference between the species composition was observed in samples from dredged areas and boundary areas was observed (ANOSIM, P=0.98), whereas at site B1–D1 a significant effect of dredging was observed (P=0.044). No effect of time was observed at the two study sites (P=0.79 and 0.29, respectively). A SIMPER analysis of the sandy bottom samples from station B1–D1 indicated that the densities of oph-

Table 4. Density of the taxa outside the mussel beds, which contribute most to the species composition dissimilarity in control bottom and dredged bottom.

Taxa	Density ± 2 S.E. (indiv. m^{-2})		
	Boundary	Dredged	P
Station B1–D1, outside mussel beds			
Philine aperta	193.3±24.1	88.9±14.3	0.006**
Cerastoderma edule	3.3±1.3	2.2±1.1	0.650
Polydora cornuta	4.4±2.5	1.1±0.8	0.540
Isopoda	0.0±0.0	5.6±2.8	–
Corbula gibba	126.7±13.4	134.4±15.0	0.722
Nepthys hombergi	22.2±1.7	32.2±5.4	0.697
Neanthes succinea	0.0±0.0	4.4±3.3	–

Differences were tested by pairwise Mann–Whitney tests and the significances are indicated as P values: *P<0.05; **P<0.01.

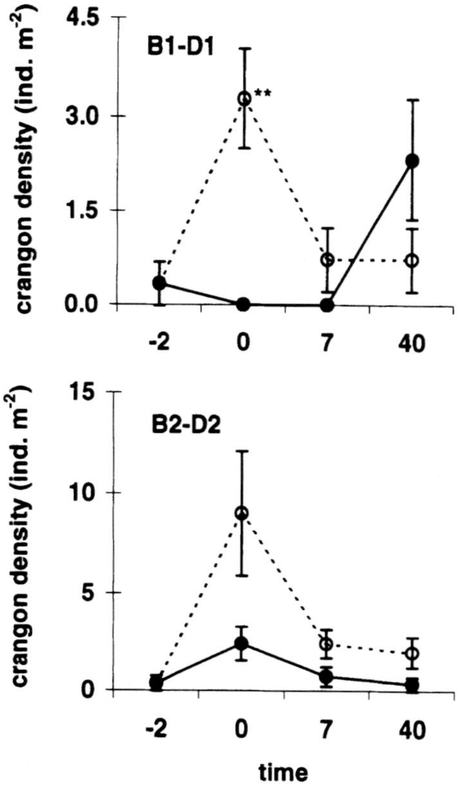

Figure 6. Density of brown shrimps, *Crangon crangon*, in boundary areas (●) and dredged areas (○) 2 days before and 0, 7 and 40 days after the experimental mussel dredging. The errorbars indicate mean± S.E.

istobranch, *P. aperta*, the bivalve *C. edule*, and the polychaetes *P. cornuta* was reduced in the dredged area, whereas an increased density of the bivalve *C. gibba* and the polychaetes *Nepthus hombergi* and *N. succinea* was observed in the dredged area (Table 4).

Density of Brown shrimps

The density of brown shrimps, *C. crangon,* was assessed before and after dredging at the two dredged study sites (Fig. 6). At the two dredged areas, an increased density of the *C. crangon* was observed just after dredging. At study site B1–D1 the mean density (±2.S.E.) was 0.3±0.7 individuals m^{-2} 2 days before dredging. Just after dredging, the density was 3.3±1.5 individuals m^{-2} in the dredging area which was significantly higher than the density in the boundary area B1 where the shrimps has disappeared completely after dredging. At the other study site B2–D2 there was a non-significant increase in shrimp density in the boundary area after dredging. The mean density (±2.S.E.) of *C. crangon* before dredging was 0.4±0.8 individuals m^{-2} increasing to 9.0±6.2 individuals m^{-2} at the dredged area and 2.5±1.5 individuals m^{-2} at the boundary area just after dredging. No trends in the *C. crangon* distribution were observed after 7 and 40 days (Fig. 6).

Mussel growth and condition index

The length and the growth of the shells during the study period are given in Table 5. The condition index (CI) of mussels at the dredged and boundary areas was estimated just after dredging and after 40 days (Table 5). The ANOVA showed significant interactions between time and dredging at both study sites (D1–B1: P=0.044; B2–D2: P=0.024). At B1–D1, Tukeys test indicated a significant reduction (P<0.05) in CI of the mussels in the dredged area from 2.2±0.2 μg mm^{-3} (mean±2 S.E.) at time 0 to 1.7±0.2 μg mm^{-3} 40 days after dredging. No change in CI was observed in the boundary area. At B2–D2, a significant difference in CI between mussels in dredged and boundary area at time 0 was observed, whereas no difference was observed after 40 days.

The specific mussel growth rate was low in all areas (Table 5). The highest growth rate was measured in area B2 (+4.7% $month^{-1}$) and the lowest in area D1 (−31.4% $month^{-1}$).

Sediment texture

The size distributions of sediment particle were similar among the boundary areas (B1–B2) and control area C2 having a median diameter of approximately 80 μm (Φ=3.65). At area C1 the median diameter was 150 μm (Φ=2.8). At study site D1, the dredging process tended to increase median diameter (88 μm, Φ=3.50), but a χ^2-test, testing the weight frequencies of the sediment from dredged and control area, indicated no significant differences (P>0.05). The organic content of the sediment varied among the experimental areas (Table 6). At the dredged areas the organic content tended to be lower than in the boundary areas, but no significant differences were observed (t-tests, P>0.05).

Discussion

A prerequisite for interpreting studies on the impact of bottom trawls and dredges in benthic ecosystems

Table 5. Shell-length, shell-growth and ash free dry weight (AFDW) of mussels in area D1–B1–D2–B2 at time 0 and 40 days after dredging. In addition, the condition index (CI= W/L^3) and the specific growth rate ($\ln(W_{40}/W_0)/t$) is given.

	Shell length mm		Shell growth (mm)	AFDW (g)		CI (g mm^{-3})		Specific growth (% month^{-1})
	L_0	L_{40}	L_{40}–L_0	W_0	W_{40}	CI_0	CI_{40}	
D1	41.0±0.9	42.6±1.1	1.6	0.173±0.080	0.115±0.024	2.2±0.3	1.7±0.2	−31.4
B1	42.2±1.1	44.6±1.1	2.4	0.162±0.019	0.161±0.019	2.2±0.2	2.1±0.2	−0.4
D2	38.1±1.0	41.9±0.8	3.8	0.133±0.042	0.129±0.049	2.0±0.2	1.9±0.1	−2.3
B2	38.3±0.7	40.1±0.5	1.8	0.112±0.013	0.119±0.010	1.7±0.2	1.9±0.2	+4.7

is proper control situations unaffected by fishery. Increased utilisation of marine resources with extensive and repeated disturbance of the ecosystems may induce long-term changes in benthic communities, which make it difficult to define baselines for unaffected scenarios (Gislason, 1994). Locations not previously exposed to fishing activities have demonstrated larger short-term ecosystem effects of experimental fishing than at established fishing grounds (Pranovi & Giovanardi, 1994, Thrush et al., 1995). Consequently, experimental investigations on the impact of trawling and dredging in habitats serving as commercial fishing grounds may fail to demonstrate or underestimate significant short-term effects of bottom disturbance. In the present study the number of species in each sample was low. In the mussel beds 8.7 species per sample was recorded whereas only 4.7 species per sample was recorded outside. Time-series of samples of epi- or infauna has been used to document long-term effects of fishing activities on benthic ecosystems and a benthic infauna-sampling programme was conducted in Limfjorden by Petersen in 1911–1952. A reanalysis of the data from Løgstør Broad, where the present experiment was conducted indicated that the species richness in the period 1911-1952 ranged between five and 12 species 0.1 m^{-2} (Hylleberg 1993). Although the mesh size used in their sampling ranged between 1.2 and 2 mm compared to 1 mm in this study, the time series indicate that there has been a change in the species composition, but the species richness has not been reduced since the beginning of this century. Other factors than the fishery may have changed the species richness during the last century. High ecosystem productivity is hypothesised to support a total higher abundance and species richness (Srivastava & Lawton, 1998). From that hypothesis, increased species richness could be expected in many coastal areas as an effect of eutrophication, but this effect may be counteracted by other effects as fishery.

The present study demonstrated that one of the adjacent boundary area was indirectly affected by the experimental dredging (Study site B2–D2). A reduced number of species and individuals was observed in the dredged area after 40 days coincident with increased numbers in the boundary areas and constant densities in the control areas. The analysis of the species composition at the study site showed that the fauna in both the dredged and boundary area changed following dredging. Furthermore, an increased density of brown shrimps, *C. crangon*, was observed in both the dredged and the boundary area after dredging indicating that they are attracted to both areas. In many impact experiments, the control areas are immediately outside the trawled or dredged area (Hall et al., 1990; Brylinsky & Gibson, 1994) and it is implicitly assumed that treated and control areas are independent. In order to avoid sedimentation of material in the control plot from the dredging process Currie & Parry (1996) dredged only during periods with a tidal current away from a control area and avoided spreading of suspended material into the area. Hall et al. (1990) investigated the effect of hydraulic dredging on the infauna by comparing the species composition in dredged and adjacent undredged areas. They observed a reduced number of species and individuals inside the dredged areas compared to the adjacent undredged areas the day after experimental dredging. Forty days after dredging the density of species and individuals were increased in the dredged area and decreased in the adjacent control area. The dimensions of the experimental plots used by Hall et al. (1990) was 50×100 m, so it is reasonable to interpret the observed patterns as an effect of suspension of benthic animals by the dredge followed by a sedimentation of animals in the adjacent areas.

A reduction in density of infauna was observed in the samples from the commercial dredge track and from the experimentally dredged areas. Smaller polychaetes such as *Harmothoe imbricata, Scoloplos armiger, Polydora* sp. and *Nereimyra* were significantly reduced in density. Other investigations have also shown that small polychaetes are vulnerable to dermersal fishing gear (Eleftheriou & Robertson, 1992; Thrush et al., 1995). The bivalve, *Arctica islandica* had a higher density in samples from dredged areas than in adjacent areas in both the commercial and experimental study. This bivalve lives in the upper 10 cm and dredging may have removed the upper layer of sediment changed the sampling profile of the grab. Other species such as *Corbula gibba* and *Neanthes succinea* show conflicting results comparing commercial and experimental areas. The effect of experimental dredging was more significant on mussel beds than on the surrounding sediment. The complex, tri-dimensional mussel bed serve as habitat of a rich associated fauna (Tsuchiya & Nishihira, 1985, 1986; Svane & Setyobudiandi, 1996). Just after the dredging process in the present study, an increased species richness was recorded in sandy substrates, whereas a reduced richness was observed in the mussel bed. These observations indicate that the dredging process destroy the mussel bed structure and homogenises the distribution of species.

A significant impact of dermersal fishing activities is an increase in density of scavengers (Gislason, 1994; Jennings & Kaiser, 1998). Invertebrates, injured by the fishing gear, and discards of non target species is reported to serve as food source for scavenging invertebrates (Pranovi & Giovanardi, 1994; Thrush et al., 1995). In the present study, the dredged areas was invaded by brown shrimps, *C. crangon*, following dredging increasing the density from 0.3 to 3–9 individuals m^{-2}. This species is reported to play a key role in structuring benthic communities (Reise, 1977; Pihl & Rosenberg, 1984; Jensen & Jensen, 1985; Pihl, 1985, 1990; Veer & Bergmann, 1987; Hedqvist-Johnson & André, 1991; Nielsson et al., 1993; Kamermans & Huitema, 1994) and investigations in shallow waters in western Sweden have demonstrated that larger shrimps prey on macrofauna such as juvenile individuals of *Mya arenaria, Cerastoderma edule, Nereis* sp. and *Corophium volutator* (Pihl & Rosenberg, 1984). The observed reductions in the density of a number of smaller polychaetes in both the observations from the commercial dredge track and the experimental dredging may be caused by a displacement of the animals to the surface making them vulnerable to brown shrimps, which reduce the density of individuals and species. This increased predation rate of smaller invertebrates may then accelerate the transport of energy to the higher trophic levels in the ecosystem.

Table 6. Organic content in percent in sediment from control, boundary and dredged areas.

	Undredged	Dredged
C1	1.5±0.0	
C2	2.6±0.6	
B1–D1	3.6±0.6	3.0± 0.5
B2–D2	3.2±0.3	3.1± 0.4

Studies on food limitation in blue mussels have demonstrated that they pay a cost for their aggregated distribution. Mussels are reported to reduce the amount of seston in the near-bottom water (Frechette & Bourget, 1985a; Asmus & Asmus, 1991; Dolmer, 2000a,b) and thereby reduce their growth rate (Frechette & Bourget, 1985b; Smaal et al., 1986; Petersen et al., 1997). Mussel growth is further reduced in the centre of mussel beds (Newell, 1990; Svane & Ompi, 1993). Therefore, when mussel density was reduced in dredged areas an increased food uptake and biomass accumulation might then be expected. However, the present experiment demonstrated that the individual biomass accumulation was lower in the dredged areas which had a lower density than in control areas. One reason for this observation could be that a proportion of the mussels may have become buried into the sediment and had to spend time and energy to move to the sediment surface. Furthermore, destruction of the mussel bed structure may also be critical for the mussels. In dense mussel beds, mussels are normally orientated with the siphons upwards at a certain height above the seabed. Reorientation of the mussels may then reduce the quality of inhaled water ex. amount of suspended sediment. Since the biomass accumulation may be correlated to filtration activity of mussels, the reduced individual biomass accumulation in dredged areas suggests that the dredging process reduced the filtration of those mussels not removed by the dredge.

Previous investigations have demonstrated that fishing with dermersal gears changes the sediment composition and the topography of the sea bed (Hall et al., 1990; Pravoni & Giovanardi, 1994; Kaiser & Spencer, 1996). Visual inspection of the seabed at the two dredged study sites showed that the topography was changed. Previous investigations of mussel dredging in Limfjorden has demonstrated that the dredge only affects the upper 0.2–2.0 cm of the sediment (Dyekjær et al., 1995), which is a very slight penetration compared to other dermersal fishing gears (de Groot, 1984; Hall et al., 1990; Bergman & Hup, 1992). The analyses of the median diameter and organic content of the sediment showed no effects of dredging. Since the analysed sediment samples was taken down to 5 cm in the sediment and assuming, that the dredge penetrated down to about 2 cm (Dyekjær et al., 1995), the samples included a large fraction of unaffected sediment. This mixing of affected and unaffected sediments may reduce our ability to demonstrate a difference between dredged and boundary sediments.

The ecosystem in Limfjorden is eutrophic because of nutrient-rich freshwater run-off from the watershed which sustains a high primary production of microalgae and a derived secondary production of blue mussels (Dolmer, 1998). Oxygen depletion in the bottom water occur almost bi- or triannually during summer periods with low mixing of the water column which results in mass mortality of benthic invertebrates (Jørgensen, 1980; Dolmer et al., 1999). During the summers of 1994 and 1997 oxygen depletion exterminated 25–50% of the mussel populations. The dimensions of these events has to be considered when evaluating the effects of mussel fishery on the ecosystem. Dolmer et al. (1999) demonstrated that two factors control the large-scale distribution of blue mussels in Limfjorden. In a year without oxygen depletion the fishery for blue mussels significantly reduced the mussel population, whereas in a year with oxygen depletion, a high mussel mortality or a total elimination was observed in the affected areas. The mussel fishers manage the fishery, so almost all mussels in an area used as fishery ground are harvested within a couple of months before the fishery activities move to other grounds. New mussels then have to be recruited to fished grounds and grow to a minimum shell length of 4.5 cm before the grounds can sustain a new fishery. Since the increase in length from recruitment to 4.5 cm takes at least 2.5 years (Dolmer, 1998), the grounds are exposed to mussel fishery in a short period followed by recruitment and growth in cycles of at least 2.5 years duration which is a low fishery intensity compared to other areas (Churchill, 1989; Rijnsdorp et al., 1997). Jennings & Kaiser (1998) pointed out that it is difficult to demonstrate any impact of fishing activities in habitats frequently disturbed by natural events. This is probably the situation in Limfjorden where the impact of dredging every third year has to be scaled against the ecosystem effect of oxygen depletion with the same recurrent frequency.

Acknowledgements

Thanks are due to Dr D. Eibye-Jacobsen for taxonomical advice and help and to Drs G. Burnell, K.T. Jensen, N. Dankers and an anonymous referee for linguistic corrections and critical comments.

References

Asmus, R. M. & H. Asmus, 1991. Mussel beds – limiting or promoting phytoplankton? J. mar. Biol. Ecol. 148: 215–232.

Bergman, M. J. N. & M. Hup, 1992. Direct effects of beamtrawling on macrofauna in a sandy sediment in the southern North Sea. ICES J. Sea. Res. 49: 5–11.

Brylinsky, M. & J. Gibson, 1994. Impacts of flounder trawls on the intertidal habitat and community of the Minas Basin, Bay of Fundy. Can. J Fish. aquat. Sci. 51: 650–661.

Churchill, J. H., 1989. The effect of commercial trawling on sediment resuspension and transport over the Middle Atlantic Bight continental shelf. Con. Shelf Res. 9: 841–864.

Clarke, K. R. & R. M. Warwick. Change in marine communities: An approach to statistical analysis and interpretation. Plymouth Marine Laboratory, 1994, 144 pp.

Currie, D. R. & G. D. Parry, 1996. Effects of scallop dredging on a soft sediment community: a large-scale experimental study. Mar. Ecol. Prog. Ser. 134: 131–150.

de Groot, S. J., 1984. The importance of bottom trawling on benthic fauna of the North Sea. Ocean Mgmt. 9: 177–190.

Dolmer, P., 1998. Seasonal and spatial variability in growth of *Mytilus edulis* L. in a brackish sound: comparisons of individual mussel growth and growth of size classes. Fish. Res. 34: 17–26.

Dolmer, P., 2000a. Algal concentration profiles above mussel beds. J. Sea Res. 43: 113–119.

Dolmer, P., 2000b. Feeding activity of mussels related to near-bed currents and phytoplankton biomass. J. Sea Res. 44: 221–231.

Dolmer, P., P. S. Kristensen & E. Hoffmann, 1999. Dredging of blue mussels (*Mytilus edulis* L) in a Danish sound: stock sizes and fishery-effects on mussel population dynamic. Fish. Res. 40: 73–80.

Dyekjær, S. M., J. K. Jensen & E. Hoffmann, 1995. Mussel dredging and effects on the marine environment. ICES C.M. 1995/E:13 ref K, 19 pp.

Eleftheriou, A. & M. R. Robertson, 1992. The effects of experimental scallop dredging on the fauna and physical environment of a shallow sandy community. Neth. J. Sea Res. 30: 289–299.

Fréchette, M. & E. Bourget, 1985a. Energy flow between the pelagic and benthic zones: factors controlling particulate organic matter available to an intertidal mussel bed. Can. J. Fish. aquat. Sci. 42: 1158–1165.

Fréchette, M. & E. Bourget, 1985b. Food limited growth of *Mytilus edulis* L. in relation to the benthic boundary layer. Can. J. Fish. aquat. Sci. 42: 1166–1170.

Gislason, H., 1994. Ecosystem effects of fishing activities in the North Sea. Marine Poll. Bull. 29: 520–527.

Hall, S. J., 1999. The effects of fishing on marine ecosystems and communities. Blackwell Science Ltd, Oxford.

Hall, S. J., D. J. Basford & M. R. Robertson, 1990. The impact of hydraulic dredging for Razor clams *Ensis* sp. on the infaunal community. Neth. J. Sea Res. 3–27: 119–125.

Hedqvist-Johnson, K. & C. André, 1991. The impact of the brown shrimp *Crangon crangon* L on soft bottom meiofauna: An experimental approach. Ophelia 34: 41–49.

Hoffmann, E. & P. Dolmer, 2000. Effect of closed areas on the distribution of fish and benthos. ICES J. Mar. Sci. 57: 1310–1314.

Hylleberg, J., 1993. Extinction and immigration of benthic fauna. The value of historical data from Limfjorden, Denmark. In: Symposium Mediterranean Seas 2000, NFR, Della Croce (ed.), Istuto Scienze Ambientali Marine, Universita di Genova: 43–73.

Jennings, S. & M. J. Kaiser, 1998. The effects of fishery on marine ecosystems. Adv. Mar. Biol. 34: 201–352.

Jensen, K. T. & J. N. Jensen, 1985. The importance of some epibenthic predators on the density of juvenile benthic macrofauna in the Danish Wadden Sea. J. exp. Mar. Biol. Ecol. 89: 157–174.

Jørgensen, B. B., 1980. Seasonal oxygen depletion in the bottom waters of a Danish fjord and its effect on the benthic community. Oikos 34: 34–76.

Kaiser, M. J. & B. E. Spencer, 1996. The effects of beam-trawl disturbance on infaunal communities in different habitats. J. anim. Ecol. 65: 348–358.

Kamermans, P. & H. J. Huitema, 1994. Shrimp (*Crangon crangon* L.) browsing upon siphon tips inhibits feeding and growth in the bivalve *Macoma balthica* (L.). J. exp. mar. Biol. Ecol. 175: 59–75.

Kristensen, P. S., 1997. Oyster and mussel fisheries in Denmark. US Dep Commer, NOAA tech Rep NMFS 129: 5–38.

Meeuwig, J. J., J. B. Rasmussen & R. H. Peters, 1998. Turbid waters and clarifying mussels. Their moderation of impirical chl:nutrient relations in estuaries in Prince Edward Island, Canada. Mar. Ecol. Prog. Ser. 171: 139–150.

Møhlenberg, F., 1995. Regulating mechanisms of phytoplankton growth and biomass in a shallow estuary. Ophelia 42: 429–256.

Newell, C. R., 1990. The effects of mussel (*Mytilus edulis* Linnaeus, 1758) position in seeded bottom patches on growth at subtidal lease site in Maine. J. Shellfish. Res. 9: 113–118.

Nilsson, P., K. Sundbäck & B. Jönsson, 1993. Effect of the brown shrimp *Crangon crangon* L. on endobenthic macrofauna, meiofauna and meiofaunal grazing rates. Neth. J. Sea Res. 31: 95–106.

Petersen, J.K., O. Schou & P. Thor, 1997. In situ growth of the ascidian *Ciona intestinalis* (L) and the blue mussel *Mytilus edulis* in an eelgrass meadow. J. exp. Mar. Biol. Ecol. 218: 1–11.

Pihl, L., 1985. Food selection and consumption of mobile epibenthic fauna in shallow marine areas. Mar. Ecol. Prog. Ser. 22: 169–179.

Pihl, L., 1990. Year-class strength in plaice (*Pleuronectes platesse* L) on the Swedish west coast. Hydrobiologia 195: 79–88.

Pihl, L. &R. Rosenberg, 1984. Food selection and consumption of the shrimp *Crangon crangon* in some shallow marine areas in western Sweden. Mar. Ecol. Prog. Ser. 15: 159–168.

Pranovi, F. & O. Giovanardi, 1994. The impact of hydraulic dredging for short-necked clams, *Tapes* spp, on an infaunal community in the lagoon of Venice. Sci. Mar. 58: 345–353.

Reise, K., 1977. Predator exclusion experiments in an intertidal mud flat. Helgoländer wiss. Meeresunters. 30: 263–271.

Riemann, B. & E. Hoffmann, 1991. Ecological consequences of dredging and bottom trawling in the Limfjord, Denmark. Mar. Ecol. Prog. Ser. 69: 171–178.

Rijnsdorp, A. D., A. M. Buys, F. Storbech & E. Visser, 1997. Microdistribution of the Dutch trawl fishery between April 1993 and March 1996. Reports of the Netherlands Institute for Fisheries Research.

Shephard, S. A., 1983. The epifauna of megaripples: species' adaptations and population responses to disturbance. Aust. J. Ecol. 8: 3–8.

Smaal, A. C., J. H. G. Verhagen, J. Coosen & H. A. Haas, 1986. Interactions between seston quantity and quality and benthic suspension feeders in the Oosterschelde, the Netherlands. Ophelia 26: 385–399.

Srivastava, D. S. & J. H. Lawton, 1998. Why more productive sites have more species: an experimental test of theory using tree-hole communities. Am. Nat. 152: 510–529.

Svane, I. & M. Ompi, 1993. Patch dynamics in beds of the blue mussel*Mytilus edulis* L: effects of site, patch size, and position within a patch. Ophelia 37: 187–202.

Svane, I. & I. Setyobudiandi, 1996. Diversity of associated fauna in beds of the blue mussel *Mytilus edulis* L: Effects of location, patch size, and position within a patch. Ophelia 45: 39–53.

Thrush, S. F., J. E. Hewitt, V. J. Cummings & P. K. Dayton, 1995. The impact of habitat disturbance by scallop dredging on marine benthic communities: what can be predicted from results of experiments? Mar. Ecol. Prog. Ser. 129: 141–150.

Tsuchiya, M. & M. Nishihira, 1986a. Islands of *Mytilus edulis* as a habitat for small intertidal animals: effect of island size on community structure. Mar. Ecol. Prog. Ser. 25: 71–81.

Tsuchiya, M. & M. Nishihira, 1986b. Islands of *Mytilus edulis* as a habitat for small intertidal animals: effects of *Mytilus* age structure on the species composition of the associated fauna and community organization. Mar. Ecol. Prog. Ser. 31: 171–178.

Van der Veer, H. & J. N. Bergman, 1987. Predation by crustaceans on the newly settled 0-group plaice *Pleuronectes platessa* population in the western Wadden Sea. Mar. Ecol. Prog. Ser. 35: 203–215.

Hydrobiologia **465:** 129–138, 2001.

The effect of scallop dredging on Irish Sea benthos: experiments using a closed area

C. Bradshaw, L.O. Veale, A.S. Hill & A.R. Brand
Port Erin Marine Laboratory, University of Liverpool, Port Erin, Isle of Man IM9 6JA, British Isles
Tel: +44-1624-831006. Fax: +44-1624-831001. E-mail: CBRADS88@hotmail.com

Key words: dredging, disturbance, benthos, closed area, *Pecten maximus*

Abstract

A 2 km^2 area off the southwest coast of the Isle of Man (Irish Sea) has been closed to commercial fishing with mobile gear since March 1989. This area was heavily fished for *Pecten maximus* (Linnaeus, 1758) prior to closure, and the seabed immediately surrounding the closed area is still one of the most heavily dredged in the Irish Sea. Two methods have been used to study the effect of scallop dredging on the benthos in this closed area and adjacent fished areas. Firstly, twice-yearly grab sampling of experimental plots inside and outside the closed area since 1995 has enabled comparisons of the benthic infauna and epifauna of experimentally dredged plots, undredged control plots and plots exposed to commercial dredging. Secondly, divers have carried out visual transect surveys of *P. maximus* numbers regularly since closure. Communities of experimentally disturbed plots have become less similar to adjacent undisturbed control areas and more similar to commercially dredged areas. At each sampling date, similarity between dredged samples was greater than between undredged samples. Since 1989, there have been increases in the mean numbers of *P. maximus* in the closed area. The age structure of the closed area *P. maximus* population is also different to that outside, with a higher mean age due to the presence of large, old individuals.

These results present strong evidence that scallop dredging alters benthic communities and suggest that the closure of areas to commercial dredging may allow the development of more heterogeneous communities and permit the populations of some species to increase. A common problem with studying fishing disturbance is the lack of good control sites and this work also demonstrates the value of closed areas to scientific studies of demersal fishing.

Introduction

Bottom fishing is known to have an impact on benthic fauna: the short term effects include damage and death to bycatch species, injury and death on the seabed due to contact with the gear (especially of large fragile animals e.g. urchins, starfish), influxes of predators and scavengers feeding on damaged or dead animals, turnover of the sediment and attached animals, and smothering of animals by resuspended sediment (see Jones, 1992; Dayton et al., 1995; Jennings & Kaiser, 1998: Lindeboom & De Groot, 1998 for reviews of the subject). Impacts vary with different seabed types (Shepard & Auster, 1991; Jones, 1992; Kaiser & Spencer, 1996), different fishing gear (e.g. Kaiser et al., 1996) and level of background disturbance (e.g. Hall et al., 1990; Hall, 1994; Kaiser, 1998). Several previous studies have taken an experimental approach (e.g. Eleftheriou & Robertson, 1992; Thrush et al., 1995; Currie & Parry, 1996), but these studies concentrated mostly on the effects of short-term, small-scale or one-off dredging events, or had inadequate control sites. In this paper, we present results from a 5-year long experimental investigation.

The scallop, *Pecten maximus* (Linnaeus, 1758), and queen scallop, *Aequipecten opercularis* (Linnaeus, 1758), have formed the basis of an important fishery in the north Irish Sea around the Isle of Man (Fig. 1) for the last 60 years. The fishery began in the late 1930s, close inshore off the SW coast and consisted of only nine small boats towing between one and three dredges. The fishery did not start expanding until

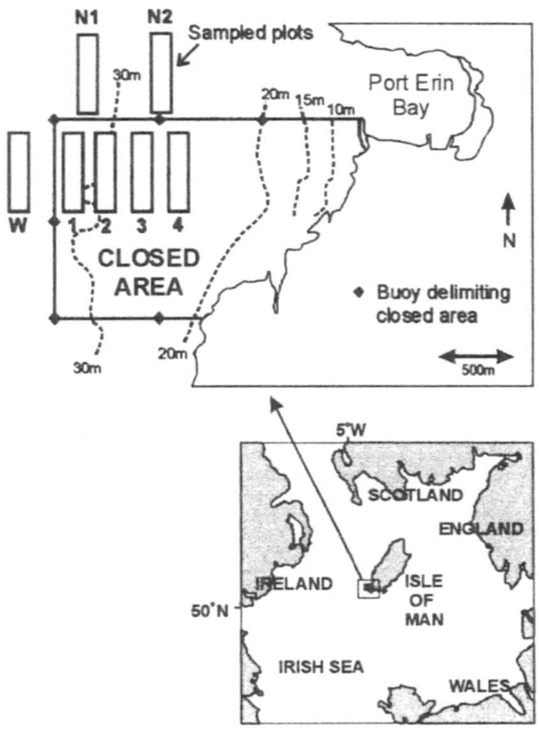

Figure 1. Map showing layout and location of the sampling plots in the closed area. Plots 2 and 4 have been undredged since 1989, plots 1 and 3 experimentally dredged every two months since 1995, and plots N1, N2 and W are exposed to commercial scallop dredging. Inset shows location of the Isle of Man in the Irish Sea.

the mid 1950s, reaching a maximum of c. 70 boats in 1983–4. The development of larger and more powerful boats with improved gear has enabled the fishery to exploit rougher fishing grounds and move further offshore. There are currently around 30 Manx boats, each fishing 12–24 dredges. Boats from Ireland, Northern Ireland, Wales, England and Scotland also now fish for scallops in the north Irish Sea (Brand et al., 1991). In 1998, the combined tonnage of queens and scallops landed on the Isle of Man was 1696t, worth £1.2 million (equivalent to 62% of the value of all fisheries on the island) (Brand & Beukers–Stewart, 2000).

Conservation legislation in operation since 1943 restricts fishing for scallops to an open season that currently runs from 1 November to 31 May inclusive; queens may be fished all year round. Other legislation includes a minimum landing size of 110 mm shell length for scallops and restrictions on boat size, gear type and gear width (Brand et al., 1991).

Scallops live on, or partly buried in, the surface sediments and are fished with 2′ or 2′6″ (61 or 76 cm) spring-toothed, Newhaven-type scallop dredges with metal belly rings, fished in gangs of 6–12 each side. The gear penetrates the upper layers of the seabed to rake out the partially buried scallops and in so doing disrupt the benthos (e.g. Caddy, 1973; Eleftheriou & Robertson, 1992; Thrush et al., 1995; Hall–Spencer & Moore, 2000a, b).

The seabed around the Isle of Man comprises a mixture of mud and sand with a variable amount of dead shell and stone, especially at the surface. Small-scale patchiness of the seabed provides a heterogeneous substrate with many habitat niches. Coarse stone and shell fragments can provide shelter for mobile animals and also support a rich sessile community (e.g. hydroids, bryozoans, tunicates), which in turn provide attachment surfaces, shelter or food for other animals. These species are thought to be especially sensitive to overturning by dredges and smothering by sediment (Dayton et al., 1995; Bradshaw et al., 2000).

An exclusion zone of nearly 2 km^2 off the south-west of the Isle of Man (Fig. 1) has been closed to commercial fishing by dredge or trawl (though not to static gear) since March 1989 under Isle of Man legislation. The area had been heavily dredged for 50 years prior to closure, and the surrounding area, known as the Bradda Inshore fishing ground, continues to be one of the most intensively fished grounds in the Irish Sea, with a mean annual effort (1982–1995) of 7919 m h (Hill et al., 1997), equivalent to 43% of the seabed being dredged each year. Maximum water depth in the closed area is 40 m at the outer edge and the seabed is a mixture of gravel, sand and mud. This closure was implemented initially to enable Port Erin Marine Laboratory (PEML) scientists to investigate the feasibility and effectiveness of scallop cultivation and stock enhancement, but subsequently provided an opportunity to monitor the numbers of benthic species after the cessation of dredging and to carry out controlled experiments into the effect of dredging on benthic communities.

This study investigates the effect of scallop dredging on infaunal and epifaunal communities using controlled experiments in, and monitoring of, the closed area. In 1994, 5 years after closure, a programme of experimental dredging began in the closed area. The abundance of *P. maximus* throughout the closed area has also been monitored by diver survey since closure.

Materials and methods

Experimental design

The NW corner of the closed area was chosen as the site for the experiment, due to its reasonably uniform water depth and sediment type. The experimental layout (Fig. 1) consists of four parallel plots of c. 500 m length and c. 100 m width, running in a north–south direction, two of which are experimentally dredged ('treatment' plots 1 and 3) and two of which are undredged (controls 2 and 4). Three identically sized plots are located outside the closed area, two to the north (N1 and N2) and one to the west (W). These latter three plots are exposed to commercial scallop dredging. The two treatment plots have been experimentally dredged approximately every 2 months since January 1995, using two sets of four spring-toothed, Newhaven-type scallop dredges, towed 10 times along each line.

Sampling

The benthic community of all seven plots has been sampled twice a year starting in 1995, in the spring (immediately before the closed season for scallops) and autumn (immediately after the closed season). On each plot, 0.1 m^2 Day grab samples were taken, in three groups of four up to 1998 and subsequently in four groups of six. Three groups of two 0.1 m^2 Day grabs were also taken for sediment analysis (grain size distribution and organic carbon content). Depth and dGPS-fixed sample positions were recorded simultaneously with the grab sampling using 'Microplot' software (Sea Information Systems, Aberdeen). In months where experimental disturbance and sampling coincided, sampling was always carried out first to avoid picking up short-term effects.

Grab samples were sieved immediately over a 1 mm mesh, covered with 35 ppt $MgCl_2$ and 4 ppt Rose Bengal for 1–3 h to relax and stain the animals before fixing in 4% formalin. All the animals were extracted from each sample by hand, preserved in 70% alcohol and subsequently identified as far as taxonomically possible (mostly to species).

An additional monitoring programme has used divers to carry out surveys of *Pecten maximus* densities since 1989 throughout the closed area and in a control (commercially fished) area 1 km to the north. Each survey consisted of a pair of divers counting scallops along two parallel 2 m $\times$ 50 m belt transects. Counts from the two divers were pooled into animal numbers per 200 m^2. The number of surveys completed each year varied from four to 20. In 2000, additional data from longer dive surveys carried out as part of another study in the same area were also used and standardised to numbers per 200 m^2 (unpublished data, Beukers–Stewart & Jenkins).

To measure the age structure of scallops in the area, samples of *Pecten* were obtained from 1 km north of the closed area using a standard commercial scallop dredge. Three 1-h tows at a speed of 2.5 knots were taken, and all commercial-sized scallops measured and aged. These data were compared with similar data from scallops obtained from an illegal fishing event in the NE corner of the closed area in April 2000 (unpublished data, Beukers–Stewart & Jenkins).

Data analysis

The naturally patchy distribution of fauna on the seabed makes variability high between replicates; groups of grabs were, therefore, pooled [into 3 groups of four grabs (pre-1998), or 4 groups of 6 grabs (March 1998)] before any analysis. Sample numbers were increased in 1998 in order to improve the power of univariate tests. Count data were $4\sqrt{}$ transformed to reduce the effect of very abundant species and presence–absence transformation used for colonial species data and combinations of count and colonial data. All data were reduced by 3% to eliminate bias from infrequently occurring species and standardised to account for different sample volumes.

Multivariate analyses were performed with PRIMER software (Plymouth Marine Laboratory, U.K.). The data were clustered using group-average clustering and the Bray-Curtis measure of sample similarity, and plotted using non-metric multi-dimensional scaling (MDS). ANOSIM (analysis of similarities) was used to test the significance of differences in composition of communities from dredged and undredged sites and SIMPER (similarity percentages) to determine which species contributed most to these differences. The Index of Multivariate Dispersion (IMD; Warwick & Clarke, 1993) was calculated to investigate levels of variability in dredged and undredged plots.

Univariate analyses were restricted to one-way and two-way ANOVA tests and subsequent multiple comparisons tests (Tukey's parametric or Dunn's non-parametric) on species number, Shannon–Wiener diversity (log_2), species richness, Pielou's Evenness and Simpson's dominance. Differences between dredged and undredged lines were tested separately for each

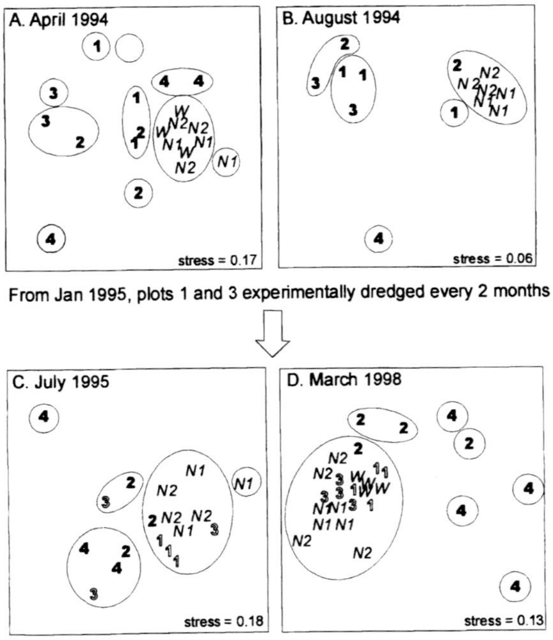

Figure 2. MDS plots of benthic community data before and after experimental dredging. For all dates, dredged plots cluster separately from, and are more tightly clustered than, undredged plots. Italicised text represents commercially dredged plots outside the closed area; bold letters, undredged plots inside the closed area; open letters, experimentally dredged plots inside the closed area (after Jan 1995). Plots are based on standardised, 3% reduced, 4$\sqrt{}$ transformed, pooled count data (see 'Methods').

Table 1. Comparison of the indices of multivariate dispersion (IMD) for undredged, commercially dredged and experimentally dredged plots. Values of relative dispersion are shown for each sampling period. Higher values indicate greater variability and therefore greater heterogeneity. IMDs were calculated from Bray–Curtis similarities of the 5% reduced dataset, which had been presence–absence transformed and standardised. Note: April and August 94 are pre-experimental dredging; plots 1–4 were undredged on those occasions

Date	Undredged	Commercially dredged	Experimentally dredged
Mar 98	1.65	0.98	0.40
Apr 97	1.49	0.68	1.28
Aug 94	1.15	0.72	–
Apr 94	1.31	0.44	–

sampling occasion. More detailed analyses were performed on the March 1998 data, due to this having a higher level of replication and being the most-recent in the time series. For these two reasons, any significant results were most likely to be found in this dataset.

Results

Experimentally dredged plots in the closed area are more similar to commercially dredged plots outside the closed area than undredged controls within the closed area

Cluster analysis of samples taken before experimental dredging began (April and August 1994) shows that the plots outside the closed area (exposed to commercial fishing) cluster separately from those inside (all undredged) (Fig. 2a,b). In January 1995, experimental dredging first took place on plots 1 and 3. The samples taken 6 months later, in July, indicate that the fauna of plots 1 and 3 were now more similar to the commercially dredged plots outside the closed area (N1, N2 and W) (Fig. 2c).

Experimental dredging continued approximately every 2 months on plots 1 and 3, and samples taken over the next three years continued to show the same trend of experimentally dredged plots (1 and 3) being more similar to commercially dredged plots (N1, N2 and W) than undredged plots in the closed area (2 and 4) (e.g. March 1998: Fig. 2d). However, no significant differences between treatments were identified using ANOSIM, and SIMPER analyses did not show any clear trends of particular species or groups of species being responsible for these groupings.

To see if these changes were reflected in changes in species number or diversity, Kruskal–Wallis one-way ANOVA on ranks were performed on various univariate indices calculated from the March 1998 data. No significant differences in total species number or species richness were found between treatments. Diversity (Shannon–Wiener) was slightly higher in the experimentally dredged than the undredged plots, and Pielou's evenness and Simpson's Dominance were both higher in the commercially dredged plots than the experimentally dredged and undredged plots. These rather confusing results suggest that condensing complex data with high levels of variability such as these into single indices can mask clear trends in community change shown by multivariate methods.

Undredged plots are more heterogeneous than dredged plots

The MDS plots of data from 1994 (April and August; Fig. 2a,b) show that undredged plots (1–4) are less

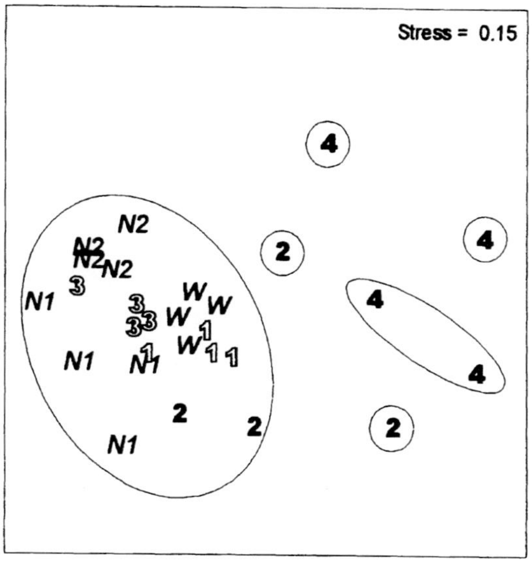

Figure 3. MDS plot of sessile benthos data from March 1998. Dredged plots (both experimentally and commercially dredged) cluster separately from, and are more tightly clustered than, undredged plots. Data has been standardised, 3% reduced, presence-absence transformed and pooled into 4 groups of six replicates for each experimental plot (see 'Methods').

tightly clustered than commercially dredged plots (N1, N2 and W). This is also the case after the experimental dredging regime began: July 1995 and March 1998 show that plots 1 and 3 are now more tightly clustered than 2 and 4 (Fig. 2c,d). Calculation of the Index of Multivariate Dispersion confirms this, with highest values for the undredged plots, followed by commercially and experimentally dredged (Table 1). These results imply that dredging reduces heterogeneity in these communities.

Sessile species

As sessile epifaunal species might be considered especially sensitive to dredging disturbance, more detailed multivariate analyses were done on the sessile species subset of the March 1998 dataset. The pattern of dredged and undredged plots clustering separately, and of dredged plots being more similar to each other, is still seen in these cluster analyses (Fig. 3). The sessile species that contribute most to similarity amongst samples in the dredged plots (experimentally and commercially dredged) are encrusting bryozoans, encrusting sponges and small ascidians. Those that make the undredged samples similar to each other are upright forms, such as the bryozoans *Cellaria* spp Ellis & Solander, 1786 and *Crisidia cornuta* (Linnaeus, 1758), and the hydroids *Halecium* spp Oken, 1815. It is these differences in morphology that contribute most to the dissimilarities between undredged-experimentally dredged and undredged-commercially dredged (Fig. 4; Table 2).

Population structure of Pecten maximus *in the closed area*

There are upward trends in numbers of the scallop *Pecten maximus* both inside and outside the closed area (Fig. 5a). However, the densities of scallops in the closed area are consistently higher than those outside (though this not statistically significant, due to high variability in numbers). There are also striking differences in the population structure of scallops in the two areas (Fig. 5b). Mean age inside the closed area is 6.5 years, compared with 5.3 years outside, due to the presence of large, old individuals (>9 years old) in the closed area, in contrast to the dominance of 4–5 year olds outside. In comparison, the mean age of scallops fished off Bradda in 1937 was 9.9 years, (n=294 scallops fished by commercial boats), with three scallops over 20 years old.

Discussion and conclusions

Community changes

From this study, it is clear that ongoing scallop dredging over a period of 5 years alters benthic communities. We have shown that is possible to experimentally drive a benthic community from one state to another through regular scallop dredging. Previous work has demonstrated that communities change after one-off experimental dredging; however, in this study we subjected the benthos to the cumulative effects of 5 years of ongoing dredging, a level similar to that occurring on commercially fished grounds.

There is evidence from this work that dredging makes benthic communities more homogeneous (see also Bradshaw et al., 2000). Similar results have been shown by (Auster et al., 1996), who recorded reduced habitat complexity due to mobile fishing gear in the Gulf of Maine. Closure of areas to commercial dredging in this study, therefore, allows the development of more heterogeneous communities. Heterogeneity of communities, and the associated increased habitat complexity, has important implications for success of larval settlement and species diversity (Roberts

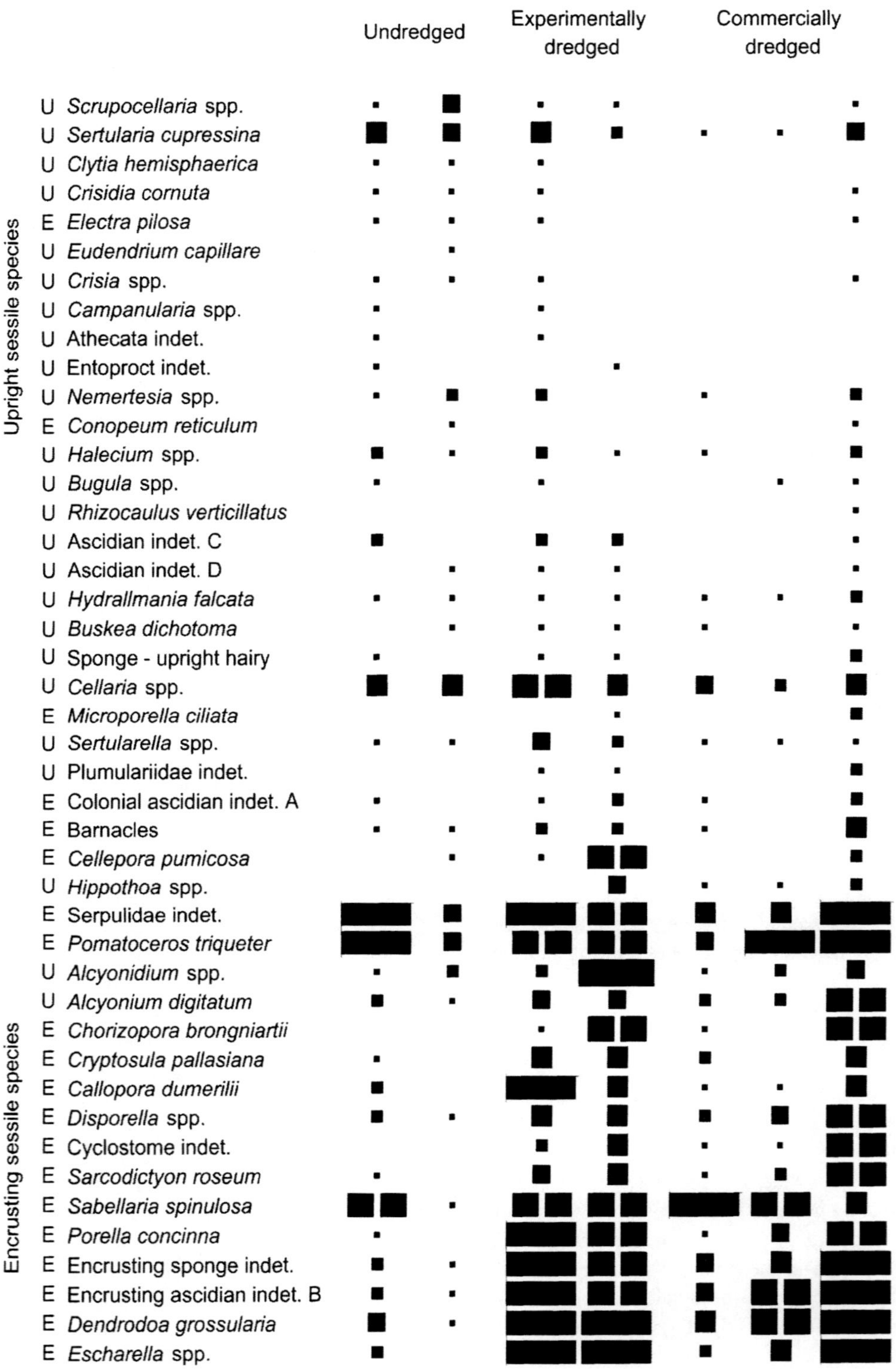

Figure 4. Shade diagram of sessile species abundances for March 1998. Species more common in the undredged plots are upright (U) species, whereas those more common in the dredged plots have an encrusting morphology (E). The data has been pooled for each plot and the relative abundance of each species estimated from the number of replicates which contained that species. Darker shading indicates higher abundance. The species have been ranked by the degree of difference between the abundances between undredged and combined commercially- and experimentally-dredged plots.

Table 2. Species identified by SIMPER analysis as contributing most to the differences between treatments. In both cases, encrusting sessile species are more abundant in the dredged plots, while upright forms flourish in the undredged plots. UD = undredged (plots 2 & 4), ED = experimentally dredged (plots 1 & 3), CD = commercially dredged (plots N1, N2, W). The analysis is based on Bray–Curtis similarities from the full dataset from March 1998, with standardisation and presence-absence transformation. 'enc.' = encrusting

Top 10 species that contributed most to the dissimilarity between undredged and experimentally dredged plots			Top 10 species that contributed most to the dissimilarity between undredged and commercially dredged plots		
Species	Life form	Relative abundance	Species	Life form	Relative abundance
Cryptosula pallasiana	enc. bryozoan	ED>UD	*Escharella* spp	enc. bryozoan	CD>UD
Cyclostome spp	enc. bryozoan	ED>UD	*Porella* spp	enc. bryozoan	CD>UD
Escharella spp	enc. bryozoan	ED>UD	*Crisidia cornuta*	upright bryozoan	UD>CD
Porella spp	enc. bryozoan	ED>UD	*Scrupocellaria* spp	upright bryozoan	UD>CD
Sarcodictyon roseum	enc. octocoral	ED>UD	*Sarcodictyon roseum*	enc. octocoral	CD>UD
Cellepora pumicosa	enc. bryozoan	ED>UD	*Halecium* spp	upright hydroid	UD>CD
Schizomavella spp	enc. bryozoan	ED>UD	*Clytia hemisphaerica*	upright hydroid	UD>CD
Chorizopora brongniartii	enc. bryozoan	ED>UD	*Cyclostome* spp	enc. bryozoan	CD>UD
Crisidia cornuta	upright bryozoan	UD>ED	*Schizomavella* spp	enc. bryozoan	CD>UD
Clytia hemisphaerica	upright hydroid	UD>ED	*Cryptosula pallasiana*	enc. bryozoan	CD>UD

& Polunin, 1991; Auster et al., 1996). The use of complex habitats may also increase the survival of mobile species (Auster et al., 1996).

Closure changes the population structure of P. maximus

Not surprisingly, numbers of the target animal, *Pecten maximus*, show a consistent significant increase in the closed area since 1989. Although numbers have also been increasing outside the closed area over this time, numbers are consistently (though not significantly) higher inside the closed area (Fig. 5a). Rice et al. (1989) found the same trend for quahog clams [*Mercenaria mercenaria* (Linnaeus, 1758)] in a closed area in Narragansett Bay, as did Murawski et al. (2000) for *Placopecten magellanicus* (Gmelin, 1791) on Georges Bank. The present study clearly shows that the average age of the scallops is greater in the closed area (Fig 5b), a trend also described by Rice et al. (1989) and Murawski et al. (2000). However, the average age of *P. maximus* in our closed area is nowhere near as high as the 1937 population measured by Tang (1941), right at the start of the scallop fishing industry on the island. This suggests that the scallop population in the closed area is slowly changing to more closely resemble a natural population.

Sessile species

Studies to date have generally indicated that sessile epifaunal species are adversely affected by bottom fishing (e.g. Bridger, 1970; Caddy, 1973; Dayton et al., 1995; Auster et al., 1996). However, this experimental study shows that the situation is more complex than that, with upright species (mainly hydroids) less abundant in dredged plots, while encrusting species (bryozoans, sponges, small ascidians) flourish in dredged areas. We hypothesise that upright species are more prone to being directly damaged and buried by overturning of their attachment substrata, whereas the stone on which a sponge or an encrusting bryozoan is growing may be completely overturned, but there need only be a small portion of the colony remaining above the sediment surface for asexual growth to extend the colony back up on to the new top or side surfaces. Kaiser et al. (1999) also showed that not all epifauna are adversely affected by bottom fishing. After experimental beam trawling, Kaiser's study could not detect any changes in numbers or sizes of serpulid worms, which are opportunistic species and may rapidly recover from fishing disturbance.

It is likely that upright forms, which are worst affected by dredging, are also important in increasing structural complexity and habitat heterogeneity of the seabed, and thus also increasing diversity (Auster et al., 1996; Collie et al., 1997; Watling & Norse, 1998). In the closed area, upright hydroids provide an attachment surface for other hydroids [e.g. *Lafoea dumosa* (Fleming, 1828), *Diphasia rosacea* (Linnaeus, 1758)], bryozoans [e.g. *Buskea dichotoma* (Hincks, 1862), *Conopeum reticulum* (Linnaeus, 1767)], barnacles [e.g. *Scalpellum scalpellum* (Linnaeus, 1767)] and sponges [e.g. *Scypha ciliata* (Fabricius, 1780)] and also for the dwelling tubes of certain worms [*Eunice pennata* (O F Müller, 1776), spirorbid]. Scallop spat have also been seen to settle on these upright species (Bradshaw & Veale, pers. obs.). Other animals use the biogenic structures for shelter [e.g. spider crabs, pea crabs *Pisidia longicornis* (Linnaeus, 1767), sea spiders] and to some they are an important food source (e.g. various nudibranch species). Thus, the destruction of biogenic substrate also impacts on other species, as observed on *Ostrea edulis* Linnaeus, 1758 reefs (Reise, 1982; Riesen & Reise, 1982; Reise & Schubert, 1987, other papers in this volume), *Sabellaria spinulosa* Leuckart, 1849 reefs (Reise & Schubert, 1987), sponge-coral communities (Van Dolah et al., 1987), mixed epifaunal communities (Kaiser et al., 2000) and maerl beds (Hall-Spencer & Moore, 2000a,b). There is some evidence that patches of high diversity or structural complexity, such as biogenic epifauna, may also enhance recruitment of a range of invertebrates and fish (Auster et al., 1996; Smith & Witman, 1999).

Ecology

The response of an area to dredging, or any other form of disturbance, depends on many variables related to life-history characteristics of all the species, as well as local hydrography, intensity, frequency and time of year of dredging (e.g. Caswell & Cohen, 1991; Auster et al., 1996; Dial & Roughgarden, 1998; Lindeboom & De Groot, 1998). We evidently need to know much more about the ecology and life histories of individual benthic species in order to fully explain or predict the effects of dredging. If dredging kills individuals of a species, the potential to recolonise is the most important factor in its survival in the community. The success of recolonisation depends on larval supply and successful recruitment (Carr & Reed, 1993). If dredging only damages a species, its regenerative powers are more important, whether through asexual reproduction (e.g. colonial bryozoans) or regrowth of damaged parts (e.g. sponges, starfish). Other factors

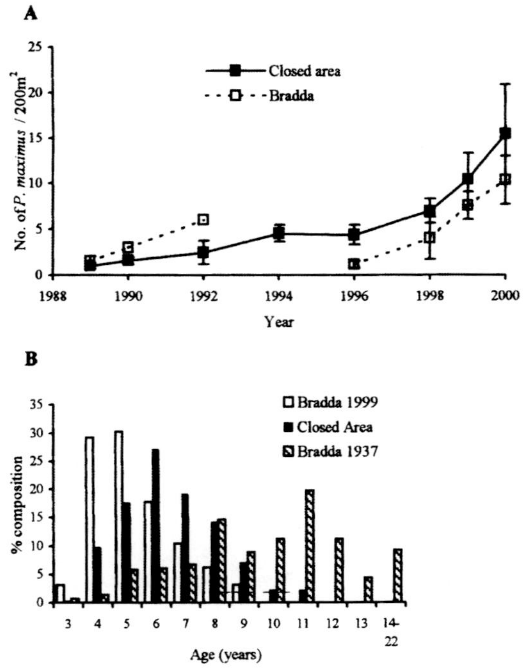

Figure 5. (**A**) Densities and (**B**) age composition of *Pecten maximus* inside and outside the closed area. Densities are expressed as numbers per 200 m^2, the area of a standard diver survey. Data in (**B**) are based on legal commercial sized (>110 mm) scallops only, and 1937 data are from Tang (1941).

such as the ability of individuals to right or rebury themselves or to shed settled sediment are also important in these non-fatal casualties of dredge disturbance. Even apparently undamaged animals have been shown to suffer increased predation after dredging (Ramsay & Kaiser, 1998). Changes in community structure through repeated disturbance will also alter other ecosystem processes such as competition and predation (Caswell & Cohen, 1991; Auster et al., 1996; Dial & Roughgarden, 1998).

Advantages of closed areas

A major problem encountered by scientists attempting to study the effects of bottom fishing has been the lack of suitable unfished control sites (e.g. Auster et al., 1996; Lindeboom, 2000). Areas may be unfished because they are intrinsically different to fished areas, if only in the absence of the target species, are near underwater obstructions (which may in themselves alter the immediate environment) or may have been fished (and therefore probably altered) in the past (Collie et al., 1997). Our work demonstrates that even a small closed area, such as in this study, is of great value to scientific studies of bottom fishing. Apart from the scientific benefits of closure, there are also potential ecological and fisheries benefits, both for the target species and other species. These may include protection of habitat (with implications for increased habitat complexity, and increased genetic and biological diversity as discussed above), increased abundance, size and age of the target species, increased productivity (and enhanced recruitment), increased catches in adjacent areas, insurance against fisheries stock collapse and formation of sources to reseed other areas (Roberts & Polunin, 1991; Dugan & Davis, 1993; Clarke, 1998).

However, closed areas cannot be seen as a panacea for all fisheries problems, and may actually create new problems of their own. For the closure of an area to be successful, there must be clear objectives (e.g. conservation of habitat, protection of particular species; Lindeboom, 2000). If a target species is to be protected, careful planning is required with regard to the biology of the species; protection of nursery grounds, spawning grounds or adult populations may have widely differing results. Among other factors to be considered are the required size of the area, the position of the area relative to other potential impacts (e.g. pollution) and the requirements of all the stakeholders in the area.

Care must be taken that fishing effort is not simply deflected elsewhere, resulting in increased fishing pressure on other stocks or other species (Horwood, 2000; Murawski, 2000); closures must fit into a wider fisheries management programme. Closures may be ineffective or unfeasible for mobile, wide-ranging species like North Sea cod (Horwood, 2000), but species with relatively static adults and mobile larvae, such as scallops, are perhaps most likely to benefit from closures through protection of brood stock. Additional measures, such as the seeding of the seabed with juvenile or adult scallops, have had mixed results, but proved successful in a number of cases [e.g. Ansell et al., 1991 (France); Ito, 1991 (Japan); Hatcher et al., 1996 (New Zealand)]. In combination with closures, this could potentially be a successful strategy for improving stocks of relatively immobile species such as these.

There is much evidence that closures can benefit habitats and fisheries and provide important scientific resources, despite the complications described above. With the current widespread problems of overfishing and habitat destruction, serious consideration should be given to the establishment of closed areas.

Acknowledgements

We wish to thank to the skipper and crew of RV *Roagan*, the numerous people who have sorted benthic samples and done diver surveys, and B. Beukers-Stewart and S. Jenkins for the use of unpublished data. The study was funded by the U.K. Ministry of Agriculture, Fisheries and Food as part of contract numbers CSA 2332 and CSA 4142 and drew also on the resources and data provided by the Isle of Man Department of Agriculture, Fisheries and Forestry scallop research project. Thanks are also due to NOAA for financial assistance to enable CB to attend the 3rd ICSR conference in Cork, which was so ably organised by G. Burnell.

References

Ansell, A. D., J. C. Dao & J. Mason, 1991. Three European scallops: *Pecten maximus*, *Chlamys (Aequipecten) opercularis* and *C. (Chlamys) varia*. In Shumway, S. E. (ed.), Scallops: Biology, Ecology and Aquaculture. Developments in Aquaculture and Fisheries Science, vol 21. Elsevier, Amsterdam: 715–751.

Auster, P. J., R. J. Malatesta, R. W. Langton, L. Watling, P. C. Valentine, C. L. S. Donaldson, E. W. Langton, A. N. Shepard & I. G. Babb, 1996. The impacts of mobile fishing gear on seafloor habitats in the Gulf of Maine (Northwest Atlantic): Implications for conservation of fish populations. Rev. Fish. Sci. 4: 185–202.

Bradshaw, C., L. O. Veale, A. S. Hill & A. R. Brand, 2000. The effects of scallop dredging on gravelly sea-bed communities. In Kaiser, M. J. & S. J. De Groot (eds), Effects of Fishing on Non-Target Species and Habitats. Gray Publishing, Tunbridge Wells: 83–104.

Brand, A. R. & B. Beukers–Stewart, 2000. Shellfish Research Report to the Isle of Man Government, Department of Agriculture, Fisheries and Forestry: 22 pp.

Brand, A. R., U. A. W. Wilson, S. J. Hawkins, E. H. Allison & N. A. Duggan, 1991. Pectinid fisheries, spat collection, and the potential for stock enhancement in the Isle of Man. ICES mar. Sci. Symp. 192: 79–86.

Bridger, J. P., 1970. Some effects of the passage of a beam trawl over the seabed. ICES Gear and Behaviour Committee Report C. M. 1970/B: 10

Caddy, J. F., 1973. Underwater observations on the tracks of dredges and trawls and some effects of dredging on a scallop ground. J. Fish Res. Bd Can. 30: 173–180.

Carr, M. H. & D. C. Reed, 1993. Conceptual issues relevant to marine harvest refuges: examples from temperate reef fishes. Can. J. Fish. aquat. Sci. 50: 2019–2028.

Caswell, H. & J. E. Cohen, 1991. Communities in patchy environments: a model of disturbance, competition and heterogeneity. In Kolasa, J. & S. T. A. Pickett (eds), Ecological Heterogeneity. Springer-Verlag, New York: 97–122.

Clarke, B. M., 1998. No Take Zones (NTZ's). A Realistic Tool for Fisheries Management? Marine Conservation Society, Ross-on-Wye: 52 pp.

Collie, J. S., G. A. Escanero & P. C. Valentine, 1997. Effects of bottom fishing on the benthic megafauna of Georges Bank. Mar. Ecol. Prog. Ser. 155: 159–172.

Currie, D. R. & G. D. Parry, 1996. Effects of scallop dredging on a soft-sediment community: a large-scale experimental study. Mar. Ecol. Prog. Ser. 134: 131–150.

Dayton, P. K., S. F. Thrush, M. Tundi Agardy & R. J. Hofman, 1995. Environmental effects of marine fishing. Aquat. Conserv. 5: 205–232.

Dial, R. & J. Roughgarden, 1998. Theory of marine communities: the intermediate disturbance hypothesis. Ecology 79: 1412–1424.

Dugan. J. E. & G. E. Davis, 1993. Applications of marine refugia to coastal fisheries management. Can. J. Fish. aquat. Sci. 50: 2029–2042.

Eleftheriou A. & M. R. Robertson, 1992. The effects of experimental scallop dredging on the fauna and physical environment of a shallow sandy community. Neth. J. Sea Res. 30: 289–299.

Hall, S. J., 1994. Physical disturbance and marine benthic communities life in unconsolidated sediments. Oceanogr. mar biol. Ann. Rev. 32: 179–239.

Hall, S. J., D. J. Raffaelli, D. J. Basford & M. R. Robertson, 1990. The importance of flatfish predation and disturbance on marine benthos: an experiment with dab *Limanda limanda* (L.). J. exp. mar. Biol. Ecol. 136: 65–76.

Hall-Spencer, J. M. & P. G. Moore, 2000a. Impacts of scallop dredging on maerl grounds. In Kaiser, M. J. & S. J. De Groot (eds), Effects of Fishing on Non-Target Species and Habitats. Gray Publishing, Tunbridge Wells: 105–117.

Hall-Spencer, J. M. & P. G. Moore, 2000b. Scallop dredging has profound, long-term impacts on maerl habitats. ICES J. mar. Sci. 57: 1407–1415.

Hatcher, B. G., R. E. Scheibling, M. A. Barbeau, A W. Hennigar, L. H. Taylor & A. J. Windust, 1996. Dispersion and mortality of a population of sea scallop (*Placopecten magellanicus*) seeded in a tidal channel. Can. J. Fish. aquat. Sci. 53: 38–54.

Hill, A. S., A. R. Brand, L. O. Veale & S. J. Hawkins, 1997. Assessment of the effects scallop dredging on benthic communities. Report to MAFF, University of Liverpool: 112 pp.

Horwood, J. W., 2000. No-take zones: a management context. In Kaiser, M. J. & S. J. De Groot (eds), Effects of Fishing on Non-Target Species and Habitat. Gray Publishing, Tunbridge Wells: 302–311.

Ito, H., 1991. Japan. In Shumway, S. E. (ed.). Scallops: Biology, Ecology and Aquaculture. Developments in aquaculture and fisheries science, vol 21. Elsevier, Amsterdam: 1017–1056.

Jennings, S. & M. J. Kaiser, 1998. The effects of fishing on marine ecosystems. Adv. mar. Biol 34: 201–351.

Jones, J. B., 1992. Environmental impact of trawling on the seabed: a review. N.Z. J. mar. Freshwat. Res. 26: 59–67.

Kaiser, M. J., 1998. Significance of bottom-fishing disturbance. Conserv. Biol. 12: 1230–1235.

Kaiser, M. J. & B. E. Spencer, 1996. The effects of beam-trawl disturbance on infaunal communities in different habitats. J. anim. Ecol. 65: 348–358.

Kaiser, M. J., A. S. Hill, K. Ramsay, B. E. Spencer, A. R. Brand, L. O. Veale, K. Prudden, E. I. S. Rees, B. W. Munday, B. Ball & S. J. Hawkins, 1996. Benthic disturbance by fishing gear in the Irish Sea: a comparison of beam trawling and scallop dredging. Aquat. Conserv. 6: 269–285.

Kaiser, M. J., K. Cheney, F. E. Spence, D. B. Edwards & K. Radford, 1999. Fishing effects in northeast Atlantic shelf seas: patterns in fishing effort, diversity and community structure VII. The effects of trawling disturbance on the fauna associated with the tubeheads of serpulid worms. Fish. Res. 40: 195–205.

Kaiser, M. J., F. E. Spence & P. J. B. Hart, 2000. Fishing gear restrictions and conservation of benthic habitat complexity. Conserv. Biol. 15(4).

Lindeboom, H. J., 2000. The need for closed areas as conservation tools. In Kaiser, M. J. & S. J. De Groot (eds), Effects of Fishing on Non-Target Species and Habitats. Gray Publishing, Tunbridge Wells: 290–301.

Lindeboom, H. J. & S. J. E. De Groot, 1998. IMPACT II The Effects of Different Types of Fisheries on the North Sea and Irish Sea Benthic Ecosystems. Netherlands Institute for Sea Research (NIOZ), Den Burg, Texel, Netherlands: 404 pp.

Murawski, S. R. Brown, H.-L. Lai, P. J. Rago & L. Hendrickson, 2000. Large-scale closed areas as a fishery-management tool in temperate marine systems: the Georges Bank experience. Bull. mar. Sci. 66: 775–798.

Ramsay, K. & M. J. Kaiser, 1998. Demersal fishing disturbance increases predation risk for whelks (*Buccinum undatum* L.). J. Sea Res. 39: 299–304.

Reise, K., 1982. Long-term changes in the macrobenthic invertebrate fauna of the Wadden Sea: are polychaetes about to take over? Neth. J. Sea Res. 16: 29–36.

Reise, K. & A. Schubert, 1987. Macrobenthic turnover in the subtidal Wadden Sea: the Norderaue revisited after 60 years. Helgoländer wiss. Meeresunters. 41: 69–82.

Riesen, W. & K. Reise, 1982. Macrobenthos of the subtidal Wadden Sea: revisited after 55 years Helgoländer wiss. Meeresunters 35: 409–423.

Rice, M. A., C. Hickox & I. Zehra, 1989. Effects of intensive fishing effort on the population structure of quahogs, *Mercenaria mercenaria* (Linnaeus 1758), in Narragansett Bay. J. Shellfish Res. 8: 345–354.

Roberts, C. M. & N. V. C. Polunin, 1991. Are marine reserves effective in the management of reef fisheries? Rev. Fish Biol. Fish. 1: 65–91.

Shepard, A. N. & P. J. Auster, 1991. Incidental (non-capture) damage to scallops caused by dragging on rock and sand substrates. In Shumway, S. E. & P. A. Sandifer (eds), An International Compendium of Scallop Biology and Culture. World Aquaculture Society, Baton Rouge: 219–230.

Smith, F. & J. D. Witman, 1999. Species diversity in subtidal landscapes: Maintenance by physical processes and larval recruitment. Ecology 80: 51–69.

Tang, S. F., 1941. The breeding of the scallop (*Pecten maximus* (L.)) with a note on growth rate. Proc. Liverpool biol. Soc. 54: 9–28.

Thrush, S. F., J. E. Hewitt, V. J. Cummings & P. K. Dayton, 1995. The impact of habitat disturbance by scallop dredging on marine benthic communities: what can be predicted from the results of experiments? Mar. Ecol. Prog. Ser. 129: 141–150.

Van Dolah, R. F., P. H. Wendt & N. Nicholson, 1987. Effects of a research trawl on a hard-bottom assemblage of sponges and corals. Fish. Res. 5: 39–54.

Warwick, R. M. & K. R. Clarke, 1993. Increased variability as a symptom of stress in marine communities. Changes In Marine Communities 172: 215–226.

Watling, L. & E. A. Norse, 1998. Disturbance of the seabed by mobile fishing gear: a comparison to forest clearcutting. Conserv. Biol. 12: 1180–1197.

Hydrobiologia **465**: 139–151, 2001.
G. Burnell (ed.), Coastal Shellfish – A Sustainable Resource.

Growth statistics of an exploited razor clam (*Ensis siliqua*) bed at Gormanstown, Co Meath, Ireland

Edward Fahy & Joanne Gaffney
Marine Institute, Fisheries Research Centre, Abbotstown, Castleknock, Dublin 15, Ireland

Key words: razor clams, *Ensis siliqua*, hydraulic dredging, Irish Sea, Ireland, fishery, growth, maturation, spawning, condition, associated benthic fauna, population density

Abstract

A clam bed of 21 km^2 was discovered at Gormanstown on the East coast of Ireland in 1997. Between July 1998 and August 1999, dredge samples were collected from vessels fishing there and these provided material for biological analysis. *E. siliqua* occurs with an inshore invertebrate fauna; occasional occurrences of large numbers of *Donax vittaus, Pharus legumen* and *Lanice conchilega* were associated with razor clams of less than 6 cm long. As harvesting proceeded smaller individuals appeared in the samples; these may have immigrated onto the bed from the periphery of high densities of larger clams. On the clam bed the animals ranged from 0+ to 19+ years old. *Ensis siliqua* grew at a slower rate on the west coast of the Irish Sea than on the coast of North Wales. Males grew slightly faster and longer than females and males were more numerous in the landings. The clams are estimated to commence maturation at approximately 4 years of age. They have a similar gonadal cycle to those in Portugal and a similar spawning period from mid-May to the end of July or early August. Condition reflects gonad development, reaching a peak in May. Harvesting by hydraulic dredging causes breakages which are higher in quarters 1 and 4 than in quarters 2 and 3. Bruised razor clams do not eliminate sand in desanding tanks and so they are rejected by processors. The total biomass of the Gormanstown bed is estimated at approximately 1500 tonnes of which 2/3 had been removed by July 1999, two years after exploitation commenced.

Introduction

In 1997, a fishery for razor clams (*Ensis* spp) commenced when a substantial concentration of the animals was discovered on the east coast, 30 km north of Dublin at Gormanstown, Co Meath. The clams were harvested by hydraulic dredging using a blade which penetrates to a depth of approximately 25 cm; water is pumped through the blade at a pressure of 2–3 BAR, creating a fluidised bed and washing the clams back into a metal cage. According to the terms of Irish fisheries legislation, all vessels which are currently licensed to fish commercially are entitled to harvest clams using these methods. Between October 1997 and August 1999, approximately 50 boats are known to have fished the Gormanstown bed although only 20 did so regularly.

Ensis siliqua (L) (as elucidated in Holme, 1951), is the dominant species in the Gormanstown clam bed. This species is widespread in the British Isles; it has a preference for fine sands (Tebble, 1966; Henderson & Richardson, 1994). The biology of this species has not been previously investigated in Ireland and the work described here was undertaken to ascertain details relevant to a rational management strategy for the species in this and other exploited clam beds.

The site

Gatherers' Registration dockets record the locations at which razor clams are harvested in accordance with EU Council Directive 492 of 1991 concerning the harvesting of live bivalve molluscs. In the majority of cases, the dockets identify the bed in question by some convenient landmark. A minority gave location by GPS and according to these the bed stretches from approximately 53°38′ 00″ to 53°40′ 00″ North and from −6°10′ 00″ to −6°12′ 00″ West. The bed extends from inside the 10 m depth contour into the inter-

Table 1. Taxa taken in the dredge samples with *Ensis siliqua*. Nomenclature of the bivalves is in accordance with Tebble (1966) and for other taxa from Marine Biological Association (1957). Animals are scored in the scale: rare-occasional-frequent-common-abundant

Taxon	Comments
Bivalves	
Acanthocardia sp (?)	rare, present as dead shell
Aporrhais pes-pelicani (L)	rare, present as dead shell
Cerastoderma edule (L)	rare, present as dead shell
Chlamys spp.	rare, present as dead shell
Donax vittatus (da Costa)	locally abundant, otherwise frequent
Dosinia exoleta (L)	rare, usually present as dead shell
Gari fervensis (Gmelin)	occasional
Mactra corallina (L)	common, in every sample
Lutraria lutraria (L)	common-abundant, if not as living then as dead shell
Mytilus edulis L	rare, present as dead shell
Ostrea edulis L	rare, dead shell
Pharus legumen (L)	locally abundant
Solen martinatus Montagu	only one specimen taken
Thracia sp	rare
Venus striatula (da Costa)	common, in every sample
Gastropods - Scaphopods	
Acteon tornatilis (L)	rare
Dentalium sp	rare
Littorina littoralis (L)	occasional, dead shell
Nassarius reticulatus (L)	occasional
Natica catena (da Costa)	very common
Polychaetes	
Aphrodite aculeata L	rare
Arenicola marina (L)	rare
Lanice conchilega (Pallas)	locally abundant.
Nereis diversicolor O F Muller	rare
Pectinaria sp	rare
Crustaceans	
Carcinus maenas (L)	occasional
Corystes cassivelaunus (Pennant)	common, in every sample
Necora sp common,	in every sample
Paguridae sp	rare
Echinoderms	
Asterias rubens L	rare
Astropecten irregularis (Pennant)	frequent
Echinocardium cordatum (Pennant)	common, in every sample
Ophiuroidea	common

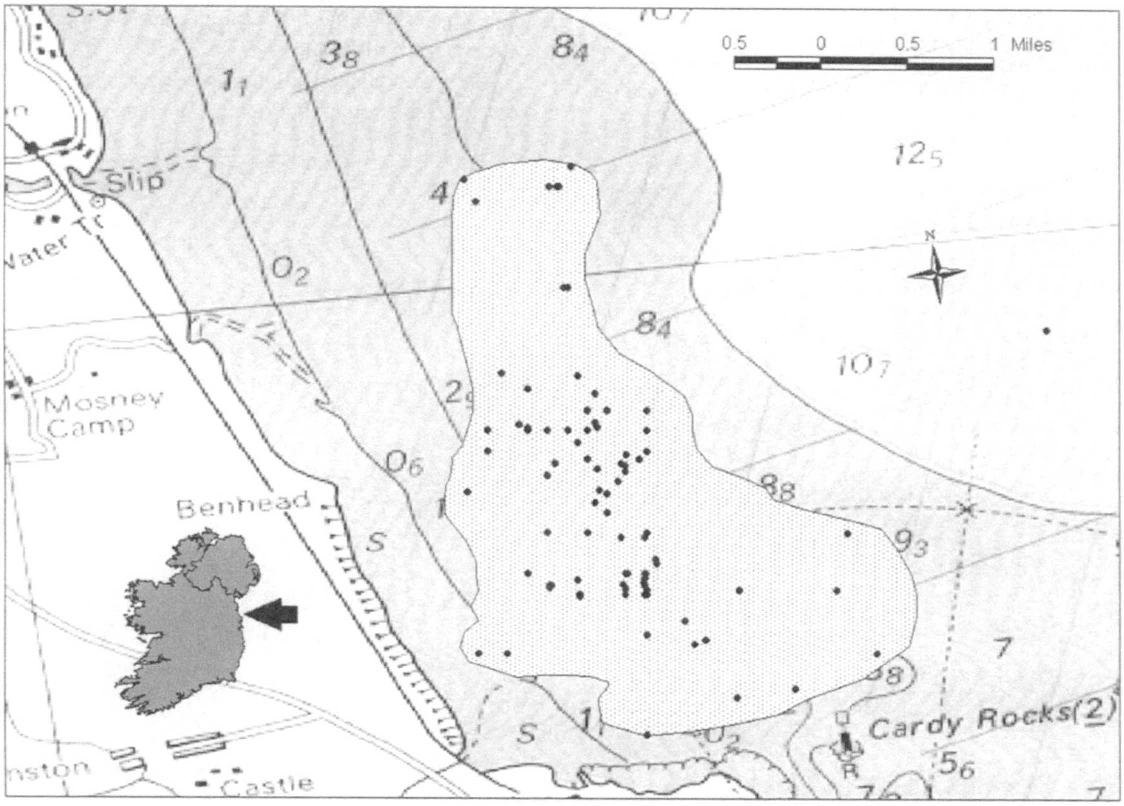

Figure 1. The Gormanstown clam bed demarcated from GPS data provided on Gatherers' Registration dockets. The data have been superimposed on the Admiralty chart: Ireland – east coast, Howth to Ardglass.

tidal. Its area has been calculated at 21 km^2 hectares (Fig. 1). Much the same area was fished in 1999 as in the previous year.

The waters overlying this area have been classified grade A (in the terminology of EU Council Directive 492 of 1991) and harvesting of bivalves there has been approved for this reason. Razor clams are widely distributed along the east coast of Ireland in lower densities but water quality is not adequate to allow harvesting everywhere.

Investigations on the Gormanstown razor clam bed commenced in July 1998 in order to monitor the progress of the fishery. Two razor species were recorded there. *Ensis siliqua* (L) accounts for almost all of those but two specimens of *Ensis ensis* (L) were identified among some 2500 individuals examined in the course of the following twelve months.

Materials and methods

Dredge samples, taken from the commercial vessels and weighing approximately 8–12 kg each provided the material for biological examination. Twenty five samples were collected during the year. Ten were gathered by boarding vessels as they were fishing and taking a fraction of the dredge contents; the remainder were provided by the vessels themselves, though the processors.

In the laboratory, all species were identified, bulked and weighed. All entire *Ensis siliqua* were measured to the nearest mm below (total length) (Plate 1) on a fish measuring board and weighed to the nearest 0.1 g after washing in salt solution to remove sand. Breakages were counted.

A condition index was obtained by regressing Ln weight on Ln length for all razors >8 cm, and then calculating weight at two total lengths, 15 and 20 cm.

From February to August 1999, gonadal tissue was taken from an average of 17 *E. siliqua* each month,

Table 2. Averaged length at age data for *Ensis siliqua*

Age	Average lt, cm	Standard deviation	Number of observations	Coefficient of variation
1	3.16	0.7	365	22.0
2	6.35	1	301	18.9
3	9.17	1.3	265	14.5
4	11.20	1	246	12.5
5	12.65	1.5	226	12.2
6	13.84	1.2	207	9.0
7	14.65	1.1	187	7.4
8	15.25	1.0	175	6.3
9	15.71	0.9	159	6.0
10	16.09	0.9	136	5.6
11	16.38	0.9	118	5.7
12	16.67	0.9	91	5.5
13	16.94	1.0	62	5.6
14	17.24	0.9	43	5.1
15	17.55	0.9	16	5.2
16	17.86	0.8	8	4.6
17	17.72	0.7	5	4.2
18	17.40	0.6	2	3.3
19	17.10		1	

sectioned and stained with eosin and hematoxylin in order to ascertain maturity; gonadal smears were taken from up to 100 animals per month in order to sex them ($n = 345$).

Annulus formation and other indications of growth intervals (such as clefts on smaller animals) on the shells were marked and the *growth length* was measured (from the tip of the shell, beneath the hinge, to the centre of the distal margin of the annulus) (Plate 1). *Total length* is approximately 4% greater than *growth length* to the shell margin. Growth data were obtained from only about 50% of *Ensis siliqua* examined, interpretation of the remainder being problematical.

Estimates were made of the length growth coefficients from the von Bertalanffy growth equation (Ricker, 1975):

$$L_t = L_\infty(1 - \exp[-k(t - t_o)]),$$

where L_t is the length at age at time t, L_∞ is the theoretical maximum length, t_0 is the theoretical age at length zero and k is the growth coefficient. The growth parameters L_∞, t_0 and k were estimated by fitting a growth curve to the observed mean length at age data.

Using the total lengths and interpreted ages, an age length key was devised for the Gormanstown clams and it was used to interpret changes in age structure of the exploited animals in 1998 and 1999.

The razor clam population at Gormanstown was reconstructed from the following elements:

Purchases } Rejects	Landings	
	} Landings, Breakages	Catches

Purchases + Rejects → Landings; Landings + Breakages → Catches

Taking each in turn:

Purchases were the weights of consignments bought by processors from fishermen and recorded in their 'books inwards'. *Rejects* (the fraction of the landing which was refused by the purchasers) consisted of clams which had become bruised in the course of harvesting so that they did not purge themselves of sand. Their occurrence was recorded by three processors whose purchases account for 65% of the total recorded; 26% of their purchases were accompanied by details of rejected clams. Purchases plus rejects constituted *landings* specific details of which were obtained from the dredge samples. Landings and *breakages* (which are discarded at sea) together made up the *catches*.

Details of removals from the clam bed were supplied by processors who made their books inwards available. Details of 1830 landings of razor clams from Gormanstown were abstracted from the books of seven buyers.

Additional corroborative data in the form of depuration dockets, export documentation and Gatherers' Registration Dockets, referred to above, supported the record of landings.

Three vessels provided data using European Communities' Logsheets which are normally used to record TAC fin fish landings. Details of one of these boats permitted calculation of the size of the Gormanstown clam bed using the methods of Leslie & Davis (1939). The vessel in question supplied a daily log giving the weight of clams harvested at Gormanstown, the number of dredging operations and their duration.

Results

The associated benthic community

Samples collected from the dredgers contained various animal species and sediment which ranged from fine sand to coarse shelly material. *Lanice conchilega, Pharus legumen* and *Donax vittatus* were occasionally

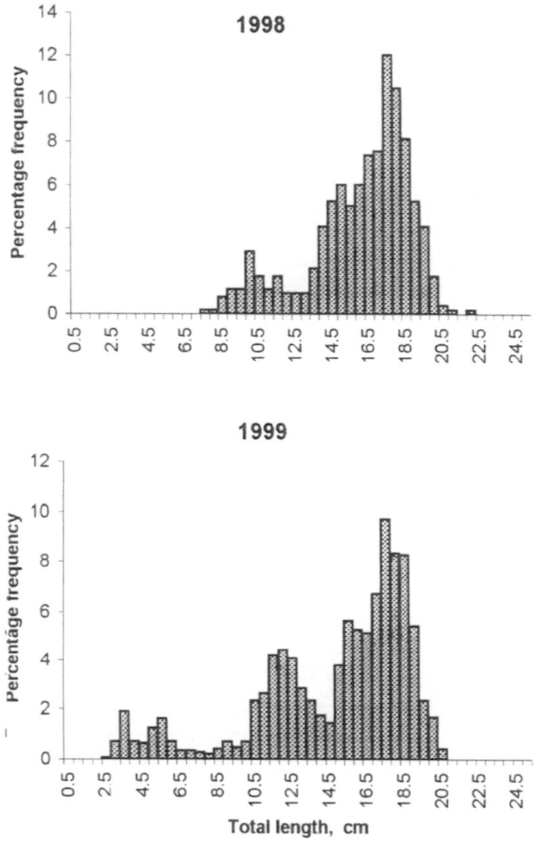

Figure 2. Percentage length–frequency distributions of *Ensis siliqua* taken in 1998 and 1999.

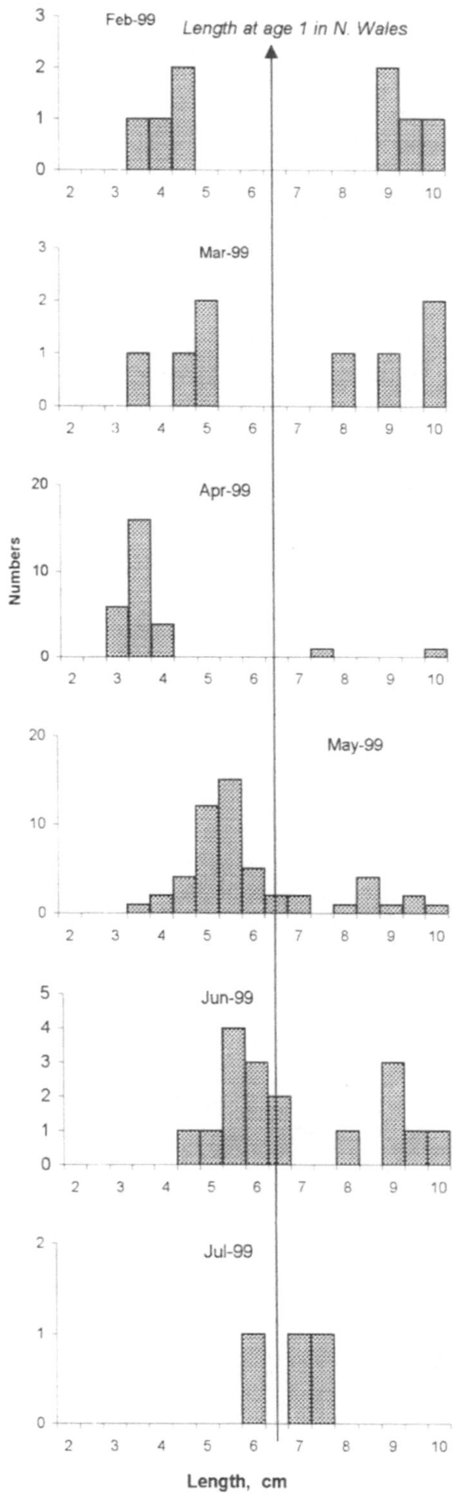

Figure 3. Length–frequency distributions of *Ensis siliqua* of less than 11 cm taken in each month from February to July 1999. The vertical arrow marks the mean length at age 1 of the species in the U.K.

found in great abundance associated with very small (<6 cm) *Ensis siliqua* in the spring and early summer months of 1999. Other species are listed in Table 1.

Age and growth

The length frequency of the samples in the two periods differed (Fig. 2): in 1998, two modal groups of razors were captured; the following year three. Fishing took place in the same places over the period, suggesting there was an immigration of smaller animals to the bed as harvesting proceeded. The possibility of marginal areas being exploited for the first time in the second year of sampling would also explain the phenomenon and cannot be discounted.

Back-calculated lengths at age of 365 animals are averaged in Table 2. The Gormanstown clams displayed greatest variability in length at age during their earliest years. When comparison was made with a growth curve in North Wales (Henderson & Richard-

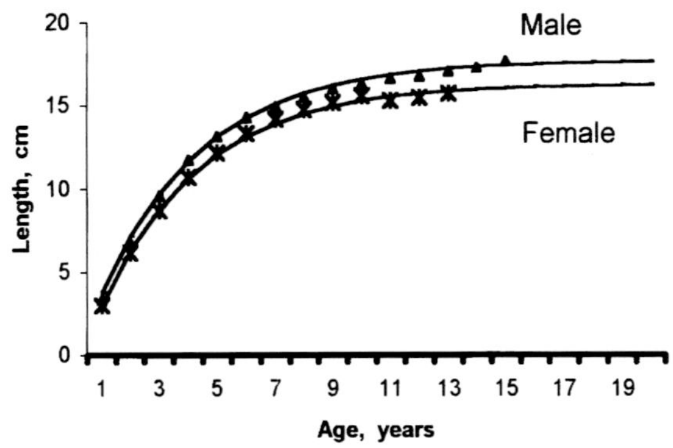

Figure 4. Von Bertalanffy growth curves and averaged lengths at age for male and female *Ensis siliqua* on the Gormanstown clam bed.

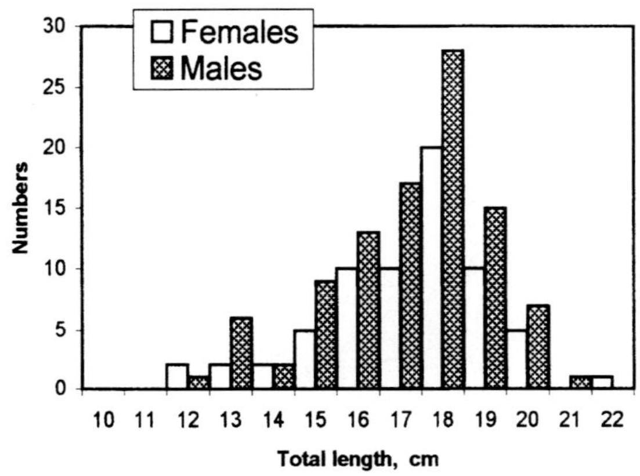

Figure 5. Length frequency distribution of male and female razor clams sampled at Gormanstown in 1999.

son, 1994) the averaged lengths of the animals in their earlier years showed some discrepancy. The Gormanstown clams are approximately half as long as the U.K. animals at age 1 although otherwise the lengths at a year's end are similar, the Irish material thus appears to be one year older: Gormanstown clams at the end of their second year's growth are of similar length to the U.K. animals at the end of their first year.

In the spring of 1999, a number of samples yielded very small *Ensis siliqua* which must have originated from a spat fall of the previous year (Fig. 3). In May 1999, the length frequency of these animals displayed a clear modal point which was still less than the mean length of age 1 clams in the British study.

Von Bertalanffy growth parameters for Gormanstown clams are provided in Table 3, only average lengths at age for which there were more than five data points being included in the calculations. Sixteen of the aged animals were identified from gonad smears as females and 25 as males. Growth parameters for these are also given in Table 3 and the growth curves are set out in Figure 4 showing that males grew at a slightly faster rate than females, the asymptote for the male curve being almost 1.5 cm longer than the female one.

This finding may go some way to explaining a phenomenon which was recorded at Gormanstown and is also known elsewhere: males are larger and more numerous than females on the razor bed (Fig. 5).

Age at length key

The age length key devised for Gormanstown using the total lengths and interpreted ages is presented in Table 4. Applying it to the data in Figure 2 converts the numbers to ages (Fig. 6); the average age in 1998 was 9.38 years from which it fell to 8.32 years one year later.

Maturation and spawning

Reference to Figures 4 and 5 indicates that it was feasible to distinguish the sexes of *Ensis siliqua* at 12 cm total length which corresponds with an age at first maturation of four years.

Six stages of maturation are recognized in the gonadal cycle of *Ensis siliqua*, including a resting stage during which the sexes are not distinguishable. This species has been described as gonochoric with the sexes undergoing synchronous development (Gaspar & Monteiro, 1998). Usually, two of the five maturity stages were present at the same time in a gonad section.

In the course of elucidating the gonadal cycle, each gonad section prepared every month was awarded two scores indicating the developmental stages observed. The scores were then amalgamated and the monthly percentage frequencies of the six stages were expressed in histogram form (Fig. 7). Spawning was judged to have taken place from mid-May to early August. At that time, the gonadal follicles had emptied and there remained some isolated oocytes and small groups of spermatozoa, the former in a state of resorption and both gametes without follicle walls. The sex of the majority of animals greater than 15 cm total length was not identifiable at that stage.

Condition

Calculating condition factor in razor clams is problematical because of the sand they retain, in spite of

Table 3. Growth parameters for *E. siliqua* from the Gormanstown bed

Parameter	All clams	Females	Males
$L\infty$ (cm)	17.82	16.28	17.71
r^2	0.9985	0.9969	0.9978
Number of points to calculate curve	16	12	14
k	−0.256	−0.287	−0.279
t_0	0.430	0.301	0.137
Number of clams examined	365	16	25

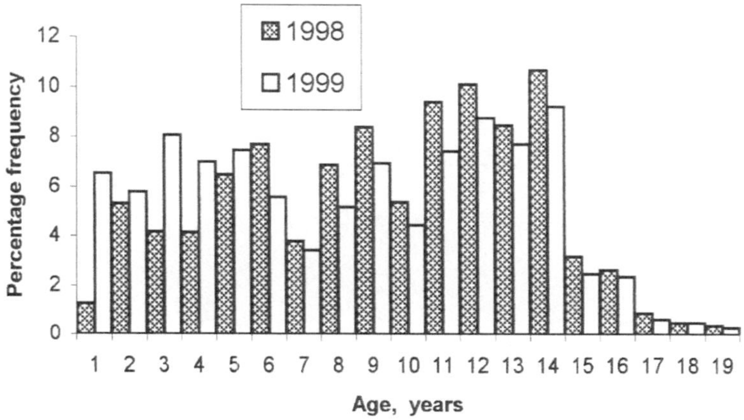

Figure 6. Age frequencies of razor clams sampled at Gormanstown in 1998 and 1999.

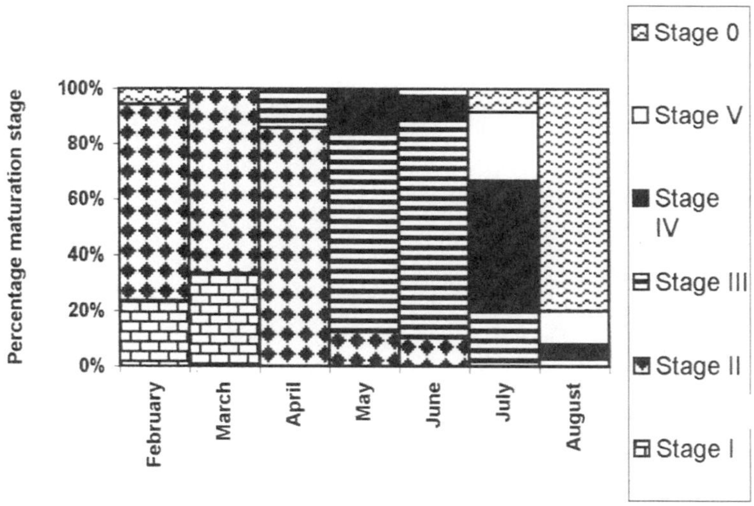

Figure 7. Percentage occurrence of six maturity stages among Gormanstown razor clams from 3rd February to 3rd August 1999.

attempts to remove it. The condition index tended to improve until May when it fell back, coinciding with the onset of spawning (Fig. 8).

Reconstructing the razor clam population

Reject rates are provided from the books inwards of three processors (Fig. 9) whose results were different – possibly because the demands of their buyers imposed

Table 4. Age length key for razors (*Ensis siliqua*), at total length

Cm	Ages																			Totals
	1	2	3	4	5	6	7	8	9	10	11	12	13	14	15	16	17	18	19	
21.0																				
20.5																				
20.0																1				1
19.5									2		2			2	2	1	1			10
19.0											1	3	2	4						10
18.5								2		1		3	4	4	1	1		1		17
18.0								1	1	2	5	2	5	3	2		1			22
17.5							1	1	2	4	2	8	4	8	2				1	33
17.0						2	0	2	4	8	8	9		3	1					37
16.5					1	0	4	2	5	3	5	3	1	3						27
16.0					1	4	2	2	3		1		1							14
15.5					1	0	1	2	4		1		2							11
15.0				1	0	3	1	2	2		2	1								12
14.5				2	0	4		2												8
14.0				0	3	1														4
13.5			1	1	2		1													5
13.0			4	4	4															12
12.5			1	4	3	4	1													13
12.0			6	4	2	1														13
11.5		1	2	2	1															6
11.0			1	1	1															3
10.5		1	3																	4
10.0	1	2																		3
9.5		1	1																	2
9.0		5																		5
8.5	1	2																		3
8.0		2																		2
7.5	1	3																		4
7.0		1																		1
6.5	3	2																		5
6.0	3																			3
5.5	10	8																		18
5.0	5	6																		11
4.5	3	1																		4
4.0	5	1																		6
3.5	15	1																		16
3.0	13																			13
2.5	3																			3
2.0	1																			1
Totals	64	37	19	19	19	19	11	16	23	18	27	29	19	27	8	3	2	1	1	362

different requirements – and did not appear to observe any seasonal pattern. In order to devise a reject rate for razor clams for which specific information was not available, all of the recorded data were averaged and a mean weight of 7.3% (st dev = 6) was applied to purchases whenever specific data were not given.

Breakages appeared to follow a seasonal pattern (Fig. 10), higher in quarters 1 and 4 than 2 and 3. Breakage rates were averaged at 30% (st dev = 11)

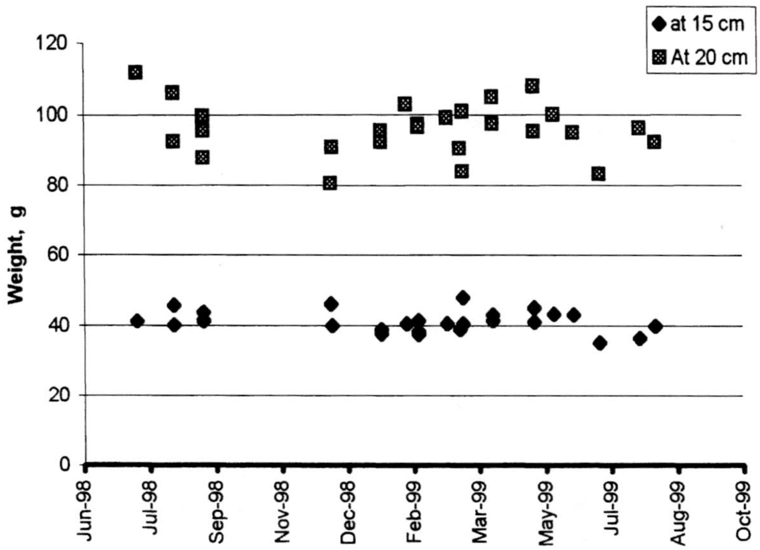

Figure 8. A condition index for *Ensis siliqua* at Gormanstown, derived by regressing LNwt on LNlt and calculating the weight at 15 and 20 cm total length. Clams included in the calculations ranged upwards from 8 cm.

in quarters 1 and 4 and 18% (st dev = 9) of the catches in quarters 2 and 3. An explanation proposes that razor clams live deeper in the sand during turbulent weather. The animals are delicate and susceptible to storm damage in the shallow sandy areas so that when they bury deeper they are more likely to be cut in two by the dredge blade.

Progress of the fishery

Amalgamating the various components of the catches and landings described above permitted calculation of the cumulative removal from the Gormanstown bed (Fig. 11) to the end of June 1999 when more than 1000 tonnes had been harvested. The rate of removal to July 1999 remained constant.

Figure 12 provides the operational and lpue record of one of the vessels whose details permit an estimate of the size of the resource using the methods of Leslie & Davis (1939). Hydraulic dredges are custom built for each vessel and the gear is constantly modified until it is considered satisfactory. Therefore the duration of each fishing operation (tow) was variable in the beginning but from September 1998 it stabilized at 0.4 h. Figure 12 (above) shows that initially effort rose and then began to decline. From this point there was a reduction in lpue measured either as kg per dredge haul or kg per hour's fishing (both are given). Substituting the date line on Figure 12 with the cumulative removals (Fig. 11) and regressing the lpue data on it provides the following estimates of the original size of the fishable stock (Table 5).

According to the first of these regressions, the size of the fishable stock was 1525 tonnes, the second gives a figure of 1595 tonnes. Assuming that the average razor clam weighs 50 g, the density of the animals would have averaged 1.45 per m^2 or, using the kg/hour figure, 1.52 per m^2.

These calculations suggest that approximately 60% of the Gormanstown bed had been harvested in July 1999, two years after dredging began there.

Discussion

The dredge cages are constructed of steel bars approximately 1.5 cm apart. These are not capable of retaining all of the youngest clams, although small organisms such as *Pharus legumen* were occasionally taken in large numbers. However, razor clams of greater than 10 cm total length were not selected hence, a comparison of aspects of the biology of *Ensis siliqua* in Ireland, Portugal and Great Britain on the basis of material gathered is feasible.

Ensis siliqua favours fine sand and is most abundant at water depths of 3–7 m (Monteiro & Gaspar, 1993) and such is the habitat it occupies at Gormanstown. Hydraulic dredging can only be conducted to a depth of 10 m at the time of writing; the boats were equipped with hoses which reached only to this

Table 5. Regressions of lpue expressed as kg per operation (per dredge tow and associated activities) and as kg per hour's fishing, on cumulative removals from the clam bed at Gormanstown (Fig. 11)

Criterion	Intercept	*X*-variable	r^2	No. obs	*P*	Significance
Kg/operation	55.196	−0.036	0.361	34	0.05<0.02	Sig.
Kg/hour	135.252	−0.422	0.422	34	0.01<0.001	V. sig.

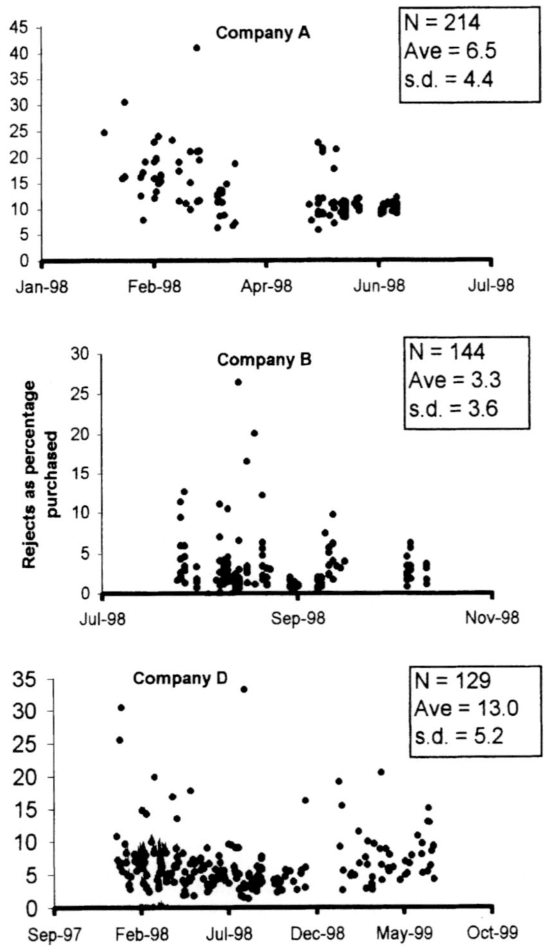

Figure 9. Reject rates of razor clams expressed as a percentage of landings to three processors.

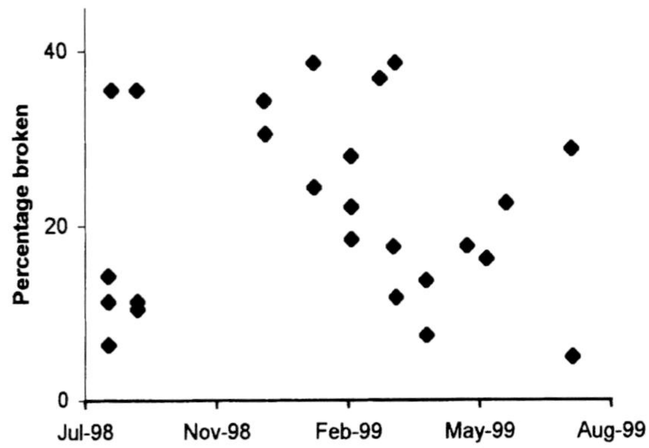

Figure 10. Percentage breakages among the catches as calculated from samples dredged from the clam bed at Gormanstown.

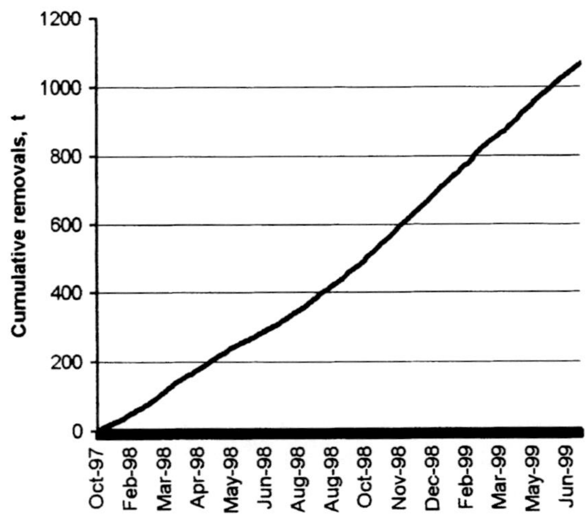

Figure 11. Cumulative removal of clams from the Gormanstown bed, the tonnage incorporating breakage and rejection rates described in the text.

depth. Divers have reported that although the Gormanstown bed extends further seawards the substratum becomes very shelly and the density of *Ensis siliqua* is dramatically reduced.

Gaspar et al. (1994) showed that sectioning the shells and using an acetate peel method of interpreting growth provided greater lengths at age than the interpretation of surface sculpture. It was also assumed to be a more accurate method of interpreting growth. However, lengths at age of the Gormanstown clams display close agreement with *E. siliqua* from North Wales (except at year 1) and there is evidence, other than from the interpretation of surface sculpture, of a smaller than reported length at the end of the first growing season.

The asymptotic length achieved by *Ensis siliqua* in Gormanstown of 17.8 cm is greater than that reached in southern Portugal and Britain where it is about 14.5 cm and 15.5 cm respectively. The animal grows faster in Portugal (Gaspar et al., 1994; Henderson & Richardson, 1994).

Gaspar & Monteiro (1998) reported that while the overall sex ratio was 1, females predominate among the smaller, younger groups and males among the larger. They suggested that greater mortality in the females explained the situation. Evidence of faster growth hence larger size achieved by males at Gormanstown may, alternatively, partially at least, explain the phenomenon.

In Vilamoura, Southern Portugal, the animal matures in its first year (Gaspar & Monteiro, 1998). Henderson & Richardson (1994) reckoned that North Wales razor clams were at least three years old and possibly older and larger than 10 cm before they mature and spawn, a finding that approximates the age at maturity at Gormanstown of four years.

Where the gonadal cycle for *Ensis siliqua* has been described there has been one spawning period per year. Lebour (1938) reported *E. siliqua* spawning in March and April in Plymouth. In Portugal, spawning has been reported commencing from May to July (Gaspar & Monteiro, 1998), a very similar situation to that described at Gormanstown.

The estimated density of 1.5 E. siliqua/m^2 must be regarded as an overall figure which includes less productive parts of the clam bed as well as areas in which the animals are tightly packed. Occasional reports by divers suggest densities might exceed 200 animals/m^2. The Gormanstown bed is not regarded as a homogenous structure, rather as a mosaic of substrates, some of them consisting of patches of fine sand with a high density of razor clams, others of shelly sand in which the clams are less densely distributed. There is also evidence of certain bivalve species being associated with the mason worm *Lanice conchilega*, occupying a nursery habitat for small *Ensis siliqua*, undoubtedly in a limited extent of the total area.

It used to be generally believed that unexploited razor clam beds consisted of large numbers of older animals. McKay (1992), for instance, described beds in North West Scotland as having individuals up to a maximum of 25 years old with relatively few smaller clams in the vicinity. He supposed that the presence of large animals could inhibit further recruitment; parent stock occupy the most favoured parts of the clam bed and consume incoming spat. The inference of this kind of age structure in a clam bed suggests that its occupants must be removed before any further recruitment occurs. Recent video evidence by Seafish Technology, Hull, U.K. suggests a different mechanism: razor clams are highly mobile, capable of jetting about using the foot for propulsion and thus being able to select more favourable substratum and growing conditions. High densities of larger and older clams might thus represent an active recent migration into those circumstances rather than a spat fall some time in the distant past. The work at Gormanstown might be interpreted to indicate that as the older and larger animals were removed smaller ones moved onto the bed. They may have migrated from discrete patches at the periphery of heavier concentrations of adult animals.

Sustaining a fishery like the Gormanstown bed is fraught with difficulties. Hall et al. (1990) observed that hydraulic dredging has short term effects and that its physical consequences are not visible after 40 days, a finding which might be expected in shallow inshore waters subject to wave disturbance. They did however remark that long lived species are likely to take much longer to recover.

Holme (1951) and Tebble (1966) both stated that large sparsely populated intertidal beds of razor clams occur around the British Isles. Henderson & Richardson (1994) reported that between 1990 and 1992 there had been a visible decline in stocks in North Wales and that no individuals greater than 17.5 cm long had re-established in the areas they examined. The vulnerability of the species to over-fishing was recognized. Gaspar & Monteiro (1998) remarked on the severe reduction in landings of *Ensis siliqua* in Portugal resulting from heavy fishing pressure from 1969 onwards.

Razor clams have been locally harvested in Ireland for many years, either being dug out or salted in the inter-tidal. Hydraulic dredging is a new phenomenon which has been embraced enthusiastically by fishermen anxious to identify alternative or new species to supplant the growing list of over-fished ones. The largest market destination for razor clams to date is Spain and the Gormanstown bed has largely supplied the Spanish market since fishing began there. *E. siliqua* is a large species, suitable for canning but it is less easy to sell after April when condition is lost at the onset of spawning and canneries cease to purchase it. A 'live' market is available during the summer months but it is not so extensive and *E. ensis*, a smaller species,

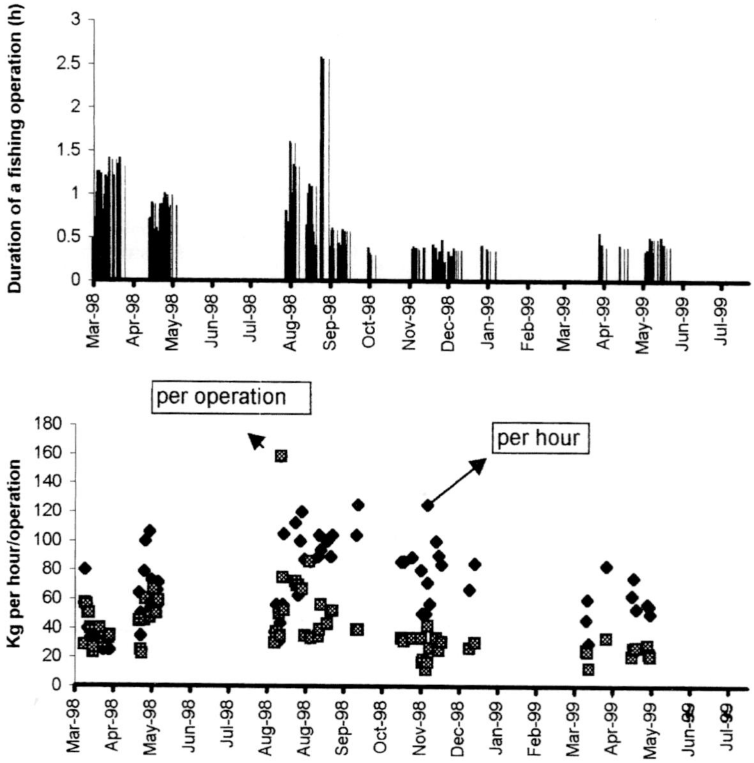

Figure 12. Logbook data used to estimate the size of the Gormanstown clam bed: duration of fishing operation (above) lpue expressed as kg per fishing operation and kg per hour (below). These data were supplied in the format of the European Communities' Logbook by a single vessel.

is more acceptable to it. The history of the Gormanstown clam bed includes episodes of over-harvesting and dumping of animals which the market has been unable to absorb. First sale prices have also fallen dramatically in response to over-supply.

The Irish industry has no experience of managing razor clams in a sustainable way and too little is known about their ecology to enable accurate pronouncements about how they should be managed. The frequency of successful spat falls is obviously a key to their renewable nature. If hydraulic dredging is to be the only means of recovering them, a period of fallowing of four to five years after fishing would seem the appropriate management strategy.

A more selective and less destructive way to harvest clams would be by diving. However, bye-law number 533 of 1996 prohibits diving for shellfish and its amendment is currently under consideration. The advantages of doing so would include the selective harvesting of certain sizes (possibly also species) of razor clams in response to market demand, the curtailment of breakages and discards and the recovery of a better product with longer survival and depuration prospects.

Many other management problems also exist, in common with the rest of the inshore sector, among them the absence of a specific licensing system for hydraulic dredging which permits anyone with a fishing licence to participate for a brief period and, in consequence, discourages individual fishermen from taking a longer term interest in a sustainable fishery for these species.

Acknowledgements

The fishery officers of the Department of the Marine and Natural Resources provided advice and assistance at sea and I am particularly grateful to Mick O'Driscoll, Kevin Concannon, Mick Lyne and John Falvey for this. The following processors supplied me with data: Bill Price of Rockabill Shellfish, Pat Rosney of Lusk, Co Dublin, Jimmy and Amelia Doyle of Rush, Co Dublin, Turlough Sheehan of Chirocon, Tom

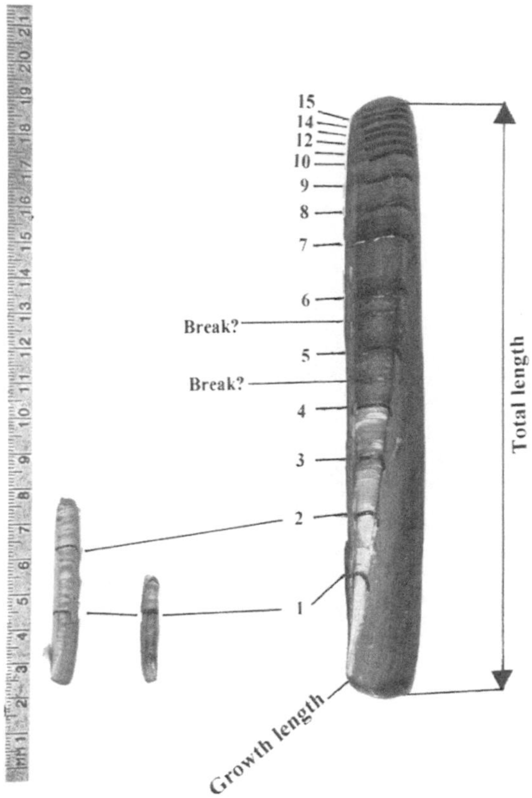

Plate 1. A shell of *Ensis siliqua* showing the total length and the origin of the growth length (right), a clam of 1+ years (centre) and a clam of 2+ years (left).

Jones of Tawnagh, Dunmore East, Co Waterford and Mark and Darren Wright of Wright's of Howth, Co Dublin. Fergal Nolan of Bord Iascaigh Mhara advised on various aspects of setting up the work. Various colleagues at the Marine Institute contributed to the project: John McArdle and his staff in the Fish Health Unit prepared histological sections of the gonads and advised on their interpretation. Ivor Marsh prepared Figure 1. Colm Duggan and Sorcha Wheatley suggested amendments to an early draft of the paper. Finally, Chris Richardson of the School of Ocean Sciences, University of Wales, Bangor discussed the early stages of the work described here with me and Martin Pyke of Seafish Technology, Hull, gave me video evidence of the mobility of an *Ensis* species.

References

Gaspar, M. B. & C. C. Monteiro, 1998. Reproductive cycles of the razor clam *Ensis siliqua* and the clam *Venus striatula* off Vilamoura, Southern Portugal. J. mar. biol. Ass. U.K. 78: 1247–1258.

Gaspar, M. B., C. A. Richardson & C. C. Monteiro, 1994. The effects of dredging on shell formation in the razor clam Ensis siliqua from Barrinha, Southern Portugal. J. mar. biol. Ass. U.K. 74: 927–938

Hall, S. J., D. J. Basford & M. R. Robertson , 1990. The impact of hydraulic dredging for razor clams *Ensis* sp. on an infaunal community. Neth. J. Sea Res. 27: 119–125

Henderson, S. M. & C. A. Richardson , 1994. A comparison of the age, growth rate and burrowing behaviour of the razor clams, *Ensis siliqua* and *E. ensis*. J. mar. biol. Ass. U.K. 74: 939–954.

Holme, N. A., 1951, The identification of British species of the genus *Ensis* Schumacher (Lamellibranchiata) J. mar. biol. Ass. U.K. 29: 639–647

Lebour, M. V., 1938 Notes on the breeding of some lamellibranchs from Plymouth and their larvae. J. mar. biol. Ass. U.K 23: 119–145.

Leslie, P. H. & D. H. S. Davis, 1939. An attempt to determine the absolute number of rats in a given area J. anim. Ecol. 8: 94–113.

Marine Biological Association, 1957, Plymouth Marine Fauna. Latineer, Trend & Co Ltd. Plymouth.

McKay, D. W., 1992. Report on a survey around Scotland of potentially exploitable burrowing bivalve molluscs. Marine Laboratory, Aberdeen Fisheries Research Services Report No 1/92

Monteiro, C. & M. Gaspar, 1993. Bivalves dolitoral oceâico algarvino: breve noticia sobre a situaçâo actual dos Principais bancos (Julio 1993) Relatório Tecnico Cientifico INIP Lisboa 65: 19.

Ricker, W. E., 1975. Computation and interpretation of biological statistics of fish populations. Bull. Fish. Res. Bd Can. 191: 382 pp.

Tebble, N., 1966, British Bivalve Seashells: a Handbook for Identification. British Museum, London.

Hydrobiologia **465:** 153–167, 2001.
G. Burnell (ed.), Coastal Shellfish – A Sustainable Resource.

The influence of dredge design on the catch of *Callista chione* (Linnaeus, 1758)

M. B. Gaspar[1], M. D. Dias[3], A. Campos[4], C.C. Monteiro[1], M. N. Santos[1], A. Chicharo[2] & L. Chicharo[2]
[1]*Instituto de Investigação das Pescas e do Mar, Centro Regional de Investigação Pesqueira do Sul. Av. 5 de Outubro, 8700-305 Olhão, Portugal*
[2]*Universidade do Algarve, Unidade de Ciências e Tecnologias dos Recursos Aquáticos*
[3]*Instituto de Investigação das Pescas e do Mar, Del. de Setúbal. Av. Jaime Rebelo 29A, 2900-409 Setúbal, Portugal*
[4]*Instituto de Investigação das Pescas e do Mar. Av de Brasilia, 1400 Lisboa, Portugal*

Key words: dredge, fishery impact, *Callista chione*, mortality, by-catch, *Pharus legumen*, *Ensis ensis*, *Solen marginatus*, *Ensis siliqua*, *Mactra glauca*, *Lutraria anguistor*, *Echinocardium cordatum*, *Laevicardium crassum*, *Spisula solida*, *Venus striatula*, *Dosinia exoleta*, *Dosinia lupinus*, *Acanthocardia tuberculata*, *Cymbium olla*, *Gibbula* sp., *Nassarius reticulatus*, *Astropecten aranciacus*, *Pagurus* sp., *Polybius henslowi*, *Calappa granulata*

Abstract

To evaluate a possible introduction of a new dredge in the fishery of *Callista chione* (Linnaeus, 1758), IPIMAR has conducted a study with the objective of comparing the efficiency of two dredges (traditional dredge and the new dredge design) and evaluating their impact on the benthic community. The experiments were carried out during March 1999 on the Southwest coast of Portugal, from a site off Troia. Three different tow durations of 5, 10 and 20 min were investigated. A total of 24 hauls were accomplished, 4 for each tow duration and dredge. The experiments were conducted by attaching a cover bag with a 20 mm mesh to the gear. After each haul, the catches in the bag and in the cover were sorted separately. All individuals retained were attributed scores on a scale of 1–4 in which 1 equates to good and 4 equates to dead. The results obtained showed that catches from the traditional dredge (TD) are composed of a great fraction of juveniles of *C. chione*, while in the new dredge (NDD) catches are composed, almost entirely, by individuals with a superior size to the minimum legal length (50 mm). This result indicates that the mesh of the bag of the TD used in the exploitation of this resource is not adequate. For the 3 different tow durations, the mean fishing yield obtained for the NDD was always superior to the TD, due to its greater efficiency in capture. The proportion of by-catch is significantly higher when the TD is used. For all 3 tow duration, the TD caused mortalities on the target species and on the macrobenthic community in the same order of magnitude as the NDD. Since the fishery of *C. chione* is managed by daily quotas per boat, when using the NDD the impact on the macrobenthic community is reduced by about 50% due to its greater efficiency of capture. Another advantage in the usage of the NDD relatively to the TD, is to allow the smallest individuals (independently of the species) to escape rapidly through the metallic bars on the grid, increasing their probability of survival.

Introduction

An important bivalve fishery takes place along the Southwest coast of Portugal. At present, in this part of the Portuguese coast, the most important commercial species is the clam *Callista chione* (Linnaeus, 1758). This species has been recorded from Great Britain, to south to the Iberian Peninsula, into the Mediterranean, along the Atlantic coast of Morocco and to the Canary Islands and the Azores (Tebble, 1966). On the Southwest coast of Portugal, this species is preferentially distributed on clean sandy bottoms from 10 to 20 m depths (Gaspar et al., 1998a).

C. chione constitutes the target of a specific fishing activity carried out by an artisanal fleet. Dredges are the gears operated by the fishery. The shape and struc-

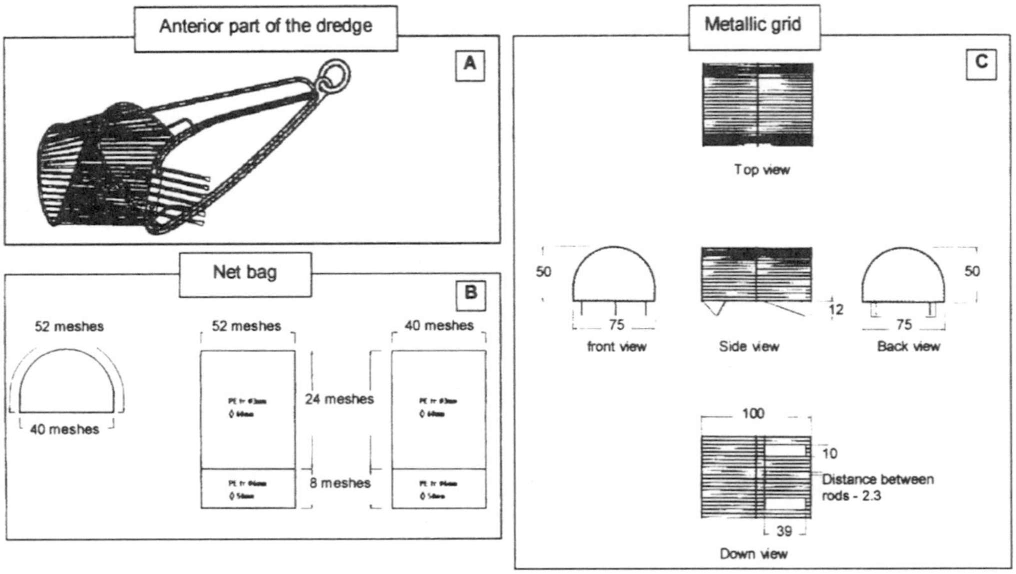

Figure 1. Diagram of the *Callista chione* dredge. Technical specifications of the net bag and the metallic grid (measures in cm). 1B is Traditional Dredge (TD); 1C is New Dredge Design (NDD).

Table 1. Gear specifications of the anterior part of the dredges used in the present study

	New dredge design	Traditional dredge
Length of the mouth	43 cm	60 cm
Space between rods	2.4 cm	2.4 cm
Number of teeth	6	6
Space between teeth	4.3 cm	4.0 cm
Tooth length	30 cm	34 cm
Tooth angle	20°	20°

ture of the Portuguese dredge has remained unchanged throughout time and consists of a small, heavy semicircular iron structure, with a net bag and a toothed lower bar at the mouth. Welded to this iron structure are three metal shafts forming a kind of hen's foot where the towing cable is attached (Fig. 1A, B).

Recently, some fishermen from the Setúbal region have been trying to introduce a new dredge typology in the *C. chione* fishery arguing that it is a more efficient fishing gear. The basic difference between the two dredges resides in the retention structure for the bivalves. In fact, in the new dredge the net bag is replaced by a semicircular metallic grid (Fig. 1A, C).

The introduction of a new dredge design in any fishery should take into consideration two aspects: first it should reduce the impact on the macrobenthic community and second it should bring something to the fishermen such as higher incomes. Thus, in the comparative study of the two dredge typologies, besides the efficiency comparison of the gears, the ecological impact should also be assessed. The fact that the dredge is a fishing gear that is towed over the bottom causes necessarily an impact on the sediment as well as on the benthic community. However, although there may be a similar impact on the sediment in both gears, the same should not happen in the benthic community even though using the same capture methods.

In this context, a comparative study was conducted with the aim of evaluating a possible introduction of the new dredge design in the fishery of *C. chione*.

Methodology

Sampling design

The comparative study was conducted during March 1999 in the Setúbal region on the western coast of Portugal, in a site off Comporta, the most important fishing ground for *C. chione*. The study was carried out onboard the 'Surpresa' (SB-867-C), a vessel belonging to the commercial dredging fleet. The samples

Table 2. Criteria used in the attribution of a score to each *taxon*

	Score	1	2	3	4
	Bivalvia	In good condition	Edge of shell chipped	Hinge broken	Crushed / dead
	Gastropoda	In good condition	Edge of shell chipped	Shell cracked or punctured	Crushed / dead
	Asteroidea	In good condition	Arms missing	Worn and arms missing	dead
	Ophiuroidea	In good condition	Arms missing	Worm and arms missing /minor disc damage	Major disc damaged / dead
	Echinoidea	In good condition	<50% spine loss	>50% spine loss / minor cracks	Crushed / dead
Crustacea	**Anomura**	In good condition	Out of shell and intact	Out of shell and damaged	Crushed / dead
	Brachyura	In good condition	Legs missing / small carapace cracks	Major carapace cracks	Crushed / dead
	Natantia	In good condition			Dead
	Osteichthyes	In good condition	Small amount of scales missing/small cuts or wound	Large amount of scales missing/severe wounds	Dead
	Polychaeta	In good condition			Sectioned

were collected on sandy bottoms between 8 and 10 m depth. The specifications of the anterior part of the gears used in the experiment are summarised in Table 1.

In order to estimate the effect of tow duration on the catch of *C. chione* and on the number of damaged individuals, three-tow durations were investigated (5, 10 and 20 min). A total of 24 tows were accomplished, 12 for each type of dredge and with consistent durations for both dredges. The duration of dredge hauls was measured from the time the winch stopped paying out the towing cable to the time when the winch was restarted. The dredges were towed at a commercial speed of 2–3 knots.

In order to determine the proportion of clams under the size of 50 mm retained by both dredges and also to determine the respective impact, by means of the number of damaged individuals, the cover method was employed. This technique consists in the covering of the net bag, in the case of the TD, or the grid, in the case of the NDD, with a second net bag with a small mesh size. This bag has a 20 mm mesh and was stretched loosely over the entire back of the dredges, enough slack being left in the cover to reduce any masking effect on selection and to allow the natural flow of water through the net (Gaspar et al., 1999).

The captures of each dredge, coming from the bag or the grid and from the cover, were processed separately in all tows. All catches were identified, weighed and measured (in the case of the bivalves). Standard shell length measurements were made along the longest linear dimension to the millimetre below. The species identification was made according to Bucquoy et al. (1882–98), Tebble (1966), FAO (1987) and Poppe & Goto (1993). The nomenclature adopted was that of FAO (1987).

To determine indirect mortality each organism caught was attributed a score, which evaluated the degree of damage suffered (Table 2). The scale in the present work used four scores, where score 1 corresponds to the individuals in perfect condition while score 4 corresponds to the dead individuals. Ten *taxa* were considered: Bivalvia, Gastropoda, Asteroidea, Ophiuroidea, Echinoidea, Anomura, Brachyura, Natantia, Osteichthyes and Polychaeta.

For all *taxa* considered, except for Bivalvia, the attribution of the respective score for each individual caught was done visually during the sampling operation on board. For Bivalvia, the attribution was done in the laboratory, during the processing of the collected samples. Underwater images were collected by divers during the tows.

In the present work, the dredge usually used to harvest bivalves was called the traditional dredge (TD), while the new dredge was denominated by new dredge design (NDD).

Data analysis

Statistical procedures followed Zar (1996) and Sokal & Rohlf (1981). Size frequency distributions were constructed from raw data and compared using a Kolmogorov–Smirnov test. To investigate the differences between the fishing yields obtained for both dredges and for the three-tow duration, a two sample *t*-test was used. Analyses of variance ANOVA or Kruskal–Wallis ANOVA for comparison of overall proportions was used to test the effect of tow duration within each dredge on the proportion of by-catch and on the proportion of damaged individuals. Multiple comparisons were performed using the Student–Newman–Keuls test. If no significant statistical differences were observed on the effect of tow duration on the proportion of both by-catch and damaged individuals, the data was pooled and *t*-test or Mann–Whitney test were used to investigate the effect of dredge design on those proportions. Prior to the application of ANOVA, Kruskal–Wallis ANOVA, *t*-test or Mann–Whitney test, data was standardised and, when expressed as a percentage, was converted to arcsine square root values. All statistical analyses but Kolmogorov–Smirnov test were performed using the SigmaStat statistical software. For all tests the level of significance was 5%.

Results

Overall catches

Table 3 shows the species that entered the dredges for each *taxon* considered. A total of 36 species entered the dredges. Bivalvia was the most common *taxon* with 22 species, followed by Gastropoda and Brachyura with 3 species each.

The analysis of Table 3 show that a higher number of species entered the NDD than in the TD (34 and 29 species, respectively). In terms of weight, *C. chione* largely dominated the total catches for the two dredges (70.05% and 58.91%, respectively for NDD and TD). In terms of numbers, the most abundant species was *C. chione*, *Venus striatula* (Da Costa, 1778) and *Laevicardium crassum* (Linnaeu, 1758) (39.3%, 34.6% and 10% for the NDD and 33.3%, 21.9% and 24.7% for the TD, respectively).

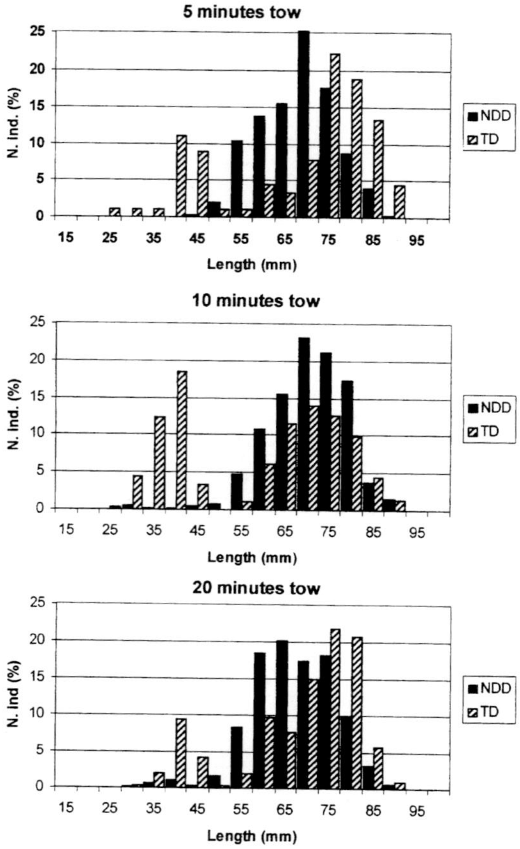

Figure 2. Callista chione length frequency distributions for the NDD and TD and for the 3 tow duration investigated.

Callista chione *catches*

Histograms of size frequency distributions for tow duration of 5, 10 and 20 min. and for the two dredges are shown in Fig. 2. Results from the Kolmogorov–Smirnov test indicate that all size distributions were significantly different ($P<0.05$).

From Figure 2 it can be seen that the catches from TD, for the 3-tow duration investigated, are made up of a significant fraction of sub-dimensioned individuals, or else, with a size inferior to 50 mm. This fraction corresponds to 23.35%, 37.8% and 16.1%, respectively, for 5, 10 and 20 minutes tow duration. In the NDD, the captures are almost totally made up of in-

Table 3. Total number of individuals caught during the experimental catches

	NDD		TD	
	Number	Weight (g)	Number	Weight (g)
Bivalvia				
Acanthocardia aculeata	53	7421	12	1720
Acanthocardia spinosa	4	629	1	321
Acanthocardia tuberculata	198	20 356	114	9663
Callista chione	2288	166 615	838	69 311
Dosinia exoleta	79	2228	32	1067
Dosinia lupinus	61	293	5	47
Ensis ensis	23	130	2	11
Ensis siliqua	5	260		
Gari fervensis	2	7		
Laevicardium crassum	575	19 676	621	21 533
Lutraria anguistor	27	161	3	6
Lutraria lutraria	2	141	2	176
Mactra coralina	10	226	17	223
Mactra glauca	86	4622	127	7439
Pandora inaequivalvis	1	2	1	2
Pharus legumen	3	14	2	18
Solen marginatus	1	25		
Spisula solida	2	7	1	3
Tellina incarnata	76	210	4	8
Thracya papyracea	4	1		
Venus fasciata	3	8	2	6
Venus striatula	2014	7632	552	1783
Gastropoda				
Cymbium olla	3	167		
Gibbula sp.	10	33	4	4
Nassarius reticulatus	3	12	1	5
Asteroidea				
Astropecten aranciacus	31	2622	19	1921
Ophiuroidea				
Ophiocomina nigra	7	19	11	10
Echinoidea				
Echinocardium cordatum	51	1340	27	354
Sphaerechinus granularis	1	5		
Anomura				
Pagurus sp.	14	41	6	25
Brachyura				
Calappa granulata	21	605	21	704
Corystes cassivelaunus	1	8		
Polybius henslowi	66	768	71	936
Natantia				
Palaemon serratus			1	52
Osteichthyes				
Dicologoglossa cuneata			2	39
Polychaeta	97	1557		274
Total	5822	237 841	2499	117 661

Table 4. Comparison of the mean number of individuals damaged (scores 2 to 4) and dead (scores 3 and 4) for each *taxon*

		New dredge design					Traditioinal dredge				
		Total	Damaged		Dead		Total	Damaged		Dead	
			N	(%)	N	(%)		n	(%)	N	(%)
5 minutes	Bivalvia	317.0	15.8	(5.0)	15.3	(4.8)	162.0	11.3	(6.9)	10.5	(6.5)
	Gastropoda	1.3	0.0	(0.0)	0.0	(0.0)	0.3	0.0	(0.0)	0.0	(0.0)
	Asteroidea	1.3	0.3	(20.0)	0.0	(0.0)	0.8	0.5	(66.7)	0.5	(66.7)
	Ophiuroidea										
	Echinoidea	5.0	4.3	(85.0)	4.0	(80.0)	1.3	1.3	(100.0)	1.3	(100.0)
	Brachyura	5.3	1.0	(19.0)	1.0	(19.0)	5.0	1.8	(35.0)	1.3	(25.0)
	Anomura	0.3	0.0	(0.0)	0.0	(0.0)	0.3	0.0	(0.0)	0.0	(0.0)
	Natantia										
	Osteichthyes						0.5	0.0	(0.0)	0.0	(0.0)
	Polychaeta	8.8	0.0	(0.0)	0.0	(0.0)	0.8	0.0	(0.0)	0.0	(0.0)
	Total	**338.8**	**21.3**	**(6.3)**	**20.3**	**(6.0)**	**170.8**	**14.8**	**(8.6)**	**13.5**	(7.9)
10 minutes	Bivalvia	399.8	33.0	(8.3)	30.0	(7.5)	242.6	12.4	(5.1)	11.2	(4.6)
	Gastropoda	1.5	0.0	(0.0)	0.0	(0.0)	0.2	0.0	(0.0)	0.0	(0.0)
	Asteroidea	3.3	1.3	(38.5)	1.0	(30.8)	1.6	0.8	(50.0)	0.4	(25)
	Ophiuroidea	0.3	0.0	(0.0)	0.0	(0.0)	2.2	2.2	(100.0)	2.2	(100.0)
	Echinoidea	4.0	2.5	(62.5)	2.5	(62.5)	4.0	3.0	(75.0)	3.0	(75.0)
	Brachyura	7.5	1.5	(20.0)	1.5	(20.0)	6.6	0.2	(3.0)	0.0	(0.0)
	Anomura	1.5	0.3	(16.7)	0.0	(0.0)	1.0	0.2	(20.0)	0.0	(0.0)
	Natantia						0.2	0.0	(0.0)	0.0	(0.0)
	Osteichthyes										
	Polychaeta	5.8	0.0	(0.0)	0.0	(0.0)	0.4	0.0	(0.0)	0.0	(0.0)
	Total	**423.5**	**38.5**	**(9.1)**	**35.0**	**(8.3)**	**258.8**	**18.8**	**(7.3)**	**16.8**	**(6.5)**
20 minutes	Bivalvia	662.5	40.3	(6.1)	36.8	(5.5)	158.0	19.3	(12.2)	16.3	(10.3)
	Gastropoda	1.3	0.0	(0.0)	0.0	(0.0)	1.0	0.0	(0.0)	0.0	(0.0)
	Asteroidea	3.3	2.3	(69.2)	0.3	(7.7)	2.7	2.3	(87.5)	0.0	(0.0)
	Ophiuroidea	1.5	0.0	(0.0)	0.0	(0.0)					
	Echinoidea	4.0	3.0	(75.0)	3.0	(75.0)	0.7	0.7	(100.0)	0.7	(100.0)
	Brachyura	9.3	2.0	(21.6)	1.8	(18.9)	13.0	3.0	(23.1)	2.7	(20.5)
	Anomura	1.8	0.3	(14.3)	0.0	(0.0)					
	Natantia										
	Osteichthyes										
	Polychaeta	9.8	0.0	(0.0)	0.0	(0.0)	4.3	0.0	(0.0)	0.0	(0.0)
	Total	**693.3**	**47.8**	**(6.9)**	**41.8**	**(6.0)**	**179.7**	**25.3**	**(14.1)**	**19.7**	**(10.9)**

dividuals with a size superior to the legal minimum (99.6%, 98.8% and 97.8% for tow duration of 5, 10 and 20 min, respectively).

The retention analysis of *C. chione* by both dredges was accomplished by isolating the fraction of the retained individuals below the minimum legal size (MLS – 50 mm) and the fraction of those with size equal or superior to the MLS. It was hoped, in this manner, to enhance the differences between the efficiency of both fishing gears in view of both groups.

Figure 3 represents the average percentage of individuals retained with a size inferior to the MLS, for each of the tow durations. The NDD only retains a small percentage of undersized individuals (<7%). In the TD, the opposite was observed. In fact, the TD retained a higher percentage of undersized individuals

Table 5. Relative vulnerability of the species to dredging

	New dredge design				Mean proportion expected to die
	1	2	3	4	
Bivalvia					
Pharus legumen	0	0	0	3	100.0
Solen marginatus	0	0	0	1	100.0
Thracya papyracea	0	0	0	4	100.0
Lutearia anguistor	1	0	2	24	96.3
Mactra glauca	7	6	2	71	84.9
Ensis siliqua	1	0	0	4	80.0
Acanthocardia aculeata	26	0	4	23	50.9
Mactra coralina	5	0	0	5	50.0
Acanthocardia spinosa	2	0	0	2	50.0
Ensis ensis	12	0	1	10	47.8
Tellina incarnata	70	0	0	6	7.9
Acanthocardia tuberculata	183	0	5	10	7.6
Callista chione	2156	4	18	110	5.6
Dosinia lupinus	51	7	0	3	4.9
Laevicardium crassum	557	3	1	14	2.6
Dosinia exoleta	69	8	0	2	2.5
Venus striatula	2011	0	0	3	0.1
Pandora inaequivalvis	1	0	0	0	0.0
Venus fasciata	3	0	0	0	0.0
Spisula solida	2	0	0	0	0.0
Lutraria lutraria	2	0	0	0	0.0
Gari fervensis	2	0	0	0	0.0
Gastropoda					
Cymbium olla	3	0	0	0	0.0
Gibbula sp.	10	0	0	0	0.0
Nassarius reticulatus	3	0	0	0	0.0
Asteroides					
Astropecten aranciacus	16	13	0	2	6.5
Ophiuroidea					
Ophiocomina nigra	7	0	0	0	0.0
Echinoidea					
Echinocardium cordatum	12	1	0	38	74.5
Sphaerechinus granularis	1	0	0	0	0.0
Anomura					
Pagurus sp.	12	2	0	0	0.0
Brachyura					
Corystes cassivelaunus	0	0	0	1	100.0
Polybius henslowi	49	0	0	17	25.8
Calappa granulata	21	0	0	0	0.0
Polychaeta	97	0	0	0	0.0

	Traditional dredge				Mean proportion expected to die
	1	2	3	4	
Bivalvia					
Pharus legumen	0	0	0	2	100.0
Lutraria anguistor	0	0	0	3	100.0
Mactra glauca	20	6	3	98	79.5
Acanthocardia aculeata	6	0	0	6	50.0
Ensis ensis	1	0	0	1	50.0
Lutraria lutraria	1	0	0	1	50.0
Mactra coralina stultorum	13	1	0	3	18.8
Acanthocardia tuberculata	102	0	0	12	10.5
Dosinia exoleta	26	4	0	2	6.3
Callista chione	804	2	0	31	3.7
Laevicardium crassum	606	5	4	6	1.6
Tellina incarnata	4	0	0	0	0.0
Pandora inaequivalvis	1	0	0	0	0.0
Dosinia lupinus	5	0	0	0	0.0
Venus fasciata	2	0	0	0	0.0
Spisula solda	1	0	0	0	0.0
Acanthocardia spinosa	1	0	0	0	0.0
Venus striatula	552	0	0	0	0.0
Gastropoda					0.0
Gibbula sp.	4	0	0	0	0.0
Nassarius reticulatus	1	0	0	0	0.0
Asteroidea					
Astropecten aranciacus	6	9	2	2	21.1
Ophiuroidea					
Ophiocomina nigra	0	0	0	11	100.0
Echinoidea					
Echinocardium cordatum	5	0	0	22	81.5
Anomura					
Pagurus sp.	5	1	0	0	0.0
Brachyura					
Polybius henslowi	57	3	2	9	15.5
Calappa granulata	17	2	0	2	9.5
Natantia					
Palaemon serratus	1	0	0	0	0.0
Osteichthyes					
Dicologoglossa cuneata	2	0	0	0	0.0
Polychaeta	18	0	0	0	0.0

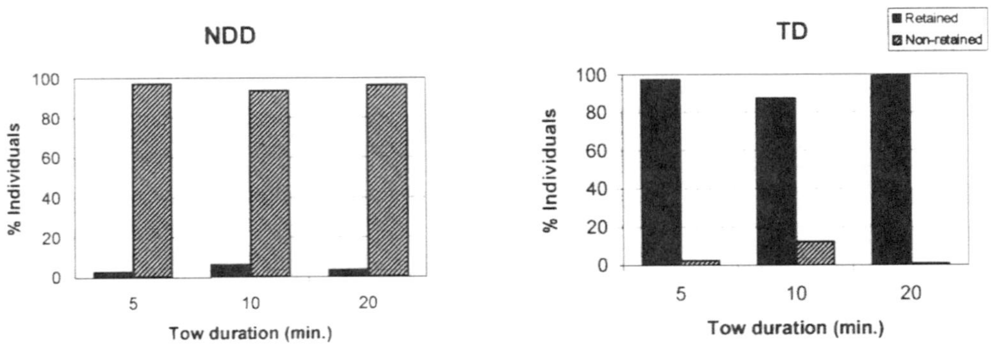

Figure 3. Proportion of individuals retained with length < 50 mm (MLS).

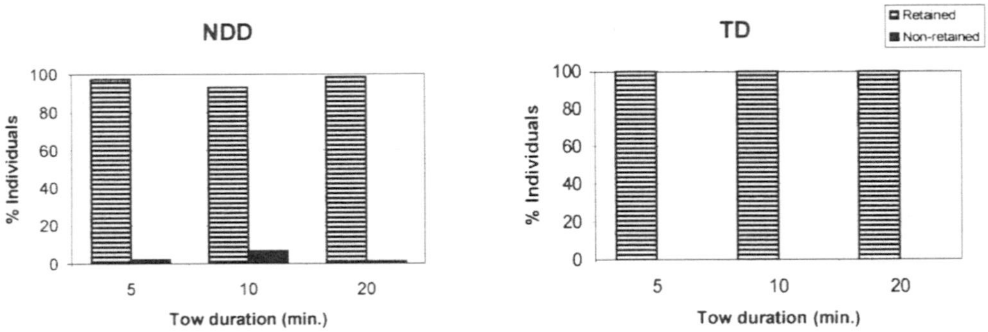

Figure 4. Proportion of individuals retained with length ≥50 mm (MLS).

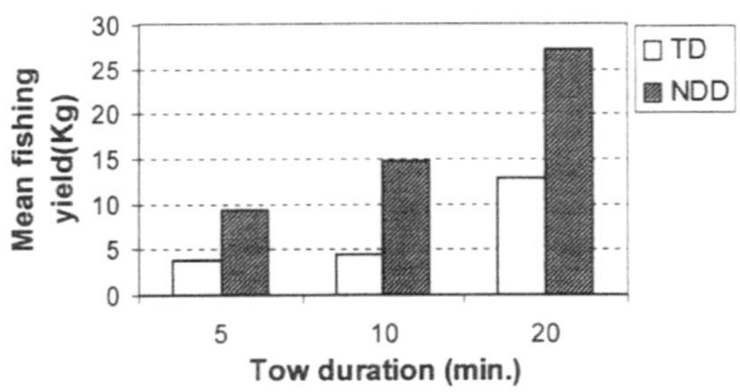

Figure 5. Comparisons of the mean fishing yield (Kg / tow duration) from NDD and TD.

(97.6%, 87.6% and 99.1%, respectively, for 5, 10 and 20 min of tow duration).

The average retention percentage of both dredges with regard to the accessible legal population (Fig. 4) was always high, exceeding 93% retention in the case of the NDD and reaching 100% retention in the case of TD.

The mean fishing yield obtained for both dredges and for 5, 10 and 20 min tow duration is shown in Fig. 5. From its analysis it can be seen that for the same tow duration the mean fishing yield obtained for the NDD is always substantially superior to that obtained by the TD. Results of the *t*-test showed significant differences between the mean fishing yield obtained for the 3 tow duration (*t*-test. 5 min tow: t=3.241, d.f.=6, P=0.018; 10 min tow: t=3.125, d.f.= 6, P=0.017; 20 min tow: t=2.871; d.f.=6, P=0.035)

By-catch

In this fishery, only the target species is retained on board because legislation allows landing of one species per boat each day. Therefore, all the other species caught are discarded. The comparison of the average percentage of by-catch, in numbers and weight, obtained for each dredge and for the 3-tow duration investigated is shown in Figure 6. For 5 and 10 min tow duration, the mean by-catch from the NDD is substantially lower than those observed for the TD, both in terms of number and weight. Surprisingly, for 20 min tow duration the mean by-catch obtained for the TD both in number and weight was similar to that obtained for the NDD. It is important to emphasise that within the beds of *C. chione* there exist extended patches of *V. striatula* that are characterised by a low

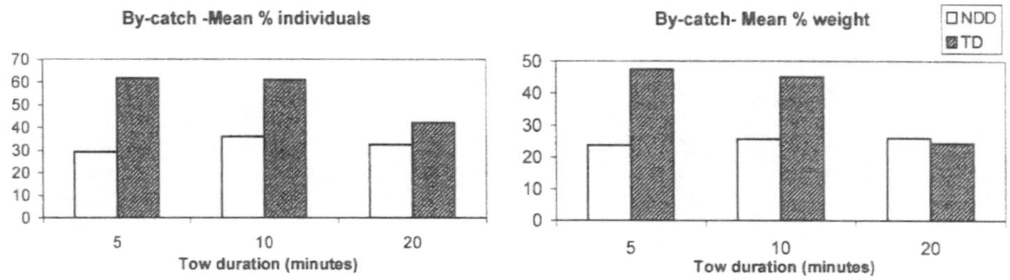

Figure 6. Comparisons of the mean proportion of by-catch in number and weight from NDD and TD, for the 3 tow duration investigated.

bio-diversity. We believe that this feature explains the reduction of the by-catch observed in the TD in the 20 min tow. In fact, for a single tow, the dredge was towed over one of these patches and caught a high number of juveniles of *V. striatula* clams that passed through the primary bag. As a consequence, the percentage of by-catch obtained for this tow was very low. If we exclude this single tow, it can be observed that the mean percentage of the by-catch remains almost constant regardless of tow duration. Thus, when using the NDD a percentage of by-catch of 31% and 25% in terms of number and weight can be expected and 61% and 48%, respectively, in number and weight, when the TD is used.

The statistical analysis carried out showed that, within each dredge, the tow duration has no effect on the percentage of the by-catch (K-W, number H=0.269, d.f.=2, P_{est}=0.874, P_{exact}=0.913; ANOVA, weight $P>0.05$). Results of the t-test showed significant differences between the mean percentage of by-catch obtained for the 2 dredges, both in number and in weight (t-test, number t=–3.411, df=22, P=0.003; weight t=–2.653, df=22, P=0.015).

Indirect mortality

Figure 7 graphically shows the mean percentage of damaged and undamaged *C. chione* clams obtained during the fishing experiments for 5, 10 and 20 min tow duration and for the dredges assayed. For both dredges, the percentage of clams damaged was very low, never exceeding 9% in the case of the NDD and 6% in the case of TD. For this species and for both dredges the percentage of damaged clams did not vary with tow duration (ANOVA, $P>0.05$). Similarly, there was no significant difference between of dredge designs in the percentage of *Callista chione* specimens damaged (t-test, t=0.971, d.f.=22, P=0.342).

Regarding the impact of this type of fishing upon the macrobenthic community (Table 4), in terms of percentage of damaged individuals (scores 2–4), it was observed that the most affected *taxa* in the area where the study was conducted was, in decreasing order of importance, the echinoidea, asteroidea and brachyura.

In a comparative analysis of the influence of both dredges in the percentage of individuals damaged, not many differences were found. It is worth to underline that for all *taxa,* with the exception for Bivalvia, the mean number of individuals caught was considerably low. Nevertheless, the TD damaged a greater percentage of sea urchins and sea stars than the NDD (Table 4). The results showed that for the NDD the average percentage of damaged individuals was 6.3%, 9.1% and 6.9% for tow duration of 5, 10 and 20 min, respectively, whilst for the TD the mean percentage of damaged individual observed was 8.6%, 7.3% and 14%, respectively, for 5, 10 and 20 min tow duration.

Regarding the influence of tow duration on the percentage of overall damaged individuals, the results of the statistical tests do not reveal significant differences in both dredges (ANOVA, $P>0.05$). The dredge design also doesn't have any effect on the percentage of damaged individuals (t-test, t=1.929, d.f.=22, P=0.067).

In the present work, a survival study was not carried out. Thus it was assumed that all individuals severely damaged (score 3) died and that all individuals lightly damaged (score 2) survived. Therefore, the impact produced by both dredges on the macrobenthic community was assessed considering only the individuals scored as 3 or 4. The analysis of Table 4 showed that for the NDD the average percentage of dead individuals was 6.0%, 8.3% and 6.0% for tow duration of 5, 10 and 20 min, respectively. For the TD, the mean percentage of damaged individual observed were 7.9%, 6.5% and 10.9%, respectively, for 5, 10 and 20 min tow duration. The comparison of mortality produced by both dredges revealed that the dredge design doesn't influence the number of dead in-

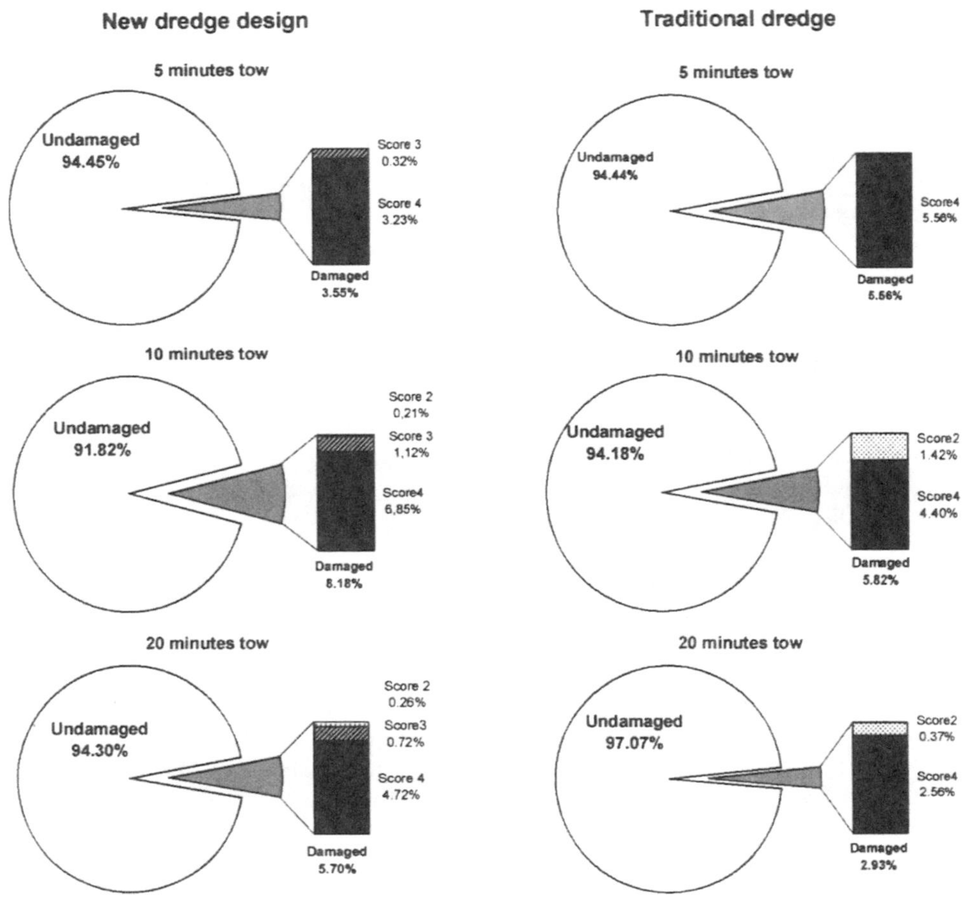

Figure 7. Relative proportion of damaged individuals of *Callista chione* for the two dredges and for the 3 tow duration investigated.

dividuals (Mann–Whitney, T=169, n_{small}=12, n_{big}=12, P=0.285).

For each dredge, data from all tows was pooled in order to evaluate the relative vulnerability of the species that comprises the by-catch to dredging (Table 5). As it was expected, the most sensitive species to this kind of fishery are the thin shelled bivalves such as *Pharus legumen* (Linnaeus, 1758), *Ensis ensis* (Linnaeus, 1758), *Solen marginatus* (Pennant, 1777), *Ensis siliqua* (Linnaeus, 1758), *Mactra glauca* Born, 1778 and *Lutraria anguistor* Phillipi, 1844, and the sea urchin *Echinocardium cordatum* (Pennant, 1777). Little or no damage occurs to the solid shelled bivalve such as *C. chione*, *L. crassum*, *Spisula solida* (Linnaeus, 1758) *V. striatula*, *Dosinia exoleta* (Linnaeus, 1758), *Dosinia lupinus* (Linnaeus, 1758), and *Acanthocardia tuberculata* (Linnaeus, 1758). The mean proportion expected to die is also very low for the gastropods *Cymbium olla* (Linnaeus, 1758), *Gibbula* sp. and *Nassarius reticulatus* (Linnaeus, 1758), for the large sea star *Astropecten aranciacus* (Linnaeus, 1758), for hermit crabs *Pagurus* sp. and for the crabs *Polybius henslowi* Leach, 1820 and *Calappa granulata* (Linnaeus, 1758).

Discussion

The catch composition of a species is conditioned by the structure of its population in the location where fishing is exercised and by the selectivity of the fishing gear. By this means, the comparative analysis of the clam composition coming from both dredges must be done, on its first approach, by its length frequency distributions retained in the net bag. In regard to the retention capacity of gears, one must particularly consider the percentages of individuals retained with size inferior to the minimum legal and, on the other hand, the percentages of individuals retained with size superior to the minimum legal.

The comparison of the length frequency distribution of *C. chione*, from the TD and the NDD, showed that the captures from the former dredge were composed of a great fraction of juveniles, while the captures from the NDD were composed, almost in its whole, of individuals with size superior to the minimum legal size. Considering the clams captured exclusively with size inferior to the minimum legal (50 mm), it was observed that the NDD retained a small percentage of these individuals, whilst the TD retained a higher percentage. The difference obtained in percentage of undersized individuals retained by both dredges is obviously related with its selectivity. Therefore, from the results obtained for the TD, it can be concluded that the mesh size that has been used in the *C. chione* fishery, 50 mm, is not the most adequate for the exploitation of this resource. In fact, the selectivity studies conducted by IPIMAR concluded that the mesh size used in the exploitation of this species must not be inferior to 70 mm.

Although the selectivity of the TD may be improved, becoming identical to the NDD, by increasing the mesh size of the bag, the way that the dredge operates during the tow must be taken in consideration. The underwater observations carried out, allowed to verify that, during the tow, the individuals that are not damaged and that pass through the parallel rods of the NDD grid are buried immediately (in the case of the infauna) or recover their activity (in the case of epifauna). In the case of the TD, during the tow, the mesh of the net bag closes preventing that most of the individuals with smaller dimension escape. In this way, the gear only turns truly selective during the operation of reversing the gear, namely, when the net bag is being washed. In the case of bivalves, the fact that individuals are retained for a long time in the bag makes them rotate a lot, increasing the stress in which they are submitted. As a consequence, the bivalve that escape through the net bag don't bury themselves immediately, being subject to predation. The predation of exposed invertebrates was reported by several authors (e.g. Caddy, 1968; Arntz & Weber, 1970; Meyer et al., 1981; Michael et al., 1990; Kaiser & Spencer, 1994; Gaspar, 1996). Meyer et al. (1981) found that 80% of the *Spisula solidissima* dislodged by hydraulic dredges reburied within 2 hours. Michael et al. (1990) have shown that almost all of the clams *Paphies donacina*, *Spisula aequilateralis*, *Dosinia anus*, *Mactra discors* and *Mactra murchiosini* were reburied after 20 min. Meyer et al., *op. cit.* stated that 3% of the undamaged clams were eaten by predators. Ramsay & Kaiser (1998) reported a high rate of predation of undamaged whelks by starfish in an area disturbed by scallop dredging. Their laboratory studies demonstrated that whelks that have been in contact in bottom gears and which have been rolled, took longer to straighten themselves and were less likely to have an escape response than whelks that had not been under the influence of the gear.

The difference of the mean fishing yield obtained in this experiment reflects a higher efficiency of capture of the NDD. This greater efficiency is a result of the fact that the sand that enters the dredge is rapidly filtered by both grids (Fig. 8). Therefore, all the clams that are found on the dredge path enter the dredge. In the case of the TD, the closing of mesh of the net during the tow prevents the sand from getting out through the mesh, blocking rapidly the dredge mouth (Fig. 9). As a consequence, part of the clams that are found on the dredge path are pulled out of the dredge's mouth decreasing, therefore, its efficiency of capture.

Despite the higher efficiency of capture of the NDD in comparison to the TD, the proportion of by-catch was significantly reduced when the NDD was used in the fishery. This is an important result since damage may occur during sorting on board of the vessels, before the animals are returned to the sea. Furthermore, if sorting times are long or conditions on deck are unfavourable, the survival of by-catch decreases (Medcof & Bourne, 1964). According to Fonds (1994), the survival chances of flatfish and roundfish discarded are almost nil. For non-commercial invertebrates, however, part of the discarded by-catch may survive. Both starfish and brittlestars autotomise their arms as an escape response to predators or as a result of gear damage, subsequently regenerating their arms. Hence the survival of both orders are expected to be high. According to Kaiser & Spencer (1995), nearly 100% of the common starfish, *Asterais rubens* and nearly 80% of brittlestars, *Ophiura ophiura*, survive to captures in the nets of beam trawls. Similarly, Bergman et al. (1990) refer that common starfish have a high chance of survival after returning to the sea. Survival among undamaged juveniles may also be low if the time spent out of water on the deck is long, especially when air temperatures are high (Gaspar & Monteiro, 1999).

Other important source of fishing mortality is damage caused to uncaught animals on the bottom. In

Figure 8. Underwater photograph of the NDD during the tow.

Figure 9. Underwater photograph of the TD during the tow.

clams, shell damage can result from animals passing under the dredge. These clams may be forced into the bottom and the mantle cavity packed with sediment which may damage the soft tissues (Caddy, 1968). Damage also occurs on the benthic organisms that are caught by the dredge, due to the mass of debris or other fauna in the net or during their escape through the meshes of the net (Bergman & Hup, 1992; Gaspar, 1996).

The study of indirect mortality carried out in the present work aimed to evaluate the influence of tow duration and the type of dredge on the number of individuals damaged both on the target species and on the macrobenthic community. Our results showed that for both dredges, tow duration didn't have any effect on the proportion of damaged individuals, neither on the target species nor on the macrobenthic community in general. Nevertheless, the influence of tow duration on the percentage of damaged individuals may be significant if the community is mainly comprised by vulnerable species. Gaspar et al. (1998b) reported that for *E. siliqua* fishery, the number of damaged individuals increases dramatically with tow duration.

The results of Gaspar et al. (1998b) on the effect of tooth length on the number of damaged razor clams (*E. siliqua*) showed that damage to these stocks is inversely proportional to catching efficiency (defined as the percentage of clams on the dredge path that entered the dredge). Although it was observed that the efficiency of capture of the NDD is higher than the TD, the proportion of damaged individuals both of the target species and other macrobenthic species was similar. This result indicates that the way that individuals are retained by the dredge doesn't influence the number of damaged individuals. It is important to point out that we have no information on the proportion of damaged individuals that have come within the influence of the dredge but that weren't caught. Nevertheless, we believe that one of the most important features of dredges is tooth length. The effect on bottom species depends on the action of fishing gear in relation to their vertical distribution. In this experiment, the tooth length used on both types of dredges was the same and, therefore, the penetration on the sediment was presumably the same, resulting on the same proportion of damaged individuals.

Our results showed that mortality induced by both dredges on the macrobenthic community was very low. Indirect fishing mortality has been found to be important in some dredge fisheries. Medcoff & Caddy (1971) reported that 80% of the uncaught clams were damaged while only 20% of the individuals caught had the shells broken. Gruffydd (1972) estimated that the Manx scallop dredge efficiency was about 15% and that death due to natural mortality and indirect fishing mortality could reach 56.5%. He also reported that from the damaged undersized scallops caught and thrown back into the sea, between 50 and 75%, would die. Meyer et al. (1981) conducted a study in order to estimate the efficiency of the hydraulic dredge used in the fishery of *S. solidissima*. These authors observed that when the dredge performance was low about 92% of the larger clams were damaged. On the other hand, when the efficiency was high the mortality decreased to 30%. Naidu (1988) estimated, on the basis of dead shell: live shell ratio, that mortality on the fishery of *Chlamys islandica* was 4–8 times higher as a result of the encounter with the fishing gear than mortality from natural causes. McLoughlin et al. (1991) found that the catching efficiency of the Australian scallop dredge was very low. They estimated that only 12% of the scallops in the tow path were caught. These authors reported that 12–22% of the initial stock in Banks Strait was landed as catch being the rest of the stock loss through direct and indirect mortality resulting from dredging. Michael et al. (1990) compared two hydraulic dredges to catch subtidal clams. Despite the high efficiency, they found that nearly 38% of the clams caught, although they hadn't shell damage, had severely or partly damaged feet. Lambert & Goudreau (1996) studied the harvesting efficiency of the New England hydraulic dredge used to catch Stimpson's surf clams (*Mactromeris polynyma*). They estimated a dredge efficiency of 90% for large clams. These authors observed that two-thirds of the clams that remained on the bottom were damaged which represents less than 10% of capturable clams. A small percentage of other species was also damaged. From the clams harvested, nearly 20% were damaged. Despite of the high efficiency of the hydraulic dredges, the percentage of damaged individuals is higher than the observed in the present study. The mortality associated to hydraulic dredges may be related with water pressure, blade height or tow speed (Kauwling & Bakus, 1979; Lambert & Goudreau, 1996).

The ecological impact of the dredge fishery on shallow sandy bottom is not as deleterious as on hard bottoms or on deeper water. In fact, on shallow waters, with a highly dynamic seabed environment resulting from natural physical disturbances, the infaunal communities are dominated by short-lived species adapted to frequent physical disturbances such as wave and tidal currents. Hence, the impact of dredging on this areas may be limited. Nevertheless, it can be expected the loss of some species in the community, usually the larger-bodied species living buried in the sand. The effects on biodiversity are most severe where natural physical disturbance is least prevalent, where storm-wave damage is negligible and biological processes tend to be slower (Watling & Norse, 1998; DeAlteris et al., 1999). Bergman & Hup (1992) studied the direct effects of beamtrawling on macrofauna in a sandy sediment in the North Sea. These authors observed that the density of a number of species of echinoderms, polychaetas and molluscs decreased 10–65% due to trawling activity. Similarly, Kaiser & Spencer (1996) concluded that beam-trawling caused significant short-term changes in the diversity and abundance of infauna species. Eleftheriou & Robertson (1992) concluded that dredging operations on sandy bottoms have a limited effect both on the physical environmental and on the smaller infauna, but a high impact on some epifauna organisms and larger infauna. From these studies and from our results, it can be concluded that the impact produced by both dredges in the area where this study was conducted is minimal. Although the indirect mortality induced by the dredges assayed was similar, from a management and ecological point of view there are advantages in using the NDD in the *C. chione* fishery. Along the Portuguese coast, this fishery is managed by daily quotas per boat, and fishermen are allowed to land only one species per day. Therefore, if we take in consideration the mean fishing yield obtained for both dredges, it can be concluded that the impact on the macrobenthic community is reduced in about 50% when using the NDD because to reach the daily quota the tow area is reduced to 50%.

As far as what has been exhibited, we consider that the New Dredge Design can be used in the exploitation of the bivalve molluscs. However, being a gear that represents a higher capture efficiency relatively to the Traditional Dredge, we underline the importance of accomplishing the maximum limits of daily capture per vessel stipulated by the current legislation. Otherwise it may contribute to the collapse of the stocks.

Conclusions

Taken into account the previous announced results, we point out the following conclusions. The capture of clams coming from the TD are composed by a great fraction of juveniles, while the NDD are composed almost in its totality by individuals with size superior to the minimum legal size. From the clams captured exclusively with size inferior to the minimum legal (50 mm), it was verified that the NDD retained a small percentage of these individuals (<7%), whilst the TD retained a very high percentage (87.6%). This result indicates that the mesh size used in this fishery isn't the most adequate. By these means, we suggest the adoption of the 70 mm-mesh size in the net bag.

For all 3 tow duration's, the mean fishing yield obtained for the NDD was always higher to the TD, due to its much greater efficiency of capture.

The proportion of by-catch is significantly higher when the TD is used.

The impact produced by both dredges on the target species and as well as on the macrobenthic community (both in terms of the percentage of damaged individuals and dead individuals) was similar and low for the same tow duration.

Considering the fishing yields obtained in the present study and that the *Callista chione* fishery is managed by daily quotas per boat, it can be concluded that the impact on the macrobenthic community is reduced to about 50% when using the NDD. The NDD captures twice the amount of the target species than that caught by the TD. Therefore, it requires less towing area in order to reach the daily quota, which results in a smaller impact on the macrobenthic community.

Another advantage in the use of the NDD relatively to the TD is to allow the small individuals (independently of the species) rapidly escape between the metallic rods of the grid, therefore increasing their probability of survival.

Acknowledgements

The authors would like to thank Dra Teresa Pina and Dra Elza Neto and two anonymous referees whose suggestions greatly improved the manuscript. The authors also thank the staff of the Delegação de Setúbal and the crew of the 'Surpresa' for their skilful handling of the boat and fishing gear. We also would like to thank the fishery organisation BIVALPESCA, namely Mr Carlos Pratas and Mr Rogério Fernandes, for their collaboration in the present work.

References

Arntz, W. E. & W. Weber, 1970. *Cyprina islandica* L. (mollusca, Bivalvia) als Nahrung von Dorschg und Kliesche in der Kieler Bucht. Ber Deut Wissen Komm Meeresforschung 21: 193–205

Bergman, M. J. N., M. Fonds, M. Hup, M. Lewis, P. Van der Puyl, A.. Stam & D. Uyl, 1990. Direct effects of beamtrawl fishing on benthic fauna in the North Sea – a pilot study. In Effects of Beamtrawl Fishery on the Bottom Fauna of the North Sea. BEON-rapport 8: 33–57.

Bergman, M. J. N. & M. Hup, 1992. Direct effects of beamtrawling on macrofauna in a sandy sediment in the southern North Sea. ICES J. mar. Sci. 49: 5–11.

Bucqoy, E., P. Dautzenberg & G. Dollfus, 1882–1898. Les mollusques marins du Rousillon. Tome 2, Pelecypodes. J. Baillière & Fils, Paris: 884 pp.

Caddy, J. F., 1968. Underwater observations on scallop (*Placopecten magellanicus* L.) behaviour and drag efficiency. J. Fish. Res. B Can., 25: 2123–2124

DeAlteris, J., L. Skrobe & C. Lipsky, 1999. The significance of seabed disturbance by mobile fishnig gear relative to natural processes: a case study in Narraganset Bay, RI. In Benaka L. (ed.), Fish Habitat: Essential Fish Habitat and Rehabilitation. American Fisheries Society: 224–237.

Eleftheriou, A. & M. R. Robertson, 1992. The effects of experimental scallop dredging on the fauna and physical environment of a shallow sandy community. Neth. J. Sea Res. 30: 289–299.

FAO, 1987. Fiches FAO identification des especes pour les besoins de la peche. (Révision 1). Mediterranee et Mer Noire. Zone de peche 37. In: Fischer, W., M. Schneider & M.-L, Bauchot (eds), Vol I, Vegetaux et Invertebres. Rome, FAO: 371–512.

Fonds, M., 1994. Mortality of fish and invertebrates in beamtrawl catches and the survival changes of discards. In: Groot S. J. & H. J. Lindeboom (eds), Environmental Impact of Bottom Gears on Benthic Fauna in Relation to Natural Resources Management and Protection of the North Sea. NIOZ Report 1994-11, RIVO-DLO Report C026/94.

Gaspar, M. B., 1996. Bivalves do litoral oceânico algarvio. Aspectos da biologia, ecologia e da pescaria dos mananciais de interesse económico: aplicação à gestão dos recursos. PhD thesis, Universidade do Algarve, Faro.

Gaspar, M. B. & C. C. Monteiro, 1999. Indirect mortality caused to juveniles of *Spisula solida* due to deck exposure. J. mar. biol. Ass. UK, 79 (3): 566–568

Gaspar, M. B., M. Castro & C. C. Monteiro, 1999. Effect of tooth spacing and mesh size on the catch of the Portuguese clam and razor clam dredge. ICES J. mar. Sci., 56: 103–110.

Gaspar, M. B., M. D. Dias & C. C. Monteiro, 1998a. A pescaria de bivalves do litoral oceânico da região de Setúbal: situação actual dos principais bancos (Junho de 1997). Rel. Tec. Cient. IPIMAR, Lisboa: 12 p.

Gaspar, M. B., M. Castro & C. C. Monteiro, 1998b. The influence of tow duration and tooth length on the number of damaged razor clams, *Ensis siliqua*. Mar. Ecol. Prog. Ser. 169: 303–305.

Gruffydd, L. L., 1972. Mortality of scallops on a Manx scallop bed due to fishing. J mar. biol. Ass. U.K. 52: 449–445.

Kaiser, M. J. & B. E. Spencer, 1994. Fish scavenging behaviour in recently trawled areas. Mar. Ecol. Prog. Ser. 126: 31–38.

Kaiser, M. J. & B. E. Spencer, 1995. Survival of by-catch from a beam trawl. Mar. Ecol. Prog. Ser., 112: 41–49.

Kaiser, M. J. & B. E. Spencer, 1996. The effects of beam-trawl disturbance on infaunal communities in different habitats. J. anim. Ecol. 65: 348–358.

Kauwling, T. J. & G. J. Bakus, 1979. Effects of hydraulic clam harvesting in the Bering Sea. Tetra Tech. Report TC3324: 183 pp.

Lambert, J. & P. Goudreau, 1996. Performance of the New England hydraulic dredge for the harvest of Stimpson's surf clams (*Mactromeris polynyma*). Can. Ind. Rep. Fish. aquat. Sci. 235: 28 p.

McLoughlin, R. J., P. Young, R. Martin & J. Parslow, 1991. The Australian scallop dredge: estimates of catching efficiency and associated indirect fishing mortality. Fish Res. 11: 1–24.

Medcof, J. & N. Bourne, 1964. Causes of mortality of the sea scallop *Placopecten magellanicus*. Proc. Natl. Shell. Assoc. 53: 33–50.

Medcof, J. & J. F. Caddy, 1971. Underwater observations on performance of clam dredges of three types. ICES C.M. 1971/B:10.

Meyer, T. L., R. A. Cooper & K. J. Pecci, 1981. The performance and environmental effects of a hydraulic clam dredge. Mar. Fish. Rev. 43 (9): 14–22.

Michael, K. P., G. P. Olsen, B. T. Hvid & H. J. Cranfield, 1990. Design and performance of two hydraulic subtidal clam dredges in the New Zealand. New Zealand Fis. Tech. Rep. 21: 16 p.

Naidu, K. S., 1988. Estimating mortality rates in the Iceland scallop, *Chlamys islandica* (O.F. Muller). J. Shell. Res. 7: 61–71.

Poppe, G.T. & Y. Goto, 1993. European Seashells. Vol II (Scaphopoda, Bivalvia e Cephalopoda). Verlag Christa Hemmen, Germany: 221 pp.

Ramsay, K. & M. J. Kaiser, 1998. Demersal fishing disturbance increases predation risk for whelks (Buccinum undatum L.). J. Sea. Res. 39: 299–304

Sokal, R. R. & F. J. Rolhf, 1981. Biometry. 2nd edn. W.H. Freeman & Company. New York: 859 pp.

Tebble, N., 1966. British Bivalve Seashells. A Handbook for Identification. British Museum. Edinburgh: 212 pp.

Watling, L. & A. Norse, 1998. Disturbance of the seabed by mobile fishing gear: a comparison with forest clear-cutting. Cons. Biology 12 (6).

Zar, J. H., 1996. Biostatistical analysis. 3rd edn. Prentice-Hall, Inc. USA: 662 pp.

Hydrobiologia **465:** 169–173, 2001.
G. Burnell (ed.), Coastal Shellfish – A Sustainable Resource.

Preliminary appraisal of an innovative hydraulic dredge with vibrating and sorting bottom on clam beds (*Chamelea gallina*)

E. Rambaldi, M.L. Bianchini[1], G. Priore, G. Prioli, N. Mietti & T. Pagliani
Consorzio Mediterraneo s.c.r.l., Via Nazionale 243, 00184 Roma, Italy
[1]*Soc. Porto Romano S.p.A. and Consiglio Nazionale Ricerche (CNR), 00100 Roma, Italy*

Key words: hydraulic dredge, clam, *Chamelea gallina*, sorting, product quality

Abstract

To solve some of the efficiency and environmental problems related to the use of the hydraulic dredge in bivalve mollusc fishing, a new experimental gear with vibrating bottom grid and other technical changes has been tested on clam (*Chamelea gallina*) beds in the Adriatic Sea. Comparative fishing surveys indicate a significantly different selectivity of the vibrating dredge, with respect to a standard gear: in fact, undersize clams are sieved out during the fishing process, and almost no juveniles are hauled. Regarding the product quality, laboratory analyses show that the amount of intervalvar sediment is significantly lower in the catch from the modified dredge, due to a sort of 'warning' device. Nevertheless, the larger number of damaged shells suggests that the vibrating grid subjects the clams to a greater mechanical stress than the standard gear. As for the environmental effects, the vibrating bottom is selective for the associated fauna too, as is shown by the higher mean weight of most the by-catch species in the experimental gear. Moreover, the riddling goes on continuously, allowing the immediate release of the sorted-out organisms, which are repositioned in the area of origin, thus avoiding a 'contagious' distribution. In conclusion, these positive preliminary results suggest real benefits of this innovative modified dredge, with wide margins for further improvement.

Introduction

The collection of bivalve molluscs, and particularly that of the striped clam *Chamelea gallina*, carried out by boats provided with hydraulic dredges, is a fishery of great economic importance on the Italian side of the Adriatic Sea (Froglia, 1989; Ferretti et al., 1990). Landing records show strong yearly fluctuations, however a clear decrease in the density of this resource is evident, associated with a reduction in the number of clams over 25 mm length, the minimum legal commercial size (Costa et al., 1987).

To solve some of the efficiency (Lombardi et al., 1996) and environmental (Vaccarella et al., 1994; Pranovi & Giovanardi, 1994; Dayton et al., 1995) problems that arise from the use of the hydraulic dredge in the clam fishery, an improved gear has been tested, modified to decrease the juvenile catch and to enhance the commercial quality of the take.

Methodology

The modified dredge cage is the usual box of steel bars (12 mm apart, parallel to the towing direction), but with the bottom partially disjointed and free to vibrate during the fishing action, via a cam system of opposing springs; the displacement is 1–2 cm at about 10 hertz. This way, the bottom grid acts as a sifter, allowing the juveniles and the undersized bivalves to escape and to resettle at once (Fig. 1).

Furthermore, the modified gear mouth presents a series of forward-oriented, low-pressure nozzles that alarm the clams, allowing an unstressed fastening of the shell prior to impact with the dredge, and the consequent reduction of intervalvar sediment.

The field experiments were carried out in March 1998, on three transects (A, B, C) north of the port of Giulianova (Italy, Central Adriatic Sea); transects were perpendicular to the coast line, 300 m apart, and two sampling positions (depth 1=3 m; depth 2=5 m) were selected on each transect.

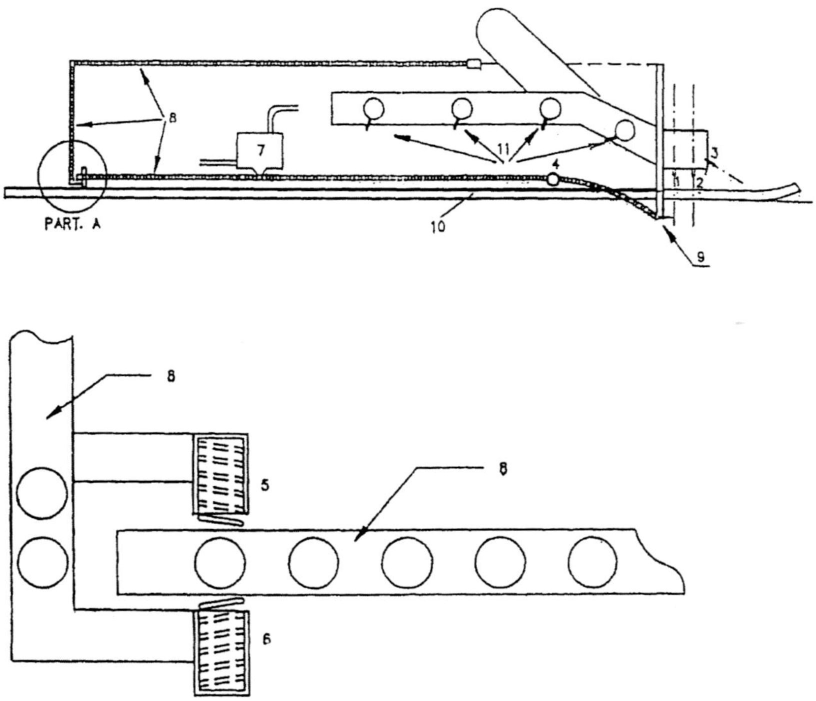

Figure 1. Longitudinal section of the vibrating dredge and detail of its bottom (1–2 breaking-through nozzles; 3 alarming nozzles; 4 hinge; 5–6 opposing springs; 7 vibrating device; 8 rigid frames; 9 sediment-penetrating blade; 10 lateral runner; 11 rinsing nozzles).

There was no significant difference between locations, so, all data were pooled to increase the sample size. Sampling was carried out by pairing two boats fishing side by side. One towed the modified vibrating gear, the other a traditional dredge, in standard use by Giulianova's fishers.

In the laboratory, samples of each transect take were subsampled as necessary, sorted, checked and measured; the intervalvar sediment was estimated (after external washing) on 1 kg of commercial-size clams, manually opened and carefully rinsed in a graduated glass cone, by measuring the volume of deposited material after 1 h; the by-catch was screened, the species were taxonomically determined and the individuals measured.

The differences in length–frequency distributions of *Chamelea gallina*, in intervalvar sediment amounts, in shell damage and in by-catch in numbers and biomass was analyzed using parametric (t-test) and non-parametric statistical tools, such as the sign test, the runs test, the Kolmogorov–Smirnov test, the Mann–Whitney test (Zar, 1984).

Table 1. Summary of measures of the *Chamelea gallina* catches, by type of gear (pooled subsamples)

	Standard	Vibrating
n	1800	1800
Total weight (g)	3410	7560
Damaged shells (%)	3.8	12.1
Clams >25 mm of shell length	283	827
Commercial catch (%)	15.7	45.9
Mean shell length (mm)	1.71	2.43
Standard deviation of shell length	0.59	0.35
Median shell length (mm)	1.48	2.48

Results and discussion

Catch analysis

The comparative fishing surveys (pooled) pointed out highly significant differences in selectivity between the two dredges (Kolmogorov–Smirnov $D_{max} = 0.633$, $D_{0.01,60} = 0.207$); even the individual trends (Box plots, not shown) on the sizes distribution of striped clam (*Chamelea gallina*) samples, taken before the on-board mechanical sorting, show that many more

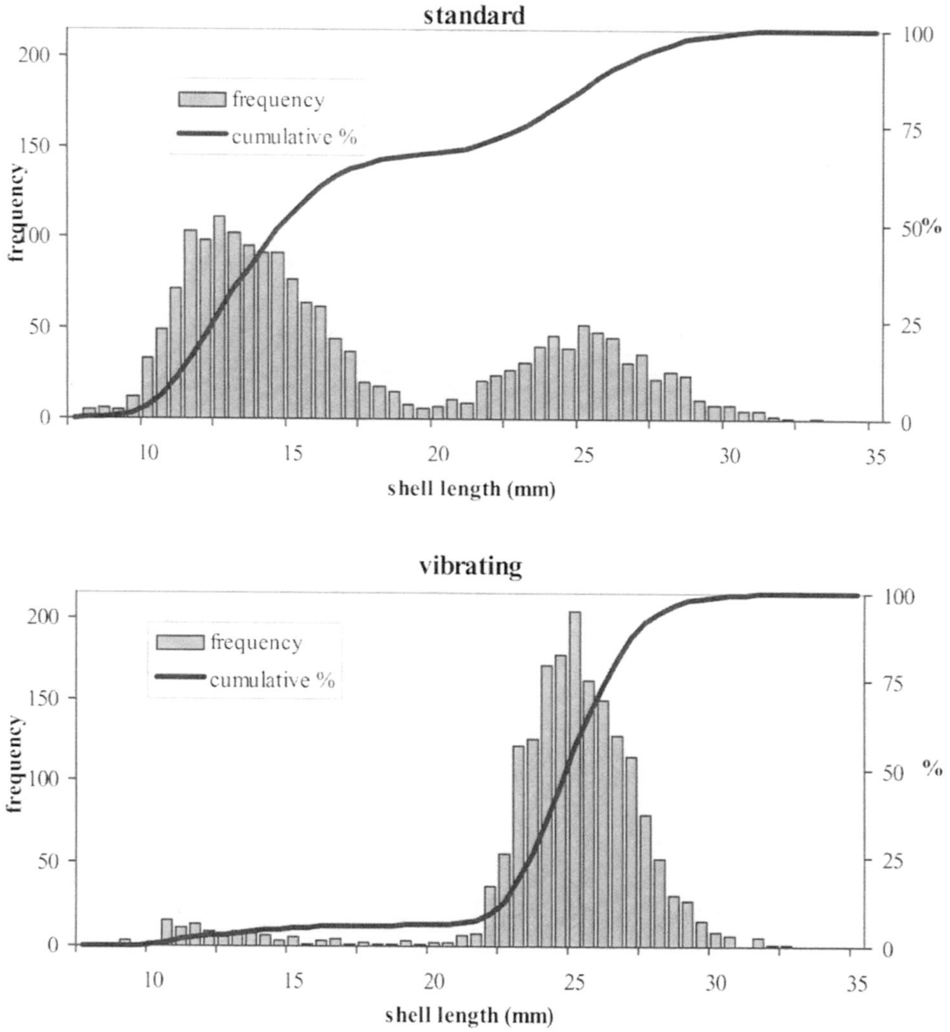

Figure 2. Length frequency distribution (absolute frequency and cumulative percent) of the clams, *Chamelea gallina*, collected with (A) the standard dredge or (B) the vibrating dredge.

undersized clams are taken with the classical gear compared to the improved one. The cumulative percent length frequency distributions of the catch (Fig. 2) show that the standard gear hauls a very low proportion (15.7%) of specimens over the legal limit (25 mm), while the figure is clearly higher (45.9%) using the vibrating dredge. Moreover, with the modified dredge, very few juveniles are taken: indeed, only 10% of the catch is of a size below 22 mm (juveniles), a proportion that reaches 70% in the traditional fishery. The main measures of the clam sample (pooled subsamples) by type of gear are summarized in Table 1.

Intervalvar sediment

The amount of bottom sediment retained inside the shell was estimated in the laboratory for commercial-size clams (Fig. 3). The volume of intervalvar sediment in the clams collected with the standard dredge showed considerable variability (mean 2.1 ml/kg of product; standard dev. 1.29), while those from the vibrating gear were more uniform (mean 1.0 ml/kg of product; standard dev. 0.13) and, with a single exception in the six hauls, were significantly lower (*t*-test $P = 0.045$; sign test $P_{0.05,5} = 0.03$). The reduction of internal mud and sand, notwithstanding the small

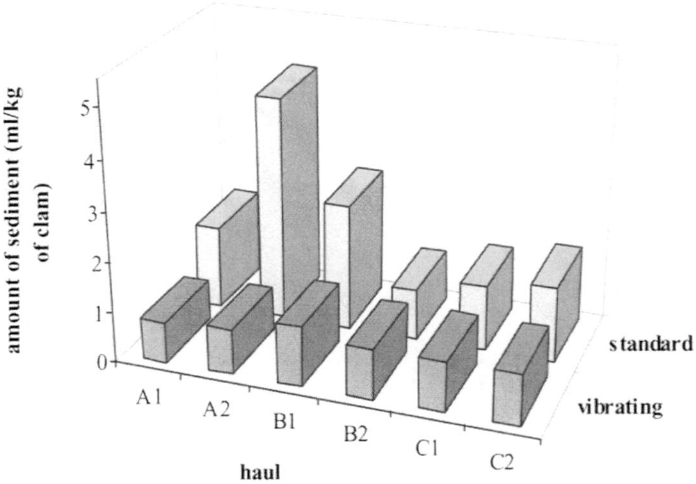

Figure 3. Sediment contamination of the clams, *Chamelea gallina*, collected with the standard dredge or the vibrating dredge (by haul; sediment in ml / product in kg). A, B and C are separate transects. 1 = 3 m depth; 2 = 5 m depth.

sample size, permits a positive evaluation of the experimental 'warning' device in relation to the quality improvement of the commercial yield.

Damaged shells

The proportion of clam shells with obvious damage (resulting from the fishing process) was calculated. A greater number of broken valves occurred in the vibrating dredge samples, with an overall value of 12.1% vs. 3.8% compared to the traditional gear (averages of 6 samples); notwithstanding the deletion of an outlying haul, the mean value for the experimental gear remains high at 8.6%. This drawback requires further examination of the vibrating apparatus, in order reduce the mechanical stress on the catch without losing its selectivity; in any event, ruptures seem to be the result of 'pinching' (due to clashing edges of box frames), and not due to 'shaking'.

By-catch

The mean weight of almost all the various associated species (Table 2) was higher in the modified gear samples (sign test $V = 2$; critical value $V_{0.05,12} = 2$); the sign test performed on the complete matrix (all species considered on every haul) demonstrates that, globally, the phenomenon is even more significant ($V = 27$; critical value $V_{0.001,89} = 31$). These results suggest that the selectivity improvement that arises from the vibrating grid acts also on the by-catch organisms. Assuming that many by-catch species are quite stout and that the 'shaking' is not too violent, it is reasonable to hypothesize a reduced environmental impact of the experimental dredge on the benthic communities living on the fishing grounds. This is based upon the fact that since the riddling goes on continuously and allows the immediate release of the sorted-out clam juveniles and discarded fauna, these will be immediately repositioned in the area of origin, and not stressed by spending long periods on deck.

Conclusions

In summary, these first trials with the vibrating dredge have already shown at least three benefits: a better size selectivity allowing the escape of undersized clams; a reduction in number of the associated by-catch; an enhancement of the quality of the commercial product.

Future research, based on a suitable number of field trials, should verify and validate the preliminary results and improve and optimize the technical aspects of the dredge, especially in order to minimize the damage to the catch; other aspects should also be examined, such as the well-being of the resettled juveniles, the comparison of the commercial yields and the economic efficiency of the new gear.

The aim of the new gear should not be to become more efficient *in toto*: the economic productivity can be raised (less labour and improved quality of the catch), but the production may even be lower, since the fishing process is probably slower and many

Table 2. Associated by-catch species in the *Chamelea gallina* fishery, by type of gear (only presence/absence is annotated for rare species): the sign test is significant ($P < 0.05$)

Species	Group	Standard dredge		Vibrating dredge		Difference in mean weight
		Mean number per haul	Mean individual weight (g)	Mean number per haul	Mean individual weight (g)	
Diogenes pugilator	(decapod)	189	2.3	110	2.5	+
Sphaeronassa mutabilis	(gastropod)	128	1.7	23	2.4	+
Hinia reticulata	(gastropod)	102	2.5	65	2.8	+
Liocarcinus vernalis	(decapod)	96	6.0	53	7.0	+
Spisula subtruncata	(bivalve)	82	1.1	18	1.3	+
Neverita josephina	(gastropod)	81	3.3	33	3.7	+
Scapharca inaequivalvis	(bivalve)	28	7.6	37	8.6	+
Tellina planata	(bivalve)	16	2.5	–		(+)
Mactra corallina	(bivalve)	15	8.1	10	8.1	
Acteon tornatilis	(gastropod)	12	0.9	–		(+)
Astropecten jonstoni	(echinoderm)	5	10.9	4	7.9	–
Donax semistriatus	(bivalve)	4	3.1	2	1.8	–
Tellina nitida	(bivalve)	3	3.4	1	3.7	(+)
Acanthocardia tuberculata	(bivalve)	+		+		
Venerupis aurea	(bivalve)	+		+		
Naticarius millepunctatus	(gastropod)	+		+		
Sipunculus nudus	(sipunculid)	+		+		
Parthenope angulifrons	(decapod)	+		+		
Ilia nucleus	(decapod)	+		+		
Macropodia rostrata	(decapod)	+		–		
Corystes cassivelaunus	(decapod)	–		+		

barely-legal (but marketable) clams are not collected at all. In summary, if the sustainability of the fishery increases and the environment benefits from the reduced by-catch (without a corresponding revenue loss for fishermen) this innovative design promises to be a worthy improvement on traditional gear.

Acknowledgements

The authors wish to thank Mr Emidio Carusi and Mr Maurizio Marconi of "Co.Ge.Vo. Abruzzo" for the technical collaboration during the tuning of the vibrating gear and the field trials.

The study was entirely funded by "Consorzio Mediterraneo", a co-operative organization active in the applied research domain through the development and the implementation of advanced projects aimed at strengthening and modernizing Italian fisheries.

References

Costa, C., M. Bianchini, P. Ceccarelli, P. Orecchia, E. Rambaldi & L. Volterra, 1987. Indagine sui molluschi bivalvi di interesse commerciale (telline, cannolicchi e vongole) delle coste della Toscana, del Lazio e della Campania: 1985–1987. Quad. Ist. Idrobiol. Acquacolt. Brunelli 7: 7–58 & 77 tables.

Dayton, P. K., S. F. Thrush, M. T. Agardy & R. J. Hofman, 1995. Environmental effects of marine fishing. Aquatic conservation: marine and freshwater ecosystems 5: 205–232.

Ferretti, M., F. Lombardo & G. Romani, 1990. Metodi di pesca dei molluschi bivalvi: vongolare tradizionali e turbosoffianti. Quaderni I.C.R.A.P. Pesca 1: 157 pp.

Froglia, C., 1989. Clam fisheries with hydraulic dredges in the Adriatic sea. In Caddy, J. F. (ed.), Marine Invertebrate Fisheries: Their Assessment and Management. Wiley & Sons, Inc.: 507–521.

Lombardi, F., E. Rambaldi & C. Brinati, 1996. Catture accessorie e rigetti nella pesca di *Ensis siliqua minor* (L.) con turbosoffiante. Quad. Ist. Idrobiol. Acquacolt. Brunelli 15: 41–157.

Pranovi, F. & O. Giovanardi, 1994. The impact of hydraulic dredging for short-necked clams, *Tapes* spp., on an infaunal community in the lagoon of Venice. Sci. mar. 58: 345–353.

Vaccarella, R., A. M. Pastorelli & G. Marano, 1994. Studio sulla efficienza delle draghe turbosoffianti e loro effetto sulle comunità bentoniche. Boll. malacologico 30: 17–28.

Zar, J. H., 1984. Biostatistical Analysis. 2nd edn. Prentice-Hall, Englewood Cliffs (N.J., U.S.A.): 718 pp.

Hydrobiologia **465:** 175–180, 2001.
G. Burnell (ed.), Coastal Shellfish – A Sustainable Resource.

A preliminary study on the effects of oyster culturing structures on birds in a sheltered Irish estuary

G. Hilgerloh, J. O' Halloran*, T.C. Kelly & G.M. Burnell
Department of Zoology & Animal Ecology, University College Cork, Cork, Ireland
*(*Corresponding author)*

Key words: shellfish aquaculture, trestles, birds, feeding behaviour, oysters, estuary, Ireland

Abstract

This study investigated the species composition, numbers and behaviour of birds in an intertidal oyster culture area in Cork Harbour . These data were compared to a nearby area free of aquaculture within the same estuary in March 1999. Species which occurred in the aquaculture free area were also observed in the trestle-area. The most abundant species were oystercatcher *Haematopus ostralegus*, redshank *Tringa totanus*, dunlin *Calidris alpina*, curlew *Numenius arquata*, black-headed gull *Larus ridibundus* and common gull *Larus canus*. Oystercatcher, curlew, black-headed gull and common gull occurred in significantly lower numbers in the trestle area, while for redshank and dunlin the differences were not significant. The percentage of birds feeding did not differ between the two areas. Oystercatcher, redshank, dunlin and curlew mostly fed in both areas. In contrast, black-headed gull and common gull generally did not feed, but surveyed the area. Whether the trestles were covered by oyster bags or not did not have any effect on the number of birds except for the dunlin. Dunlin were significantly more frequent beneath the trestles with bags compared with those without bags. In general, the percentage of birds feeding did not differ between areas. Interspecies differences occurred with regard to the position occupied by birds at the trestles. Oystercatcher, redshank and curlew spent more time underneath the trestles. Dunlin, black-headed gulls and common gulls did not differ in numbers underneath or on top of the trestles. These preliminary observations at a single time period give some insight as to the potential interactions between shellfish aquaculture and intertidal birds.

Introduction

Irish estuaries are important feeding areas for birds, especially during migration and in winter. Because of its mild climate, Ireland is the wintering area of many waders and ducks breeding in Scotland or in Iceland (Hutchinson, 1989). In Ireland, 56 sites qualify for international recognition for at least one wintering bird species (Shepherd, 1993).

Irish estuaries are also used as stopover sites for migrating birds as Ireland is situated at the margin of migration routes passing from Scandinavia or Siberia to southern Europe and Africa (Shepherd, 1993). Birds originating in Iceland and heading towards similar wintering areas also use Ireland. Certain species are divided between resident overwintering populations and transient migratory populations that overwinter elsewhere. This is the case for example, with the oystercatcher *Haematopus ostralegus*, dunlin *Calidris alpina* and redshank *Tringa totanus* (Hutchinson, 1989), though the resident populations are quite small (Gibbons et al., 1993).

Irish estuaries are also important for shellfish aquaculture which has increased in the last decade. Of special concern are sites which were recognized as especially valuable for birds and which were protected as 'Special Protected Areas' by the European Birds directive 79/409/EEC and by the Habitat directive (92/43/EEC). At some of the sites, aquaculture is permitted although its effects on foraging and resting of inter-tidal birds have not been sufficiently studied.

In Ireland, coastal shellfish culture consists mainly of the farming of oysters (*Crassostrea gigas*), clams (*Ruditapes philippinarum*) and mussels (*Mytilus edulis*). The cultivation methods differ considerably between these species.

Oysters are cultivated in plastic bags which are fixed at about 50 cm above the ground on top of structures called trestles. They are installed at the low water mark and consist usually of double rows which are separated by lanes for a tractor to facilitate husbandry. Oysters are also cultivated on the sea-bottom, from where mature individuals are harvested by dredging. Clams are ongrown intertidally in the substrate using net covered parks while mussels are suspended from floating longlines.

In an intertidal shellfish aquaculture area disturbances can be caused by either husbandry practices or aquaculture structures which cover the feeding area (Kaiser et al., 1998). In addition, the abundance of invertebrate prey species and their biomass may be influenced by aquaculture activities which will subsequently influence the feeding quality of the habitat (Spencer et al., 1996; De Grave et al., 1998). A number of studies have demonstrated clear evidence of organic enrichment by intertidal oyster cultivation causing reduced macrofaunal diversity but increased biomass (Castel et al., 1989; Nugues et al., 1996). However, at a site exposed to strong wave action these effects could not be found (De Grave et al., 1998).

The effect of oyster cultures on estuarine birds does not appear to have been studied in detail. The aim of this study was to undertake a preliminary assessment of the behaviour of birds in an estuary used for oyster culture.

We addressed the following questions:

- whether the avian species composition inside and outside of the oyster culture area was different;
- whether there were quantitative differences in bird behaviour inside and outside of the trestle area;
- whether numbers of birds were different inside and outside the oyster culture area;
- whether it would make a difference whether the oyster structures were covered by oyster bags or not.

Methods

The study was carried out in the Saleen estuary of Johnsbrook, situated in the northeast of Cork Harbour, SW Ireland. It is a sheltered estuary, and forms part of a site which is of international importance (Smiddy et al., 1995). Oyster cultivation is practiced in the mid- to low intertidal region of the estuary. Oysters are grown in mesh bags fastened onto metal frames (trestles) 40 cm in height, 90 cm in width and 3 m in length. The trestles are arranged in long double and triple rows separated by access corridors for tractors. Twinned rows of trestles are positioned perpendicular to the water line. Each bag contains approximately 700 juvenile oysters which grow from 5 to 50 g during the nursery period.

Table 1. The number of scans, period of observation, and time of low water when data were collected at Johnsbrook, Cork Harbour, March 1999

Date	No. of scans	Period	Duration of observations (min)	Low water
March 2	15	11:00–13:20	140	12:08
March 3	16	11:30–14:00	150	12:44
March 5	19	12:10–15:10	180	13:47
March 7	14	13:30–15:40	130	14:49

The aquaculture area covered a mudflat surface of one ha. In March, only the nine double rows of trestles, which were closest to the land as well as two double rows at the water side, were covered by bags: an area of 4500 m^2. The trestles without bags had algae attached to them, mainly *Fucus vesiculosus* and also *Enteromorpha* sp. and *Ulva lactuca*. A reference site without aquaculture structures was established adjacent to the trestle area. The two sites were of comparable size and were located parallel to the low water mark. However, the trestle area always contained water in some shallow depressions. Observations on birds were performed on March 2, 3, 5 and 7 1999 from a road which runs along the northern border of the estuary. A tripod telescope 22–60 $\times$ 80 and 10 $\times$ 40 binoculars were used to collect data. Focal data showing the behaviour of single birds was captured on video.

Birds at the coastline of the study area were counted once before the behavioural observations started (when the site was still covered by water) and once after they had ended (when the study site was covered by water again).

The behaviour of the birds was described as follows:

Feeding: Searching, handling; and

Not feeding: preening, washing, looking, sleeping, swimming, aggression, flying, walking.

Scan samples were taken every 10 min and the behaviour of the birds registered during low tide (Martin

& Bateson, 1986). A total of 64 scans were made as follows: 15 (March 2), 16 (March 3), 19 (March 5) and 14 (March 7) (Table 1) at different stages of the tidal cycle. The data were pooled and percentage of birds involved in different activities at different times was assessed across the tidal cycle. Single oystercatchers and curlews were observed for at least 5 min (focal scan records) to study the feeding rate. Data across all days were pooled for analysis.

Every second observation was included in each analysis to avoid auto correlation. Thus, we always had two samples of 20-min time intervals for each experiment for the same time period. Bonferroni correction of probabilities were applied to many of the tests applied to the trestle area data.

Differences between groups of data were tested by the Mann–Whitney U-test using the SPSS-program (Brosius & Brosius, 1995).

Results

The following species were present in the reference area and trestle area: cormorant *Phalacocorax carbo,* grey heron *Ardea cinerea,* Wigeon *Anas penelope,* oystercatcher *Haematopus ostralegus,* dunlin *Calidris alpina,* curlew *Numenius arquata,* redshank *Tringa totanus,* greenshank *Tringa nebularia,* turnstone *Arenaria interpres,* mediterranean gull *Larus melanocephalus,* black-headed gull *Larus ridibundus,* common gull *Larus canus,* lesser black-backed gull *Larus fuscus,* jackdaw *Corvus monedula*, hooded crow *Corvus corone cornix.*

The most common species were oystercatcher, curlew, redshank, dunlin, common gull and black-headed gull, which were studied in more detail.

Two groups of species could be distinguished with regard to their behaviour. The first group included: oystercatcher and curlew and the second group common and black-headed gull. In the first group, most birds (median 100%) spent their time feeding, in the second group this was 0%. Gulls were mainly looking around at this site (Table 2).

At the reference site, the median number of oystercatcher, redshank and common gull was highest of all studied species and that of dunlin, curlew and black-headed gull the lowest. In the trestle area birds were present in lower numbers (Table 3).

Only in black-headed gull, common gull, curlew and oystercatcher were the differences between sites significant (Mann–Whitney U-test, $P<0.05$, black-headed gull U = 152, n = 64; common gull U = 145, n = 64; oystercatcher U = 848.5, n = 66; curlew, U = 230, n = 66). In contrast the number of redshank and dunlin were not significantly different (Mann–Whitney U-test, $P>0.05$, dunlin U = 500, n = 64; redshank U = 469.5, n = 64). Similarly, in the black-headed gull, common gull, curlew and oystercatcher, the number of birds feeding per ha differed significantly between reference and trestle area (Mann–Whitney U test, $P<0.05$, oystercatcher U = 859.5, n = 66; curlew U = 250.5, n = 64; black-headed gull U = 152, n = 64; common gull U = 145,

Table 2. A comparison of the percentage of birds feeding in the two areas (numbers are given in brackets)

Species	Trestle area (median, n)	Reference area (median, n)
Oystercatcher *Haematopus ostralegus*	100 (22)	100 (33)
Dunlin *Calidris alpina*	100 (13)	100 (11)
Curlew *Numenius arquata*	100 (32)	100 (22)
Redshank *Tringa totanus*	100 (25)	100 (23)
Common gull *Larus canus*	0 (8)	0 (21)
Black-headed gull *Larus ridibundus*	0 (12)	0 (27)

Table 3. Density (numbers/ha) of birds in the reference and in the trestle area

Species	Trestle area		Reference area	
	Median	Range	Median	Range
Oystercatcher *Haematopus ostralegus*	1.95	11.71	8.14	36.06
Dunlin *Calidris alpina*	0	83.92	0	129.1
Curlew *Numenius arquata*	0	3.9	2.33	8.14
Redshank *Tringa totanus*	2.93	17.56	4.07	40.71
Black-headed gull *Larus ridibundus*	0	0.98	2.33	15.12
Common gull *Larus canus*	0	2.93	4.65	47.69

Table 4. Density of feeding birds (numbers/ha) in the study areas

Species	Trestle area		Reference area	
	Median	Range	Median	Range
Oystercatcher *Haematopus ostralegus*	1.95	11.71	6.98	34.89
Dunlin *Calidris alpina*	0.00	8.78	0.00	129.1
Curlew *Numenius arquata*	0.00	3.90	1.75	6.98
Redshank *Tringa totanus*	2.44	17.56	2.91	39.54
Common gull *Larus canus*	0.00	0.98	0.00	2.33
Black-headed gull *Larus ridibundus*	0.00	0.00	0.00	2.33

n = 64), while those of dunlin (U = 459.5, n = 64) and redshank (U = 492, n = 64) did not (Table 4).

There was no significant difference between the percentage of feeding birds of each species in both the comparison and trestle area (Mann–Whitney U-test, $P>0.05$, oystercatcher U = 250.5; n = 25; dunlin U = 67, n = 25; curlew U = 84.5, n = 32; redshank U = 191, n = 48, P=0.02; black-headed gull U = 57, n = 30; common gull U = 42, n = 31).

The feeding rate of the oystercatchers did not differ between trestle and reference area (Mann–Whitney U-test, $P>0.05$, U = 50.5, n = 24). However, within the reference site, the feeding rate was significantly higher when it was exposed for less than 3 h (140–150 min) compared with 3 h or more (180–200 min) (Mann–Whitney U-test, $P<0.05$, U = 47, n = 32).

There were no significant differences in the numbers of birds on top of the trestles between trestle areas with bags and those without bags (Mann–Whitney U-test, $P>0.01$, oystercatcher U = 512; dunlin U = 511; curlew U = 512; redshank U = 512; black-headed gull U = 496; common gull U = 512, n = 64).

There were no significant differences in the numbers of birds beneath the trestles between trestle areas with bags and those without bags (Mann–Whitney test, $P>0.01$, oystercatcher U = 435; curlew U = 450; redshank U = 427; black-headed gull U = 466; common gull U = 512, n = 64 in all species) except for the dunlins ($P<0.01$, U = 368, n = 64) (Table 5). Significantly more dunlins were in the trestle area with oyster bags.

Table 5. Bird density (numbers/ha) in the trestle area with and without bags (this includes birds underneath and on top of trestles)

Species	Trestles with bags		Trestles without bags	
	Median	Range	Median	Range
Oystercatcher *Haematopus ostralegus*	0	17.51	0	10.57
Dunlin *Calidris alpina*	0	137.89	0	79.24
Curlew *Numenius arquata*	0	8.76	0	1.76
Redshank *Tringa totanus*	0	24.08	0	14.09
Common gull *Larus canus*	0	2.19	0	3.52
Black-headed gull *Larus ridibundus*	0	2.19	0	1.76

Table 6. Density of birds (numbers/ha) underneath and on top of the trestles in trestle area (with and without bags)

Species	Under trestles		On trestles	
	Median	Range	Median	Range
Oystercatcher *Haematopus ostralegus*	0.00	17.51	0.00	0.00
Dunlin *Calidris alpina*	0.00	78.79	0.00	137.88
Curlew *Numenius arquata*	0.00	8.76	0.00	0.00
Redshank *Tringa totanus*	2.86	24.08	0.00	0.00
Common gull *Larus canus*	0.00	3.52	0.00	3.52
Black-headed gull *Larus ridibundus*	0.00	3.52	0.00	3.52

The percentage of feeding oystercatchers, curlews and redshanks underneath the trestles did not differ between trestle area with and without bags (Mann–Whitney U-test, $P>0.01$, oystercatcher U = 99, n = 31; curlew U = 14, n = 11; redshank U = 176, n = 38).

In both the trestle area with and without bags (Table 5), significantly more oystercatchers and redshanks stayed underneath the trestles whereas curlews were only observed in the trestle area with bags (Mann–Whitney U-test, $P<0.01$, in trestle area with bags: oystercatcher U = 224; curlew U = 400; redshank U = 192; without bags: oystercatcher U = 304; redshank U = 224, n = 64 in all species).

In both trestle areas, numbers of dunlins, black-headed gulls and common gulls staying underneath or on top of the trestles (Table 6) did not differ significantly (Mann–Whitney U-test, $P<0.01$, in trestle area with bags: dunlins U=408; black-headed gulls U=49; common gulls U = 512, n = 64 in all species; in trestle area without bags: dunlins U = 480; black-headed gulls U = 464; common gulls U = 512, n = 64 in all species). The results from the analysis of the second sample in the same time period (see methods) did not differ from those described here.

Discussion

This preliminary study suggests that the oyster structures did not affect the feeding behaviour of the birds. The six species with most data apparently went into the trestle area and the structures did not appear to have an effect on their behaviour. The usage of the areas differed between the species with wader species mostly feeding in both areas, outside and inside the trestle area. In contrast, the gull species were mostly loafing both outside and inside the trestle area.

The feeding rate of oystercatchers did not differ between the reference and the trestle area. However in the reference site, the feeding rate was higher on days with a shorter tidal exposure of the area than on other days. This compensation for the shorter period of food availability had been demonstrated previously in experiments performed by Swennen et al. (1989).

Most species occurred in lower numbers in the trestle area than in the reference area. Only redshank and dunlin did not differ significantly in numbers between the areas. The differences observed in the distribution of the other species cannot only be explained by the presence of the trestles, since not all environmental parameters were identical in both areas. Bird density is related to the density and availability of their prey (Wanink & Zwarts, 1993). Differences in the diversity and abundance of invertebrates can be caused not only by aquaculture structures and activities (De Grave et al., 1998) but also by naturally occurring differences in the sediment, water current and inundation time (Yates et al., 1993). Thus, redshank and dunlin might have located food more easily in the trestle area, whereas oystercatchers and curlews feed on benthic invertebrates which were probably more available in the shallow water in the reference site. Oystercatchers are known to feed on ragworms (*Hedistis* sp.), which were more abundant in the reference area (Hilgerloh unpubl.).

All species except dunlin occurred in similar numbers in the trestle area with and without oyster bags. Dunlin occurred more often beneath the trestles in the area with bags but it is not known whether it was the direct effect of the physical structure that influenced bird numbers or some indirect effect like food supply.

Redshanks, curlews and oystercatchers mostly stayed underneath trestles, while gulls and dunlin were also observed on top of the trestles. The gulls apparently preferred to rest outside of the trestle area but they occasionally used the trestles as a perch, from which they kleptoparasitized oystercatchers feeding underneath. In the trestle-free area gulls were also frequently observed kleptoparasitizing oystercatchers, whereas in other parts of the estuary they were searching for their own food. Kleptoparasitism occurs also between oystercatchers, if they occur in densities of more than 50 birds per ha (Triplet et al., 1999).

The trestles offered additional food for some species which did not occur in high numbers during the observations. Wigeon, for example, fed on green algae growing attached to the trestles but only when the water level was deep enough to allow them to swim and before the trestles were covered by the rising tide. In other estuaries brent geese *Branta bernicla* have shown similar behaviour (Hilgerloh, pers. comm.). In addition, hooded crows appeared to be feeding on small gastropods living on the attached algae.

The conclusion from this preliminary study is that while trestles may influence the presence or absence of some bird species, no clear pattern has emerged. It is also clear that the feeding ecology and behaviour of some species may be influenced directly by the presence of the trestles or indirectly by their effect on food availability. We therefore recommend that further work be carried out to quantify and elucidate the mechanisms involved.

Acknowledgements

We are grateful to N. Robertson for the help in the field. The study was performed while G. H. had a Marie Curie Fellowship for experienced scientists.

References

Brosius, G. & F. Brosius, 1995. Base System and Professional Statistics. Thomson, Bonn, 1002 pp.

Castel, J., Labourge, J.P., Escaravage, V., Auby, I. & Garcia, M. E. 1989. Influence of seagrass and oyster parks on the abundance and biomass patterns of meio- and macrobenthos in tidal flats. Est. Coastal Shelf Sci. 28: 71–85.

De Grave, S., Moore, S. J. Burnell, G. 1998. Changes in benthic macrofauna associated with intertidal oyster, *Crassostrea gigas* (Thunberg) culture. J. Shellfish Res.17: 1137–1142.

Gibbons, D. W., Reid, J. B., Chapman, R. A. 1993. The new atlas of breeding birds in Britain and Ireland: 1988–1991. T & A. D. Poyser, London.

Hutchinson, C.D. 1989. Birds in Ireland. T. & A.D. Poyser Ltd., Calton, Staffordshire, U.K.

Kaiser, M. J., Laing, I., Utting, S. D., Burnell, G. M. 1998. Environmental impacts of bivalve mariculture. J. Shellfish Res. 17: 59–66.

Martin, P. & P. Bateson, 1986. Measuring Behaviour. Cambridge University Press, Cambridge: 1200 pp.

Nugues, M. M., Kaiser, M. J., Spencer, B. E. & Edwards, D. B. 1996. Benthic community changes associated with intertidal oyster cultivation. Aquacult. Res. 27: 913–924.

Shepherd, R. (1993) Ireland's Wetland Wealth. The Irish Wildbird Conservancy. Dublin, Ireland.

Smiddy, P., O' Halloran, J., Coveney, J. A. Leonard, P. G. Shorten, M. 1995. Winter waterfowl populations of Cork Harbour: an update. Irish Birds 5: 285–294.

Spencer, B. E. Kaiser, M. J., Edwards, D. B. 1996. The effect of Manila clam cultivation on an intertidal benthic community: the early cultivation phase. Aquacult. Res. 27: 261–276.

Spencer, B. E., Kaiser, M. J., Edwards, D. B. 1997. Ecological effects of intertidal Manila clam cultivation: observations at the end of the cultivation phase. J. appl. Ecol. 34: 444–452.

Swennen, C., Leopold, M. F. de Bruijn, L. L. M. 1989. Time-stressed oystercatchers, *Haematopus ostralegus*, can increase their intake rate. Anim. Behav. 38: 8–22.

Triplet, P., Stillman, R. A., Goss-Custard, J. D. 1999. Prey abundance and the strength of interfence in a foraging shorebird. J. anim. Ecol. 68: 254–265.

Yates, M.G., Goss-Custard, J. D., McGrorty, S., Lakhani, K. H., Le V. Dit Durell, S. E. A., Clarke, R. T., Wispin, W. E., Moy, I., Zates, T., Plant, R. A., Frost, A. J. 1993. Sediment characteristics, invertebrate densities and shorebird densities on the inner banks of the Wash. J. appl. Ecol. 30: 599–614.

Wanink, J. H. & Zwarts, L. 1993. Environmental effects on the growth of intertidal invertebrates and some implications for foraging waders. Neth. J. Sea Res. 3: 407–418.

Hydrobiologia **465:** 181–186, 2001.
G. Burnell (ed.), *Coastal Shellfish – A Sustainable Resource.*

Living with bonamiasis: Irish research since 1987

Sarah C. Culloty & Maire F. Mulcahy
Department of Zoology and Animal Ecology, University College Cork, Cork, Ireland

Key words: bonamiasis, *Bonamia ostreae*, *Ostrea edulis*, Ireland, aquaculture

Abstract

The main oyster species produced in Ireland up to the 1980's was the European flat oyster *Ostrea edulis*. However, since then, production of this species has been severely affected by the presence of the protistan *Bonamia ostreae,* which was diagnosed in a population of flat oysters on the south coast following heavy mortalities. Research has been ongoing since the first diagnosis in Ireland and has concentrated on aspects of the biology of both the host and the parasite. In recent years research has concentrated on screening populations of oysters to identify any with reduced susceptibility to the parasite.

Bonamiasis in Ireland

In Ireland the native oyster, *Ostrea edulis* L., has been harvested for many centuries, and is commercially the most valuable oyster species produced. In recent years harvesting of this species has mainly occurred in a number of natural fisheries managed by co-operatives and in a number of privately managed beds. These beds are concentrated along the south, west and north coasts (Fig. 1). In 1997 total production from aquaculture in Ireland was approximately 40 000 tonnes valued at £60 000 000. Shellfish accounted for approximately a third of this value with *O. edulis* being the highest valued species produced.

The greatest threats to shellfish production in Europe in recent years have been over-fishing, mismanagement of beds, and losses caused by the presence of a number of pathogenic agents. Due to Ireland's island status, the Irish oyster industry avoided a number of the severe disease problems that affected mainland Europe in the 1960s and the 1970s, when losses occurred due to protistans such as *Marteilia refringens* (Grizel et al., 1974) and *Bonamia ostreae* (Pichot et al., 1980). Ireland's disease-free status ended in 1987 with the discovery of *B. ostreae* in native oysters in Rossmore, Cork Harbour (Fig. 1). Since the 1970s a private company has been developing and managing flat oyster production in Rossmore: restoring long derelict natural oyster beds, breeding in spatting ponds for re-stocking, and exporting 30–40 t of oysters annually to a number of European destinations. In 1984 unusually high mortality was observed on the company's oyster beds. Mortality rose steadily over the next three years, and *B. ostreae* was observed in 90% of 4-year-old oysters in 1987 (McManus, 1988; McArdle et al., 1991). From 1987 to 1990, commercial output was non-existent. Examination of Rossmore oysters, frozen in 1986, indicated a significant prevalence of *B. ostreae* at that time, which lead to the conclusion that the parasite had contributed to the mortalities that were occurring since 1984 (Rogan et al., 1991). It is impossible to determine exactly when the initial introduction of this pathogen occurred.

In 1988 *B. ostreae* was diagnosed in a consignment of wild oysters from Clew Bay (Fig. 1) exported to France. No obvious mortalities were observed there at the time, and it was a number of years before follow-up surveys of Clew Bay oysters confirmed the presence of *B. ostreae* at low levels (McArdle et al., 1991).

In 1989 *B. ostreae* was diagnosed in Galway Bay (Fig. 1). The oyster beds in Galway Bay are natural, but oysters are normally dredged from the bay, and relayed to the inner inlets, where conditions for feeding and breeding are most favourable. High mortality (70–80%) was recorded, but in localised areas only, with no mortality apparent elsewhere in the Bay. In 1991 bonamiosis was observed in five new sites within the bay, though with low prevalence and no observed

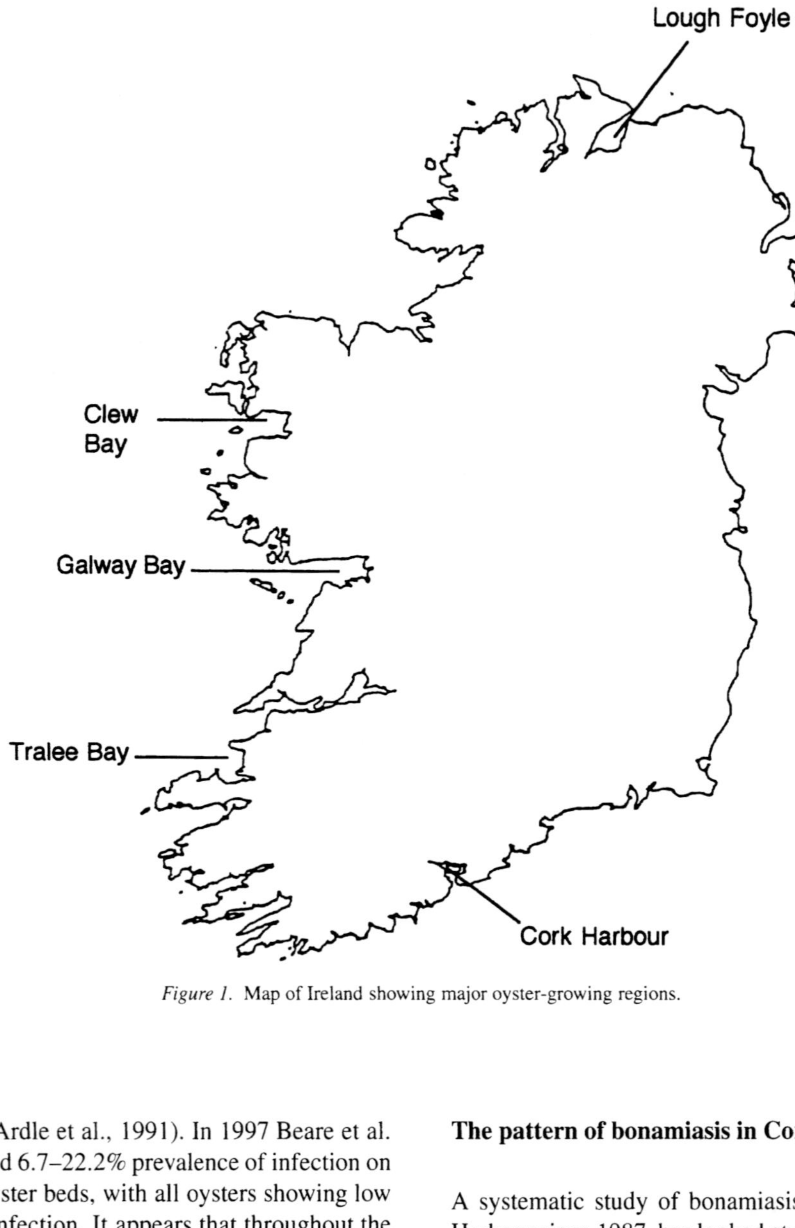

Figure 1. Map of Ireland showing major oyster-growing regions.

mortality (McArdle et al., 1991). In 1997 Beare et al. (1998) recorded 6.7–22.2% prevalence of infection on the affected oyster beds, with all oysters showing low intensities of infection. It appears that throughout the 1990s the disease has extended its range over the beds. Prevalence of *B. ostreae* has, however, remained relatively low, probably due to a combination of factors including heavy harvesting and disease-related mortality, which in turn have kept the density of oysters on the beds low. *B. ostreae* has not been diagnosed to date in the remaining Irish oyster-growing areas, including Tralee and Lough Foyle (Fig. 1).

The pattern of bonamiasis in Cork Harbour

A systematic study of bonamiasis, ongoing in Cork Harbour since 1987, has looked at a number of aspects of the disease including pathology, life cycle of the parasite, and seasonal and annual prevalence of infection in different age groups. Infection was observed in all age groups but levels became significant after the oysters spawned for the first time, when generally 2 years old. From then on older oysters exhibited higher levels of infection. The disease is equally prevalent in males and females (Culloty, 1992; Culloty & Mulcahy, 1996). No relationship with female gonadal development as had been postulated by Van Banning (1990) was observed. Continual cycles of infection and mortality occurred throughout the year, without obvious seasonality.

Using the method developed by Mialhe et al. (1988), *B. ostreae* cells isolated from infected oysters were injected into healthy oysters in the laboratory and progression of the disease followed from 2 to 42 days post-infection (p.i.) by heart smears, haemolymph smears, screening of tissue sections, and indirect immunofluorescence. A latent period occurred following inoculation during which the parasite could not be detected with available diagnostic methods, and while an inflammatory reaction was observed 14 days p.i., the first positive heart smears were not found until 28 days p.i. By 42 days the parasite was widespread in blood and tissues. Rigorous examination of serial tissue sections during the latent period using the histological-based diagnostic methods available did not reveal any early or developmental stages of the parasite.

Anaesthetics for *O. edulis*

One of the early difficulties in investigating bonamiasis in Ireland was that manipulation or inoculation of the oysters required either physically prising apart the valves or else drilling through them. A range of possible anaesthetics was tested and assessed for effectiveness with *O. edulis*, and 3.5% magnesium chloride ($MgCl_2{\cdot}6H_2O$) was found to be most effective (Culloty & Mulcahy, 1992). This allows animals to be anaesthetised and recover completely within a matter of hours, reducing stress for the oyster and for the investigator! However, what constitutes stress for an oyster and the parameters that might allow for its assessment have not yet been established.

The possible development of oyster resistance to *Bonamia*

Following the massive mortality that occurred due to bonamiasis in oysters in Cork Harbour in the mid-1980s, investigations were begun into methods to minimise infection and mortality due to the parasite. It appeared from experiences in other areas that elimination of the parasite was not an option. A decision was taken to breed only from 4-year-old survivors of the disease, in the knowledge that a degree of host resistance develops over time in many host-parasite relationships (Ford & Haskin, 1987; Ford, 1988). This breeding programme has been continued over the past 10 years, with a number of field- and laboratory-based trials being undertaken to assess its success. These trials were initially undertaken in 1990 and 1991; however, no resistance in the Cork population was evident at this stage (Culloty, 1992)

From 1996 to 1998 a further comparison was made between the relative resistance of oysters from Cork Harbour and those from two naïve oyster populations: Tralee and Lough Foyle. In both laboratory and field, the Cork Harbour oyster population showed lower prevalence of *B. ostreae* and lower cumulative mortality than the two naïve populations (Culloty & Mulcahy, 1998).

Encouraged by these results, a further assessment has been carried out from 1998–2000, using Irish and continental European oyster populations, as part of an EU CRAFT programme. Seven oyster strains, from Scotland, Ireland, Holland and France, have been relayed in Cork Harbour, in Brittany, and in Lake Grevelingen in Holland. Bonamiasis occurs in all three sites. To date, prevalence of the parasite, class of infection, mortality and cumulative mortality have been followed for over one year. Preliminary data suggests that the Cork Harbour oysters are showing lowest mortalities in Cork Harbour; however, the Dutch oysters are showing lowest mortalities in Lake Grevelingen. These results raise the possibility that not only is the host adapting to its parasite, but the parasite may be adapting also; and that the most successful oysters may be those which have been bred in the presence of the parasite from its native waters (Culloty et al., 1999a). Laboratory-based trials, which will deliver a quantitative *B. ostreae* challenge to the different European oyster populations, should help to clarify whether resistance is indeed developing, and if so whether it is to *B. ostreae* strains generally, or to the local strain in particular.

Hosts and possible vectors of *B. ostreae*

The EU Directive 91/67/EEC restricts oyster movements from areas known to have bonamiasis. However, while *Ostrea* species may not be moved from bonamiasis–enzootic to unaffected areas, the critical question is whether other bivalve species can or cannot be carriers or vectors of *B. ostreae*. It is known for example that *Crassostrea gigas* (Thunberg) blood cells phagocytose *B. ostreae* (Renault et al., 1995). Could *Crassostrea* or other species act as vectors? Furthermore, the life-cycle of *B. ostreae* is not known. Could the life-cycle be indirect, and could other species be intermediate hosts?

A study was undertaken within an EU-funded contract to examine whether (i) a range of cultured bivalves could be infected either naturally in the field or experimentally in the laboratory and (ii), having been exposed to *B. ostreae* in either the field or the lab, could any one of these bivalve species function as vectors and transmit the disease by co-habitation to healthy *O. edulis*. The bivalve species were: *C. gigas*, *Ruditapes decussatus* (L.), *Ruditapes philippinarum* (Adams & Reeve), *Mytilus edulis* (L.), *Mytilus galloprovincialis* (Lamarck), and *O. edulis* as control (Culloty et al., 1999b). The results showed that, as in previous studies, *O. edulis* became infected both naturally in the field, and experimentally by inoculation in the laboratory, and that they in turn could infect naïve *O. edulis* by co-habitation. However, *B. ostreae* was not detected in any of the other species in either laboratory or field; nor did these species, even following inoculation with *B. ostreae*, infect *O. edulis* by cohabitation. Thus it appears that the bivalve species screened, other than *O. edulis*, neither carry nor transmit *B. ostreae*. It must be emphasised that, until the complete life cycle of *B. ostreae* is established and all stages of *B. ostreae* are detectable, these conclusions cannot be 100% certain.

Diagnosis of bonamiasis

A number of histological and immunological diagnostic techniques have been developed for diagnosis and identification of *B. ostreae* including histology, heart smears and blood smears (Bachère et al., 1982), immunofluorescence (Boulo et al., 1989), and electron microscopy (Hine, 1991; Hine & Wesney, 1992, 1994). The diagnostic methods required by the EU and by the OIE are heart smears and/or histology. In a study on bonamiasis in oysters from Cork Harbour in 1995 heart smears of 2- and 3-year-old oysters were screened (O' Neill et al., 1998). The 2-year-old oysters had 7.5%, and the 3-year-old oysters 22.5% prevalence of *B. ostreae*. However, haemolymph smears of the same oysters revealed infections in two additional 3-year-olds. Even though 103 other oysters were examined subsequently by both diagnostic methods and there was no discrepancy between the results, the 1995 results raise a question over adequacy of the diagnostic criteria. When it is critical to establish the presence or absence of *B. ostreae*, e.g., in the case of proposed oyster transfers to an uninfected site, it would seem that heart smears should be used in conjunction with either haemolymph smears or histology. The work on developing a polymerase chain reaction (PCR) assay for *B. ostreae* reported by Carnegie et al. (1999) raises hopes of a method which can provide greater sensitivity of detection, as well as an invaluable tool with which to further explore the life cycle and biology of *B. ostreae*.

Defence mechanisms of *O. edulis* against *B. ostreae*

Bonamiasis results from infection of the host *O. edulis* by the parasite *B. ostreae*. Surprisingly little has been documented on the defence mechanisms of either the host or the parasite. A number of aspects of haemolymph composition and function in *O. edulis* are being examined (Cronin et al., 1999). In 1990–91 protein levels were established throughout the year, but they did not seem to vary with the intensity of bonamiasis (Culloty, 1992). Amylase, aspartate aminotransferase (ASAT) and lysozyme levels were also established. Amylase and ASAT showed seasonal variation, but variation did not appear to be related to intensity of infection.

Blood cell counts show very wide variation, much of which relates to the open circulatory system and the ability of the animal to mobilise blood cells from blood spaces and sinuses. Higher haemocyte numbers have been observed in oysters infected with *B. ostreae* (Beare, 1997), and such oysters had a higher percentage of hyalinocytes and a lower percentage of granulocytes than uninfected oysters. Also, in infected oysters the phagocytic rate and phagocytic index were significantly lower, perhaps resulting from counter defences of the parasite.

Future work

The question of whether *O. edulis* develops resistance or merely tolerance to *B. ostreae* needs to be answered if bonamiasis is to be managed effectively, and studies are continuing to address this. The development of a PCR assay for *B. ostreae* will permit investigations of the complete life cycle and biology of the parasite, including the latent period and the examination of possible alternate hosts, which are not possible at present. This knowledge is essential for prevention and control, and may suggest means of treatment.

The development of cell cultures of *O. edulis* tissues would allow *in vitro* study of parasite biology and interaction with host cells. Much work has been undertaken to develop reliable cell cultures in our laboratory and elsewhere. To date no molluscan continuous cell line is available, which severely limits research on *B. ostreae* and other molluscan pathogens. Knowledge of the immune mechanisms of *O. edulis* is required, as well as understanding of the offence and defence mechanisms of *B. ostreae* in relation to its host. Research is also required to investigate the nature and the role of stress in bonamiasis. From general principles it can be expected to be significant.

We know from studies in Ireland and abroad that no methods of eradication are available for *B. ostreae*. Once the parasite is brought into an area where *O. edulis* is grown, it is impossible to eradicate. Modified husbandry techniques may reduce the impact of the disease through reduction of stress on the oysters. However, commercial stocking densities and handling all exacerbate infection. Restocking of infected areas that have been fallow ultimately result in the return of increased infection levels. All these facts point to maintaining the *Bonamia*-free oyster growing areas. Continued emphasis must be placed on informing farmers of the dangers of illegal transfers. To date it would appear that the only solution would be the production of a stock that has reduced susceptibility to the parasite. This requires long-term approaches, and organised management practices. To date results suggest that this approach can lead to reduced mortalities and losses due to the disease. More work is required to determine if the selected strains are truly showing some 'resistance', or if some adaptation has occurred between host and parasite in their particular environment. Whether the selected strians will prove to be resistant on transfer to other areas remains to be observed: strains of parasites and environmental factors may vary from one growing area to another. All these factors have to be investigated before it is determined that 'resistant' strains are available and before transfers could occur to other areas affected by the disease. Production of a resistant stock could play a part in rehabilitating oyster growing areas throughout Europe which have been affected by huge losses.

Acknowledgements

The authors would like to thank Atlantic Shellfish Ltd., Cork for their continuing support. Work outlined in this review has been supported by: EU FAIR 94C 85/13, EU AIR AQ1-272, EU FAIR CT98-9101, FORBAIRT HE/96/113.

References

Bachère, E., J. L. Durand & G. Tigé, 1982. *Bonamia ostreae* (Pichot et al., 1979) parasite de l'huître plate: Comparison de deux methodes de diagnostic. ICES cm 1982/F28.

Beare, W. E., 1997. An investigation of the prevalence of *Bonamia ostreae* and its effect on the biology of the native flat oyster, *Ostrea edulis* L. in Galway Bay, Ireland. M.Sc. Thesis, National University of Ireland, Cork.

Beare, W. E., S. C. Culloty & G. Burnell, 1998. Some observations on spatial and temporal variation in prevalence of infection of *Bonamia ostreae* (Pichot et al., 1980) in the native flat oyster *Ostrea edulis* (L.) in Galway Bay, Ireland. Bull. eur. Ass. Fish Pathol. 18(2): 39–42.

Boulo, V., E. Mialhe, H. Rogier, F. Paolucci & H. Grizel, 1989. Immunodiagnosis of *Bonamia ostreae* (Ascetospora) infection of *O. edulis* L. and subcellular identification of epitopes by monoclonal antibodies. J. Fish Dis. 12: 257–262.

Carnegie, R. B., B. J. Barber, D. L. Distel & S. C. Culloty, 1999. Development of PCR and in situ hybridisation assays for detection of *Bonamia ostreae* in flat oysters *Ostrea edulis*. Third International Conference on Shellfish Restoration ICSR-Europe 1999: 'Coastal shellfish a sustainable resource'. Cork, Ireland 28 Sept.–2 Oct. 1999. Abstract only.

Cronin, M., S. C. Culloty & M. F. Mulcahy, 1999. A study of some of the hemolymph constituents of the European oyster *Ostrea edulis* and the effect of infection with *Bonamia ostreae*. European Association of Fish Pathologists, 9th International Conference, Sept. 19–24, Rhodes, Greece. Abstract only.

Culloty, S. C., 1992. Bonamiasis in the flat oyster *Ostrea edulis* L. Ph.D. Thesis, University College Cork, National University of Ireland.

Culloty, S. C & M. F. Mulcahy, 1992. An evaluation of anaesthetics for *Ostrea edulis* L. Aquaculture 107: 249–252

Culloty, S. C. & M. F. Mulcahy, 1996. Season-, age- and sex-related variation in the prevalence of bonamiasis in flat oysters (*Ostrea edulis* L.) on the south coast of Ireland. Aquaculture 144: 53–63.

Culloty, S. C. & M. F. Mulcahy, 1998. Results of trials to determine the relative resistance of an Irish population of flat oysters to the parasite *Bonamia ostreae*. Third Int. Symp on Aquatic Animal Health, Aug. 30–Sept. 3, Baltimore, MD USA. Abstract only.

Culloty, S. C., M. Cronin & M. F. Mulcahy, 1999a. Relative resistance of a number of European oyster populations (*Ostrea edulis*) to the parasite *Bonamia ostreae*. European Association of Fish Pathologists 9th International Conference Sept. 19–24, Rhodes, Greece. Abstract only.

Culloty, S. C., B. Novoa, M. Pernas, M. Longshaw, M. F. Mulcahy, S. W. Feist & A. Figueras, 1999b. Susceptibility of a number of bivalve species to the protozoan parasite *Bonamia ostreae* and their ability to act as vectors for this parasite. Dis. Aquat. Org. 37: 73–80.

Ford, S. E., 1988. Host–parasite interactions in eastern oysters selected for resistance to *H. nelsoni* (MSX) disease: survival mechanisms against a natural pathogen. Am. Fish. Soc. Spec. Publ. 18: 206–204.

Ford, S. E. & H. H. Haskin, 1987. Infection and mortality patterns in strains of oysters *Crassostrea virginica* selected for resistance to the parasite *Haplosporidium nelsoni*. J. Parasitol. 73(2): 368–376.

Grizel, H., M. Comps, J. R. Bonami, F. Cousserans, J. L. Duthoit & M. A. Pennec, 1974. Recherche sur l'agent de la maladie de la glande digestive de *Ostrea edulis* Linné. Bull. Inst. Pêches Marit. Maroc 240: 7–30.

Hine, P. M., 1991. Ultrastructural observations on the annual pattern of *Bonamia* sp. in flat oysters *Tiostrea chilensis*. Dis. Aquat Org. 11: 163–171.

Hine, P. M. & B. Wesney, 1992. Interrelationships of cytoplasmic structures in *Bonamia* sp. (Haplosporidia) infecting oysters *Tiostrea chilensis*: an interpretation. Dis. Aquat. Org. 14: 59–68.

Hine, P. M. & B. Wesney, 1994. The functional cytology of *Bonamia* sp. (Haplosporidia) infecting oysters *Tiostrea chilensis*: an ultrachemical study. Dis. Aquat. Org. 20: 207–217.

McArdle, J. F., F. McKiernan, H. Foley & D. Hugh-Jones, 1991. The current status of *Bonamia* disease in Ireland. Aquaculture 93: 273–278.

McManus, J., 1988. A study of the *Ostrea edulis* L. population in the North Channel, Cork Harbour. M.Sc. Thesis, National University of Ireland, Cork.

Mialhe, E., E. Bachère, D. Chagot & H. Grizel, 1988. Isolation and purification of the protozoan *Bonamia ostreae* (Pichot et al. 1980), a parasite affecting the flat oyster *Ostrea edulis* L. Aquaculture 71: 293–299.

O'Neill, G., S. C. Culloty & M. F. Mulcahy, 1998. The effectiveness of two routine diagnostic techniques for the detection of the protozoan parasite *Bonamia ostreae* (Pichot et al., 1980). Bull. eur. Ass. Fish Pathol. 18(4): 117–120.

Pichot, Y., M. Comps, G. Tigé, H. Grizel & M. A. Rabouin, 1980. Recherches sur *Bonamia ostreae* gen. n. sp. parasite nouveau de l'huître plate *Ostrea edulis* L. Rev. Trav. Inst. Pêches Marit. 43(1): 131–140.

Renault, T., N. Cochennec, & H. Grizel, 1995. *Bonamia ostreae* parasite of the European flat oyster *Ostrea edulis* does not experimentally infect the Japanese oyster *C. gigas*. Bull. eur. Ass. Fish Pathol. 15(3): 78–80.

Rogan, E., S. C. Culloty, T. F. Cross & M. F. Mulcahy, 1991. The detection of *Bonamia ostreae* (Pichot et al., 1980) in frozen oysters (*Ostrea edulis*. L.) and the effect of the parasite on condition. Aquaculture 97: 311–315.

Van Banning, P., 1990. The life-cycle of the oyster pathogen *Bonamia ostreae* with a presumptive phase in the ovarian tissue of the European flat oyster, *Ostrea edulis*. Aquaculture, 84: 189–192.

Hydrobiologia **465**: 187–193, 2001.
G. Burnell (ed.), Coastal Shellfish – A Sustainable Resource.

A satellite and field portrait of a *Karenia mikimotoi* bloom off the south coast of Ireland, August 1998

R. Raine[1], S. O'Boyle[1], T. O'Higgins[1], M. White[1], J. Patching[1], Bronwyn Cahill[2] & T. McMahon[3]
[1]*Martin Ryan Institute, National University of Ireland, Galway, Ireland*
E-mail: robin.raine@nuigalway.ie
[2]*The Irish Marine Institute, Marine Data Centre, 80 Harcourt St., Dublin 2, Ireland*
[3]*The Irish Marine Institute, Fisheries Research Centre, Abbottstown, Castleknock, Dublin 15, Ireland*

Key words: phytoplankton, blooms, *Karenia mikimotoi*, *Gyrodinium aureolum*, Irish coast

Abstract

An extensive surface bloom of the dinoflagellate *Karenia mikimotoi* occurred off southwestern Ireland during August, 1998. The bloom was evident both from remotely sensed satellite ocean colour data and as visibly discoloured water, from the mouth of Bantry Bay around towards Cork, extending some 60 km offshore. The timing of the bloom co-incided with a field survey in the area. This paper compares the surface distributions of chlorophyll and *K. mikimotoi* concentrations with satellite ocean colour and thermal infra-red sea surface temperature images, from which may be derived the origins of the bloom. It would appear that weak coastal upwelling transported a thermocline population of *K. mikimotoi* up to the surface in the region of the Fastnet Rock, where it was wind-dispersed eastwards across the northern Celtic Sea.

Introduction

Extensive blooms of *Karenia mikimotoi* (Miyake & Kominami ex Oda) Hansen & Moestrup, once known as *Gyrodinium aureolum*, are a common feature of the summer phytoplankton ecology around northwestern Europe (Gentien, 1998). They are often associated with mortalities of benthic invertebrates and other fauna. Blooms of this organism have impacted on aquaculture in that they have co-incided with mass mortalities of farmed finfish and shellfish. Around Ireland, a bloom along the south and southeastern coast in 1976 was the direct cause of substantial lugworm (*Arenicola marina* L.) mortalities (Ottway et al., 1979). A subsequent bloom around the southwest in 1979 resulted in mass mortalities of farmed rainbow trout (Jenkinson & Connors, 1980). Another bloom in 1992 along the west coast co-incided with substantial mortalities of farmed clams (*Tapes semidecussatta* L.) (Joyce, 1995). *K. mikimotoi* blooms have occurred along the south, southwest, west and north coasts of Ireland (e.g. Ottway et al., 1979; Raine et al., 1993; Raine & McMahon, 1998; Gowen et al., 1998; O'Boyle et al., 2000). In view of their potentially deleterious effect on aquaculture, substantial effort has been carried out in investigating the mechanisms through which blooms arise, particularly around the economically important coastal zone of southwestern Ireland. Here, Raine & McMahon (1998) have proposed an advective model in which potentially harmful phytoplankton populations are transported along the south coast and into the southwest under a particular sequence in wind direction. This paper describes a bloom of *K. mikimotoi* in this region in the summer of 1998, apparent from both satellite ocean colour and contemporaneous field observations, and demonstrates its mode of development and proliferation.

Methods

Field measurements were carried out from the R/V Celtic Voyager along a north-south transect of 21 stations spaced equidistantly 3 nautical miles (ca. 5.5

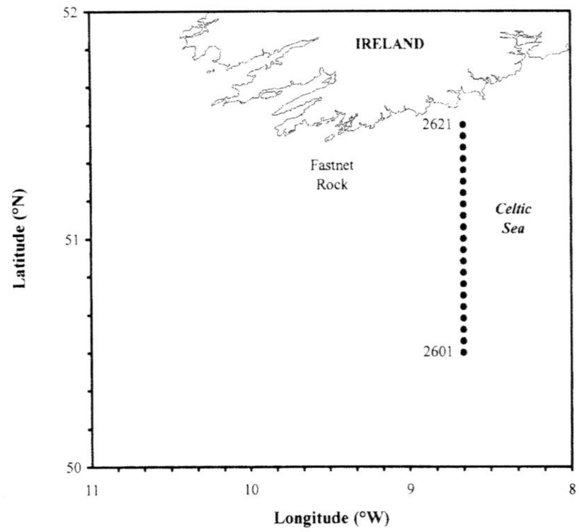

Figure 1. Positions of stations sampled on 5th August, 1998 off the south coast of Ireland. Stations are numbered sequentially from 2601 to 2621.

km) apart between 50° 30′ and 51° 30′ N at 08° 40′ W on the 5th August 1998 (Fig. 1). Temperature and salinity were profiled at each station using a SeaBird SBE19 CTD with an *in situ* fluorometer (SeaTech, Oregon) incorporated into the instrument array. Hydrocast sampling was carried out from depths selected from the vertical profiles, ideally from the surface, within the surface mixed layer, at any sub-surface chlorophyll maximum associated with the pycnocline, and the bottom mixed layer. Thus calibration measurements of temperature and salinity as well as samples for the measurements of nutrients (Grasshoff, 1976), chlorophyll (Tett, 1987a) and phytoplankton were obtained. Quantitative phytoplankton data were obtained by counting samples, preserved in Lugol's Iodine, after sedimentation using an inverted microscope (Hasle, 1978).

Whilst underway, continuous measurements of near-surface (2 m) temperature, salinity and *in vivo* fluorescence supplemented the CTD data through linkage of a thermosalinograph (Seabird SBE20) and fluorometer (Turner Model 10) into a continuous flow seawater system. Data output from these instruments were interfaced with output from the ship's position fixing equipment.

Results

Variations in sea surface temperature (SST) and chlorophyll fluorescence along the N–S transect at 08° 40′ W are shown in Figure 2. The main feature evident is a broad maximum in chlorophyll fluorescence between 50° 50′ and 51° 10 ′ N. Measurements of surface chlorophyll (Fig. 2) showed that maximum concentrations were of the order of 9 mg m^{-3}. Maximum chlorophyll fluorescence was accompanied by a decrease in surface temperature from 16 °C on either side to 14.5–15 ° within it. Elevated levels of surface chlorophyll reduced water transparency. Secchi depths of 10–14 m on either side of the bloom were 6–7 m within the zone of enhanced chlorophyll (Table 1), implying shallower euphotic depths of the order of 20 m. Only traces of nutrient (nitrate) were evident at the surface of a small number of stations along the transect (Table 1).

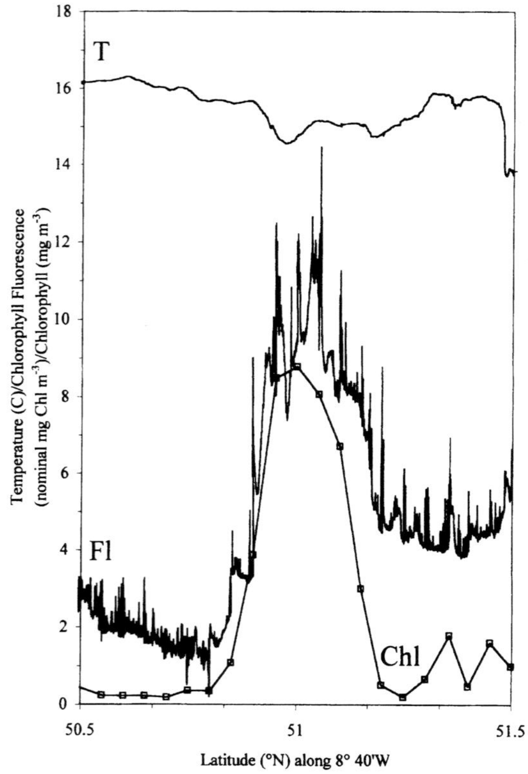

Figure 2. Continuous records of surface (2 m) chlorophyll fluorescence (Fl, nominal mg Chl m^{-3}) and water temperature (T, °C) along the ship's track on 5th August 1998 together with surface concentrations of chlorophyll *a* (Chl, mg m^{-3}) measured from discrete samples.

Profiles obtained from CTD measurements showed that the water column was thermally stratified at virtually all sampling stations along this transect (Fig. 3a). Temperature gradients of 4.5° occurred over a sharp thermocline which was located at 40 m at the outer

Table 1. Locations of sampling stations, water transparency, and surface measurements of nitrate and *Karenia mikimotoi* concentrations. For comparative purposes, the total concentrations of phytoplankton (diatoms + dinoflagellates) is shown

Station number	Latitude at 08° 40′ W (°N)	Secchi depth (m)	NO_3 (0 m) (mmol m^{-3})	*K. mikimotoi* (0 m) (cells ml^{-1})	Total phytoplankton (0 m) (cells ml^{-1})
2601	50° 30.0′	–	0.18	0	40
2602	50° 32.9′	–	0.12	3	46
2603	50° 36.0′	–	<0.10	0	23
2604	50° 38.9′	13.0	0.12	1	20
2605	50° 42.0′	14.0	<0.10	1	21
2606	50° 45.0′	12.0	<0.10	2	27
2607	50° 48.0′	11.0	0.14	15	90
2608	51° 51.0′	10.0	0.15	110	160
2609	51° 54.0′	7.0	<0.10	470	540
2610	51° 57.0′	6.5	<0.10	460	510
2611	51° 00.0′	6.5	<0.10	530	590
2612	51° 03.0′	6.0	<0.10	310	360
2613	51° 06.0′	7.0	<0.10	130	170
2614	51° 09.0′	7.0	<0.10	100	170
2615	51° 11.9′	11.0	<0.10	65	120
2616	51° 15.0′	-	<0.10	10	36
2617	51° 18.0′	14.0	<0.10	3	17
2618	51° 21.3′	9.0	<0.10	91	120
2619	51° 24.0′	10.5	<0.10	10	28
2620	51° 27.0′	9.5	<0.10	22	58
2621	51° 29.9′	8.5	<0.10	1	170

stations but shallowed to 20 m inshore of the region of the surface chlorophyll maximum. Similar uplifts in the halocline and pycnocline were also evident along the transect (Fig. 3b,c). Thermal stratification was not so well developed at the two inshore stations, possibly due to an increased mixing effect of tidal stirring resulting from the shallower water depths at these locations.

Microscopic analysis of phytoplankton samples showed that the surface chlorophyll levels were elevated due to a virtually monospecific bloom of *Karenia mikimotoi* (Table 1). Surface concentrations within the range 300–600 cells ml^{-1} were recorded at those stations (2609–2612) where the surface chlorophyll was 7–9 mg m^{-3}. Offshore of the bloom zone, the surface phytoplankton comprised mainly *Proboscia (= Rhizosolenia) alata* (Brightwell) Sandström in concentrations 5–20 cells ml^{-1} and small *Gymnodinium* spp. <20 μm diameter at concentrations up to 10 cells ml^{-1}. A mixture of *P. alata*, small *Gymnodinium* and *K. mikimotoi* comprised >90% (by cell numbers) of the flora inshore of the bloom zone, with the exception of the innermost station 2621 where diatoms such as *Chaetoceros* spp. (80 cells ml^{-1}), *Leptocylindrus danicus* Cleve (25 cells ml^{-1}), *Rhizosolenia* spp. (37 cells ml^{-1}) and *Pseudo-nitzschia* spp. (30 cells ml^{-1}) predominated.

An ocean colour derived satellite image of chlorophyll taken approximately 16 h before the field measurements is shown in Figure 4, along with a thermal infra-red sea surface temperature (SST) image taken on the same day (within 5 h). A region of elevated surface chlorophyll concentration may be seen extending from the southwest corner of Ireland southeastwards across the northern Celtic Sea, with dimensions of the order 150 km long and 40 km wide. Its position coincided with the region of elevated surface chlorophyll shown in Figure 2, which is also evident from the *in situ* fluorescence data (Fig. 3d). Similarly, an area of cooler surface water can be seen from the SST image, which extends eastwards as far as longitude 8° 40′ W.

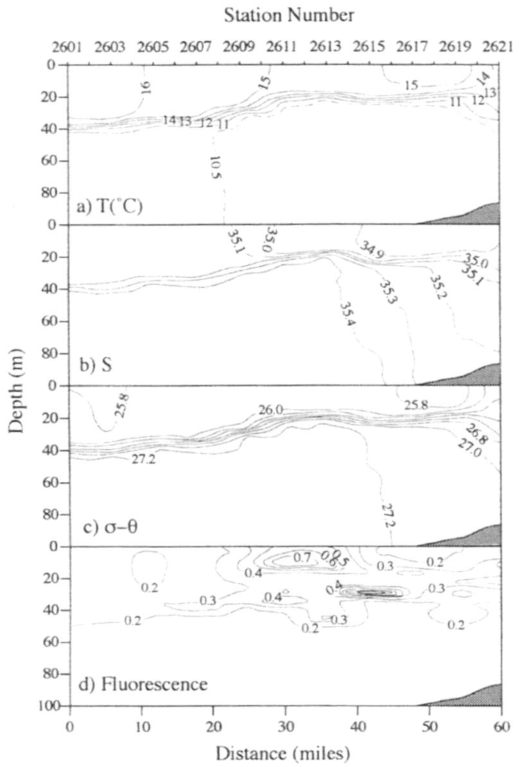

Figure 3. Distributions of (a) temperature, (b) salinity, (c) sigma-theta and (d) *in situ* chlorophyll fluorescence between 50° 30′ and 51° 30′ N along the transect of stations at 08° 40′ W depicted in Figure 1.

This is also evident in the surface temperature and CTD records (Figs 2 and 3a).

Winds recorded at the meteorological station at Valentia, approximately 60 km from the study region (Fig. 1), show west to northwest winds to have occurred between 25th and 31st July (Fig. 5). Subsequently, winds were predominantly from the southwest quarter throughout August.

Discussion

The coastal and shelf region around the southwest of Ireland is oceanographically quite complex. A common feature evident from thermal infra-red satellite imagery in summer is a region of cool surface water in the vicinity of the Fastnet Rock, similar to that observed in Figure 4. This is the result of coastal upwelling, which can occasionally result in surface water of <11 °C in summer close to the shoreline. In this situation dense diatom populations may be found (Raine et al., 1990a). At the other extreme, the water column can be thermally stratified, with surface water temperatures of ca. 16–17 °C, with a sharp thermocline situated at 30 m depth (Raine et al., 1990b), supporting an intense sub-surface chlorophyll maximum. Quite often this sub-surface maximum comprises a population of *K. mikimotoi* (Raine & McMahon, 1998). In between these two extremes one can often find weak coastal fronts at the boundary of upwelled and stratified waters (Holligan, 1981). This is most likely the situation in July and August 1979 where Roden et al. (1981) show an extensive population of *K. mikimotoi* associated with a thermal front. They describe the onshore advection of the population as the coastal upwelled water subsides and is replaced by warmer surface waters from a thermally stratified region further offshore. A population of *K. mikimotoi* associated with the thermocline is most likely the origin of this bloom.

The influence of surface wind stress on water circulation and phytoplankton dynamics around southwestern Ireland is not limited to the promotion of coastal upwelling. Raine & McMahon (1998) describe its effect on the residual clockwise circulation of coastal water which flows westwards along the south coast and around the southwestern corner. Under (prevailing) wind conditions from the southwestern quarter, the coastal circulation is constrained, and warm and highly saline oceanic water may be found very close to the shore. Relaxation of the wind forcing from this direction promotes the coastal flow, which can achieve residual (non-tidal) speeds of the order 25 cm s^{-1}. When this occurs thermocline phytoplankton populations, which include *K. mikimotoi*, are advected around to the mouths of the aquaculturally important bays of the southwest. The phytoplankton can then be advected into the bays via a two-layer oscillatory wind-driven flow mechanism (Raine et al., 1993; Edwards et al., 1996).

Perusal of the satellite archive shows coastal upwelling and enhanced surface chlorophyll to begin developing around southwestern Ireland in mid July, 1998. This patch of surface chlorophyll developed southeastwards across the Celtic Sea, culminating in the situation visible on the 4th August (Fig. 4), which persisted into September. Given the extent of our understanding of the oceanography of southwestern Ireland, it is possible to deduce events leading up to the situation on the 4/5th August. Wind data recorded at Valentia showed west to northwest winds to have blown between 25th and 31st July (Fig. 5). This is an

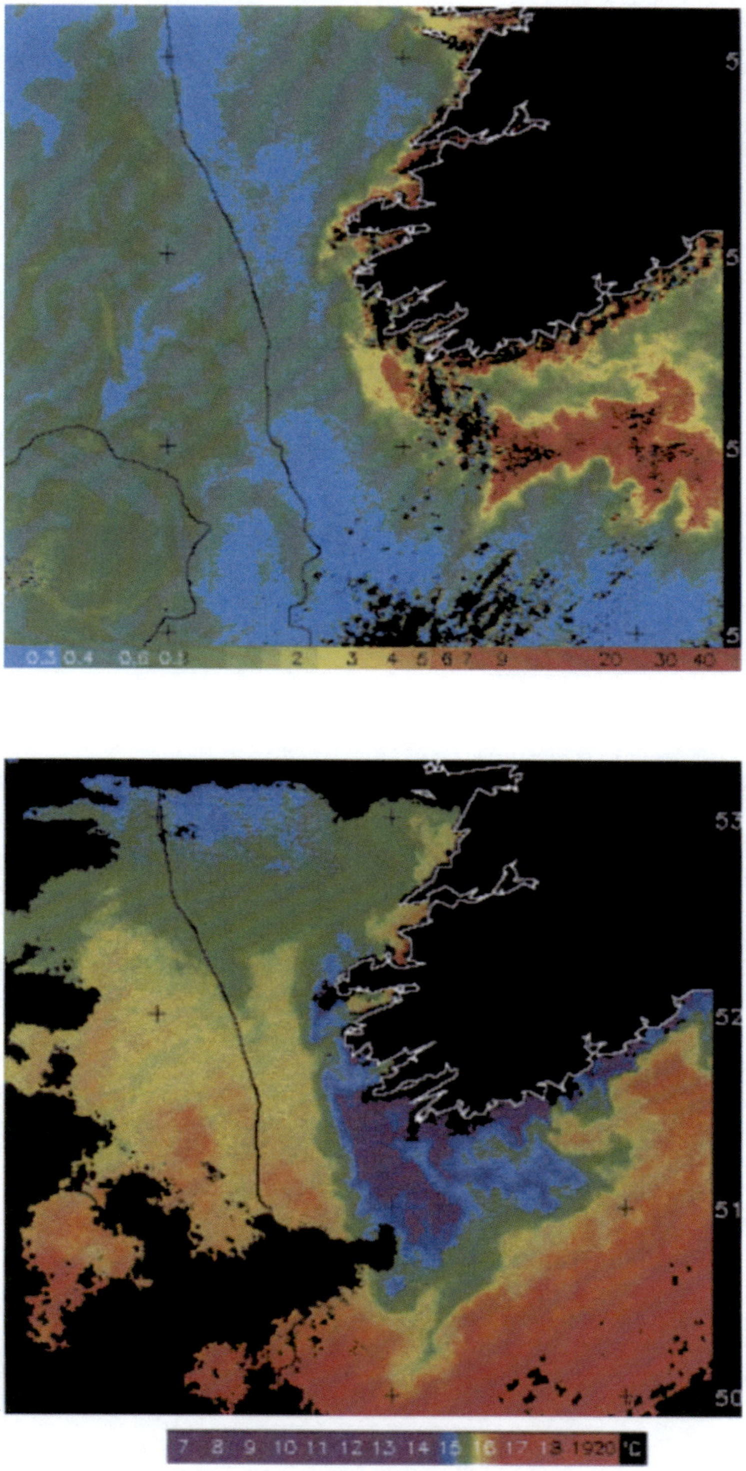

Figure 4. Satellite observations of sea surface chlorophyll and temperature around southwestern Ireland taken on 4th August 1998. Upper frame: SeaWiFS ocean colour derived chlorophyll distribution (1339 UTC). Lower frame: AVHRR SST image taken at 1748h UTC. Scale bars of chlorophyll concentration (mg m^{-3}) and temperature (C) are indicated at the bottom of each image.

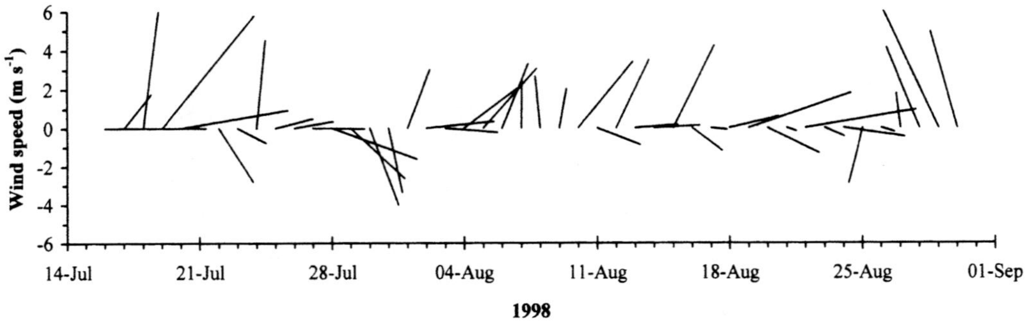

Figure 5. Stick plot of mean daily wind speed and direction measured at Valentia Island from mid-July to end of August 1998. Vectors are scaled to show true direction.

upwelling favourable wind direction, and would have caused uplift of water around the southwest, drawing any phytoplankton at the thermocline towards the surface. Once at the surface, the phytoplankton would have been blown back across the surface of the Celtic Sea under the northwesterly wind conditions. The cells would respond to the change in their environment of increased illumination by actively growing at the surface, utilising their internal nutrient pool, or Cell Quota (Droop, 1983), previously accumulated during their existence at the thermocline. It is pertinent to compare this situation with the model of Tett (1987b), who shows that it is theoretically possible to invoke an exceptional dinoflagellate (*G. aureolum*) surface bloom via initial development of the population at the thermocline followed by transport towards the surface. The difference between the model situation and that described here is that whilst Tett (1987b) invokes a 10-day flotation phase for the cells to approach the surface, a weak upwelling has provided the vertical transport mechanism off southwestern Ireland.

In conclusion, it would appear that the bloom of *K. mikimotoi* observed in 1998 is one which has arisen through physical oceanographic processes which have been driven by surface wind stress. Raine & McMahon (1998) have shown a different, though linked, mechanism of how these blooms arise in the bays of southwestern Ireland. In both cases, the origin of the blooms has been from sub-surface maxima of chlorophyll (phytoplankton) associated with the thermocline of waters in the northern Celtic Sea, and in both cases the ultimate driving variable has been wind direction. The same basic physical transport mechanisms have been shown to control the occurrence of other potentially harmful or toxic phytoplankton such as *Dinophysis* spp in the region (McMahon et al., 1998). Wind-forcing, as a function of wind speed and direction, can currently be forecasted with an accuracy equal to, or better (longer) than, the time scale of a few days. Given the compatibility of this time-scale to that within which these harmful events arise (Raine et al., 1993; McMahon et al., 1998), the desirable goal of forecastinging harmful algal events around southwestern Ireland is becoming realistic.

Acknowledgements

The authors are grateful to Steve Groom (Remote Sensing Services, Plymouth) for the provision of the satellite images taken on 4th August, 1998. Nutrient analysis was carried out at the Central Marine Services Unit at the Martin Ryan Institute, National University of Ireland, Galway. Many thanks, as always, to Michael Gillooly, captain Philip Baugh and the crew of the R/V Celtic Voyager for their co-operation throughout the study and assistance with fieldwork. This work was partly supported by EC MAST III contract MAS3-CT97-0085. The are authors are grateful to the ICSR conference organisers for their support in enabling publication of the satellite images.

References

Droop, M. R., 1983. Twenty five years of Algal Growth Kinetics: A Personal View. Bot. Mar. 26: 99–112.

Edwards, A., K. Jones, J. M. Graham, C. R. Griffiths, N. MacDougall, J. W. Patching, J. M. Richard & R. Raine, 1996. Transient coastal upwelling and water circulation in Bantry Bay, a ria on the SW coast of Ireland. Estuar. coast. shelf Sci. 42: 213–230.

Gentien, P., 1998. Bloom dynamics and ecophysiology of the *Gymnodinium mikimotoi* species complex. In Anderson, D., A. D. Cembella & G. M. Hallegraeff (eds), Physiological Ecology of Harmful Algal Blooms. NATO ASI Series G41: 155–173.

Gowen, R. J. , R. Raine, M. Dickey-Collis & M. White, 1998. Plankton distributions in relation to physical oceanographic features on the southern Malin Shelf, August 1996. ICES J. mar. Sci. 55: 1095–1111.

Grasshoff, K., 1976. Methods of Seawater Analysis. Verlag Chemie, Weinheim: 317 pp.

Hasle, G. R., 1978. The inverted microscope method. In Sournia, A. (ed.), Phytoplankton Manual. UNESCO, Paris: 88–96.

Holligan, P. M., 1981. Biological implications of fronts on the northwest European continental shelf. Phil. Trans. r. Soc., Lond. A302: 547–562.

Jenkinson, I. R. & P. P. Connors, 1980. The occurrence of the red tide organism, Gyrodinium aureolum Hulbert (Dinophyceae), round the south and west of Ireland in August and September, 1979. J. Sherkin Island 1: 127–146.

Joyce, B. A., 1995 The distribution of phytoplankton in southwestern Irish coastal waters in relation to physical oceanography. Ph.D. Thesis, National University of Ireland, Ireland: 135 pp.

McMahon, T., R. Raine & J. Silke, 1998. Oceanographic control of harmful phytoplankton blooms around southwestern Ireland. In Reguera, B., J. Blanco, M. L. Fernandez & T. Wyatt (eds), Harmful Algae. IOC Unesco: 128–130.

O'Boyle, S., G. Nolan & R. Raine, 2000. Harmful phytoplankton events caused by variability in the Irish coastal current along the west of Ireland. In Hallegraaf, G. M. et al. (eds), Proceedings of the 9th International Conference on Harmful Algal Blooms, Tasmania, February 2000, in press.

Ottway, B., M. Parker, D. McGrath & M. Crowley, 1979. Observations on a bloom of *Gyrodinium aureolum* Hulbert on the south coast of Ireland, summer 1976, associated with mortalities of littoral and sub-littoral organisms. Ir. Fish. Invest. Ser. B, No. 18.

Raine, R., B. Joyce, J. Richard, Y. Pazos, M. Moloney, K. Jones & J. W. Patching, 1993. The development of an exceptional bloom of the dinoflagellate *Gyrodinium aureolum* on the southwest Irish coast. ICES J. mar. Sci. 50: 461–469.

Raine, R., T. G. McMahon, J. O'Mahony, M. Moloney & C. M. Roden, 1990. Water circulation and phytoplankton populations in two estuaries on the West coast of Ireland. In Chambers, P. & C. Chambers (eds), Estuarine Ecotoxicology. JAPA, Wicklow: 18–27.

Raine, R., J. O'Mahony, T. G. McMahon & C. M. Roden, 1990. Hydrography and production of waters off southwest Ireland. Estuar. coast. shelf Sci. 30: 579–592.

Raine, R. & T. McMahon, 1998. Physical dynamics on the continental shelf off southwestern Ireland and their influence on coastal phytoplankton blooms. J. Continental Shelf Res. 18: 883–914.

Roden, C. M., H. J. Lennon, E. Mooney, P. Leahy & W. Lart, 1981. Red tides, water stratification and phytoplankton species succession around Sherkin Island, south-west Ireland. J. Sherkin Island 1: 50–68.

Tett, P., 1987a. Plankton. In Baker, J. M. & W. J. Wolff (eds), Biological Surveys of Estuaries and Coasts. Cambridge University Press, Cambridge: 280–341.

Tett, P., 1987b. The ecophysiology of exceptional blooms. Rapp. P.-v. Reun. Cons. int. Explor. Mer 187: 47–60.

Hydrobiologia **465:** 195–208, 2001.
G. Burnell (ed.), Coastal Shellfish – A Sustainable Resource.

The assessing of Pacific oyster (*Crassostrea gigas*) rearing performances by the IFREMER/REMORA network: method and first results (1993–98) in Brittany (France)

P.G. Fleury, E. Goyard, J. Mazurié, S. Claude, J.F. Bouget, A. Langlade & Y. Le Coguic
IFREMER, Regional Shellfish Laboratotry of Brittany, 12 rue des résistants, F-56470 La Trinité-sur-mer, France

Key words: Pacific oyster, *Crassostrea gigas*, mortality, growth, quality, *Polydora*

Abstract

Every year since 1993, from February to December, the French network IFREMER/REMORA has monitored mortality, growth and quality criteria of two oyster year-classes ('juveniles' of the first-year class and 'adults' of the second-year class), distributed among various stations of French oyster farming areas. This has provided a standard and simple annual assessment of the rearing results, generally tallied with professional observations. Each year, data series have been obtained, from which mean values (references), chronological trends or spatial differences can be analysed. Results in Brittany, from 1993 to 1998, show that 80% of oysters of 30 g in February grow to a range of 55–85 g at the end of the year, with a mortality rate lower than 20%. Mortality occurs mainly in spring for adults and in summer for juveniles. Although annual fluctuations exist, connected to climatic and hydrological influences, differences remain high between the monitored stations. This allows characterization of sites. Some chronological trends appeared significant, but the period of time remained too short for any prediction. On the other hand, REMORA data, especially when compared to climatological or hydrological data series, may support various types of studies: description of oyster quality, biological indicators in coastal water monitoring, explanatory models of the oyster-farming ecosystems. Lastly, unusual mortalities (in 1995), lack of growth (1998) and problems with quality (infestation by the worm *Polydora*) could be pointed out and quantified. Moreover, unusual results could be observed, which might not be explained only by annual variations or geographical specificities. This gives to REMORA network the role of watchdog and the possibility of giving advice for collective oyster management.

Introduction

The Pacific oyster (*Crassostrea gigas*) has been cultivated for many years throughout the world, and its performance has previously been the subject of many experimental studies. Agius et al. (1978) compared growth trials of *Crassostrea gigas* (and *Ostrea edulis*) in different places of Malta, in the Mediterranean, together with sea water variables, with the aim of testing whether oyster farming was possible or not in Malta. Brown & Hartwitch (1988a,b) in British Columbia (Canada) studied the survival, the growth and a condition index of *Crassostrea gigas* in 10 stations over a period of 14 months. They obtained a large range of results as much in survival rates (31.5–57.3% for Year 0 oysters and 80–90%, except one result, for Year 1 animals) as in growth rates (with a final weight ranging from 47.9 to 150.9 g for the Year 1 class). These large ranges in growth, connected with sea water data (mainly temperature, chlorophyll and salinity) allowed them to classify these experimental sites into three groups. Here again the aim was to determine which sites were the more suitable for oyster ongrowing.

In France, where the Pacific oyster production amounts 140 000 mT/year, various regional studies have been carried out on oyster growth, especially by scientists of IFREMER (the French Research Institute for the Exploitation of the Sea): Berthomé et al. (1986) in Marennes-Oléron, monitored the same batch of oysters during 3.5 years; final average individual weight was 80 g which appeared rather low. They pinpointed important seasonal variations in growth with better performances in spring and a cessation of growth during winter and at spawning (July). Maurer

(1989) compared oyster performances in different stations of Arcachon and found a larger variation in growth between sites than between years. Oysters of 20–30 g grew up to 57–95 g within a year. Haure & Baud (1990) studied some experimental batches in Bourgneuf bay. Mortality and growth rates were low and rather similar from one place to another one.

All these studies, correlated to various surveys of water variables, aimed to know which were the limiting factors of growth. They lasted a while and did not give any long-term references or trends. They did however point out the high variability of growth results depending on sites and years, and demonstrated the interest in standardized references for spatial and yearly comparisons.

In New Brunswick (Canada), Boghen & Allard (1990) tried to set up such a monitoring programme in collaboration with a team of oysterfarmers, in order to have a better understanding of various environmental problems encountered by the industry. Therefore, in addition to oyster annual performances, they routinely examined various water variables. Their long-term objective was to move to a permanent monitoring programme providing a data bank for the oyster industry in the New Brunswick. Such an assessment, built on both oyster performances and water data, turned to be rather heavy for a single team of researchers and farmers.

Another pluriannual monitoring programme of oyster growth was set up in Brittany (France) by Le Bec & Mazurié (1992). They aimed to record only oyster performances in a standardized procedure in order to allow comparisons. These researchers linked explanations of oyster performances to specific studies aiming to connect REMORA results with other IFREMER programmes of water data survey (especially temperature, salinity, particulate matters, phytoplankton).

After a trial period of 4 years (1989–92) in Brittany, a first analysis of the data was reported by Littaye-Mariette & Mazurié (1993) and a common methodology, the most simple as possible, was set up in 1993 for a national French network called REMORA (REseau MOllusques des Rendements Aquacoles). This network has been conducted by six IFREMER regional laboratories, for two year-classes ('adults' of the second-year class, since 1993; and 'juveniles' of the first-year class, since 1995). Specific annual IFREMER reports (Fleury et al., 1999) provide national and regional results to producers, managers and scientists.

This paper presents the initial trends observed in Brittany (France) during a 6-year period (1993–98).

Materials and methods: a standard annual monitoring

The REMORA protocol allows a standard annual monitoring of two year-classes of oysters. The standardisation concerns the location of the stations, the animal batches, the husbandry techniques, the annual sampling schedule and the laboratory analysis.

The stations

The national REMORA network comprises 39 'national' stations, distributed over the main oyster production centres (i.e., those producing over 1000 t/year). Eleven national stations are located in Brittany. If at all possible the stations are placed at the same tidal level in order to be comparable and in the centre of the farming areas in order to be as much representative as possible of the whole farming area.

In addition to these fixed national sites, 'regional' stations were added occasionally in smaller oyster areas. In Brittany, the number of these regional points varied around 20 every year, with a maximum of 27 in 1997 (Fig. 1).

The husbandry

Two year-classes of animals are monitored every year: 'juveniles' (8 months old at the beginning of the annual experiment) and 'adults' (20 months). In order to reduce the variation from year to year, oysters are issued each year from a same single origin (Arcachon) and sorted at a fixed size of 30 $\pm$ 5 g for the 'adults', and around 0.5 g for the 'juveniles'.

The standard REMORA husbandry uses traditional oyster bags, except in the Mediterranean Sea, where it follows the professional rearing technique of sticking oysters on suspended ropes. Initial densities, in February, are of 230 'adults' per bag (with a mesh of 14 mm). For 'juveniles' half-size bags are used (with a mesh of 6 mm) with an initial density of 400 animals, which is reduced to 200 by June (with a mesh of 9 mm).

The annual sampling schedule

The two batches of oysters were deployed in February of each year, and collected in December. Two seasonal

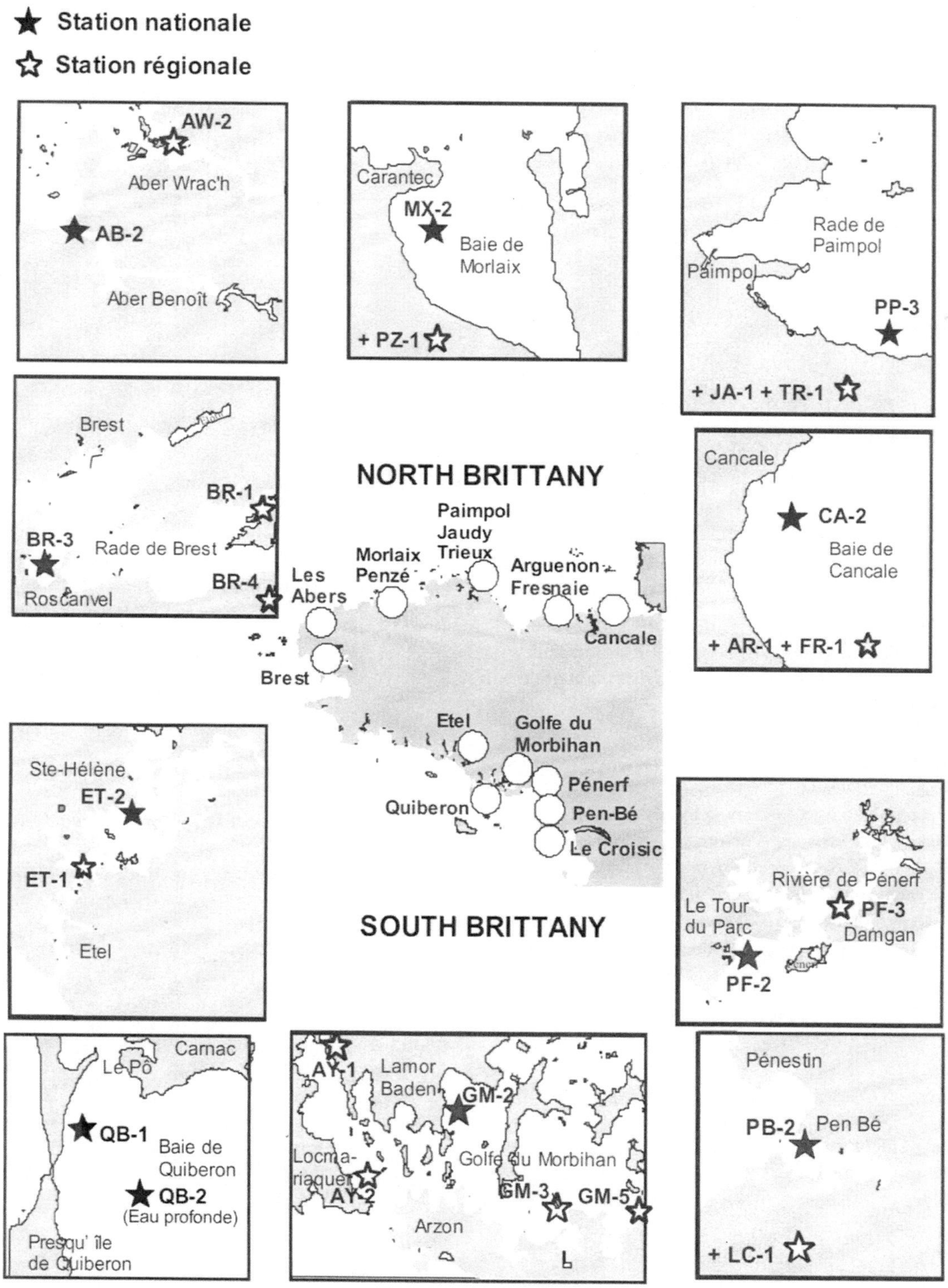

Figure 1. Location of the REMORA stations in Brittany.

intermediate samplings (30 oysters/site in June and September) were carried out at the same time in the 'national' stations (i.e., within the same 5-day period).

The analysis

The measurements aims to evaluate (1) the annual survival rate, (2) the growth, (3) the biomass production (survival × growth), (4) the quality of animals (mainly for 'adults': shell morphology, various condition indices and level of infestation by the worm *Polydora* sp.).

Various indices have been used to describe oyster morphology (Imaï & Sakaï, 1961) and flesh ratio in oysters: the one mostly used by the French oyster industry is the 'Afnor' index defined by the French standardization Association (Afnor, 1985). This commercial index sorts oysters into three classes: 'Spéciales' (Afnor index over 9), 'Fines' (index between 6.5 and 9) and not classified (index under 6.5).

$$\text{Afnor 85} = \frac{\text{Wet weight of drained flesh (g)} \times 100}{\text{Total weight (g)}}$$

Other condition indices may be calculated from REMORA data, such as the index of Lawrence and Scott (1982), defined as the ratio (×1000) of the dry matter weight to the inter-valves weight (weight of the whole animal except shell), and the index of Walne & Mahn (1975) defined as the ratio (×1000) of the dry matter weight to the shell weight. These two latter indices appeared very close to the Afnor index and were not considered worth describing in this paper.

The infestation by the worm *Polydora* (*P. ciliata* and *P. hoplura*) is estimated from a classification of oyster valves into five classes (Catherine et al., 1996). A Polydora index (PI) is calculated from this classification ranging from 0 (absence of parasite on all valves) to 1 (complete infestation of all oysters of the sample).

$$\begin{aligned}\text{PI} =& (0\ N_0) + (0.25\ N_1) + (0.5\ N_2) \\ &+ (0.75\ N_3) + N_4\end{aligned}$$

with N_0, N_1, N_2, N_3, N_4 being respectively the numbers of oysters in each class from 0 to 4.

The maturity of animals is also roughly estimated, on a macroscopic basis and clasified in three classes (empty, matured, very matured).

Table 1 gives the main variables (measured) and parameters (calculated) from the different samplings.

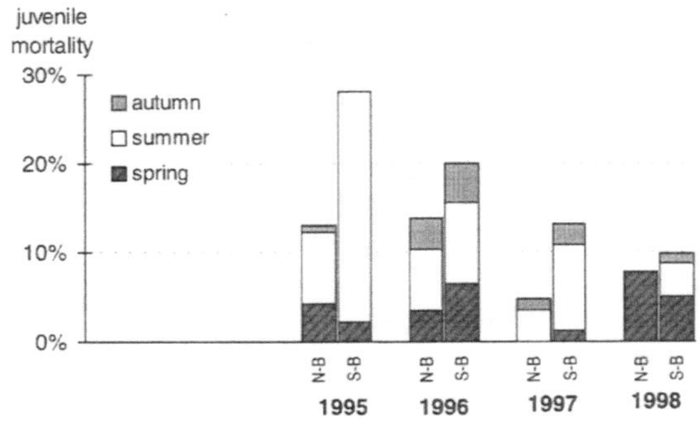

Figure 2. Seasonal mortality of the "juveniles" in national stations of North Brittany (N-B) and of South Brittany (S-B).

Statistical treatment

In order to see if there were significant differences between the means each year for the main calculated parameters, a two-factor Anova (site and year) was performed for each of them in North Brittany and in South Brittany; only national stations (references) were taken into account because the number and place of other stations (regional) were too variable from year to year.

t-Tests were also carried out on the average linear trend of these main parameters in both North and South Brittany. Mortality rates were analyzed through the transformed variable:

$$Y = \text{Arcsin (square root (mortality rate))}.$$

In the case of the Afnor index which proved to decrease very significantly in both North and South Brittany, the slopes were also compared with the trend of the Afnor index for the whole network (France).

Results

From the 6 years of data collected on 'adult' oysters, some means and trends were observed. With regard to the 'juveniles' only 4 years were collected, and only the variations in mortality were analyzed. However, for juveniles, mortality is the most important parameter for the oyster farmers.

Mortality of 'adult' and 'juvenile' oysters

Tables 2 ('juveniles') and 3 ('adults') present variations in annual mortalities in all REMORA stations of Brittany, compared to the references of national

Table 1. Main data collected from the different REMORA samplings (A, Adults; J, Juveniles)

		Initial	Samplings		Final
		Feb	June	Sept	Dec
Global measurements on batchs	Total weight (kg)	A + J			A + J
	Number of alive oysters	A + J	A + J	A + J	A + J
	Number of dead oysters	A + J	A + J	A + J	A + J
Individual measurements on 30 oysters	Length (mm)	A			A
	Width (mm)	A			A
	Thickness (mm)	A			A
	Individual weight (g)	A + J	A + J	A + J	A + J
	Shell weight (g)	A	A + J	A + J	A + J
	Drained flesh weight (g)	A	A + J	A + J	A + J
	Dry flesh weight (g)	A	A + J	A + J	A + J
	Maturity		A + J	A + J	A + J
	Infestation by *Polydora*	A	A + J	A + J	A + J
Calculated parameters	Survival rate		A + J	A + J	A + J
	Weight growth		A + J	A + J	A + J
	Growth rate		A + J	A + J	A + J
	Biomass production rate				A
	Coef. of length, of width, of thickness	A			A
	Flesh index (Afnor)	A			A + J
	Dry matter index (Lawrence & Scott)	A			A + J
	Polydora index	A	A + J	A + J	A + J

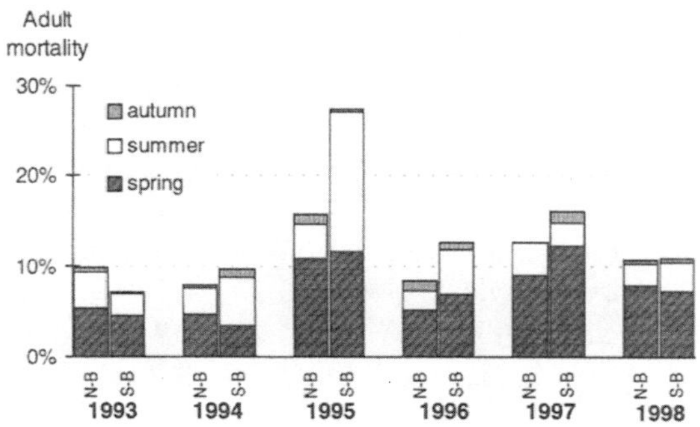

Figure 3. Seasonal mortality of the "adults" in national stations of North Brittany (N-B) and of South Brittany (S-B).

REMORA stations of Brittany and of the whole national network (France). The averages were stable: between 9 and 14%, except a peak in 1995 (24% in Brittany). But the range between stations was very large in both year classes: from about 0 to over 70% for both classes. The mortality occured mainly in summer for the 'juveniles' (Fig. 2) and in spring for the 'adults' (Fig. 3).

The year 1995 was a clear exception with high summer mortalities in most South Brittany sites (with 35% for 'adults' and 25% for 'juveniles'as an average). These mortalities may be partly explained by toxic algae blooms of *Gymnodinium* which were observed in Brittany in summer 1995 (Belin & Raffin, 1997).

A clear contrast could be observed in mortality between North and South Brittany (Table 4): in North Brittany the two-factor Anova did not point out any difference between sites or between years; on the contrary, significant differences between sites and between years were observed in juvenile and adult mortality in South Brittany, and chronological trends significantly differed from zero.

Growth of the 'adult' oysters

The final mean weight of 'adult' oysters at the final sampling in December varied from year to year between 60.9 and 80.2 g (from an initial weight of 30 g in February), with large differences between stations (see Table 4 for statistics and Table 5 for detailed data).

Table 2. Annual percentage of site mortality of 'juvenile' oysters, from 1995 to 1998, by site, region and for the whole (France) network (bold characters, national stations

North Brittany	1995	1996	1997	1998
Cancale 1	*nd*	*nd*	*nd*	*nd*
Cancale 2	**18**	**12**	**4**	**4**
Arguenon 1	*nd*	*nd*	12	9
Fresnaie 1	*nd*	*nd*	0	3
Paimpol 2	*nd*	*nd*	*nd*	*nd*
Paimpol 3	**25**	**5**	**3**	**5**
Trieux 1	*nd*	*nd*	13	4
Jaudy 1	*nd*	*nd*	0	3
Morlaix 2	**nd**	**21**	**12**	**71**
Morlaix 3	7	5	0	*nd*
Morlaix 4	17	4	4	*nd*
Penzé 1	*nd*	*nd*	0	8
Aber Wrac'h 2	*nd*	*nd*	0	15
Aber Benoit 1	3	13	0	*nd*
Aber Benoit 2	**9**	**5**	**8**	**12**
Aber Benoit 3	6	*nd*	8	*nd*
Brest 1	*nd*	*nd*	*nd*	*nd*
Brest 2	*nd*	*nd*	*nd*	*nd*
Brest 3	**1**	**25**	**1**	**6**
Brest 4	*nd*	*nd*	*nd*	*nd*
South Brittany				
Etel 1	0	6	0	4
Etel 2	**5**	**3**	**6**	**1**
Etel 3	21	30	0	*nd*
Quiberon 1	**25**	**13**	**10**	**7**
Quiberon 2	**46**	**39**	**38**	**13**
Auray 1	*nd*	*nd*	39	12
Auray 2	42	*nd*	*nd*	4
Golfe du Morbihan 2	**25**	**10**	**2**	**12**
Golfe du Morbihan 3	*nd*	16	16	10
Golfe du Morbihan 5	*nd*	*nd*	2	6
Pénerf 2	**34**	**28**	**16**	**8**
Pénerf 3	39	14	49	11
Pen Bé 1	21	34	24	21
Pen Bé 2	**14**	**25**	**12**	**20**
Pen Bé 4	24	5	12	10
Le Croisic 1	*nd*	*nd*	0	11
North Brittany	11	11	4	13
South Brittany	25	18	15	10
Brittany (national stations)	**20**	**17**	**10**	**15**
France (whole national network)	**18**	**13**	**12**	**12**

The record year in Brittany was 1994 (mean weight = 80.2 g) with oysters wheighing over 100 g in Quiberon (106 g) and Pen-Bé (107 g). The worst year for growth was 1998, when oysters in several stations of South Brittany did not reach 50 g in average (e.g., an annual increase of less than 20 g).

With regard to the chronological evolution a significative difference between years could be noticed in North Brittany but without any clear trend; on the contrary highly significative differences were pointed out in South Brittany with a significant decreasing trend from 1994 to 1998 (Table 4).

With regard to growth in each farming area, three different types of sites could be observed:

- Areas of good growth. Aber Benoit, Quiberon and Pen-Bé; in fact, Quiberon 2 and Pen-Bé decreased from excellent (over 100 g) to medium (60–70 g) which is mostly responsible for the average decline of growth results in South Brittany. Some good performances were observed also in Arguenon and Le Croisic, but these stations have been monitored since 1997 only.
- Areas consistent with the average. Paimpol, Etel, and downstream stations of the Golfe du Morbihan.
- Areas regularly under the average. Cancale and Fresnaye, the bay of Brest, Penthièvre, South of the Golfe du Morbihan and Pénerf.

Differences could also be observed between stations with more or less constant growth from year to year and from point to point (Cancale, Paimpol, Etel) and more variable areas with a spatial gradient from upstream (better) to downstream (not as good): Morlaix, Aber Benoit, Golfe du Morbihan, Pénerf, Pen-Bé.

The biomass production rate

The biomass production rate takes into account both the survival rate and the annual growth. In the oyster industry, from 18 to 30 months, the reference figure is 2 (e.g., a 6-kg bag of 200 oysters of 30 g = 6 kg is said to produce a total final weight of 12 kg). For REMORA's oysters in Brittany, the average of this coefficient was over 2 in 1993 (2.1) and in 1994 (2.4). Since 1995, it has been just under 2, which has to be compared with high mortalities in 1995 and low growth in 1997 and 1998 (Table 6).

Concerning the biomass production, the statistical analysis revealed significant differences between sites

Table 3. Annual percentage of site mortality of 'adult' oysters, from 1993 to 1998, by site, region and for the whole (France) network (**bold characters, national stations**)

North Brittany	1993	1994	1995	1996	1997	1998
Cancale 1	9	11	15	4	13	*nd*
Cancale 2	**3**	**3**	**19**	**6**	**19**	**12**
Arguenon 1	*nd*	*nd*	*nd*	*nd*	18	11
Fresnaie 1	*nd*	*nd*	*nd*	*nd*	11	3
Paimpol 2	10	5	9	5	21	*nd*
Paimpol 3	**8**	**6**	**10**	**9**	**5**	**14**
Trieux 1	*nd*	*nd*	*nd*	*nd*	15	2
Jaudy 1	*nd*	*nd*	*nd*	*nd*	10	4
Morlaix 2	**11**	**6**	**nd**	**6**	**8**	**16**
Morlaix 3	9	7	15	7	7	*nd*
Morlaix 4	9	8	11	5	10	*nd*
Penzé 1	*nd*	*nd*	*nd*	*nd*	10	9
Aber Wrac'h 2	*nd*	*nd*	*nd*	*nd*	6	15
Aber Benoit 1	10	13	12	6	11	*nd*
Aber Benoit 2	**26**	**24**	**13**	**9**	**22**	**6**
Aber Benoit 3	45	11	18	13	26	*nd*
Brest 1	8	30	15	3	*nd*	7
Brest 2	*nd*	*nd*	48	12	24	14
Brest 3	**7**	**7**	**25**	**16**	**12**	**9**
Brest 4	*nd*	*nd*	27	5	11	4
South Brittany						
Etel 1	10	7	21	11	13	5
Etel 2	**3**	**5**	**11**	**9**	**8**	**15**
Etel 3	8	7	16	11	11	*nd*
Quiberon 1	**5**	**6**	**19**	**15**	**nd**	**7**
Quiberon 2	**6**	**5**	**15**	**22**	**36**	**7**
Auray 1	*nd*	*nd*	*nd*	*nd*	9	8
Auray 2	9	11	12	3	10	3
Golfe du Morbihan 2	**5**	**9**	**32**	**8**	**9**	**7**
Golfe du Morbihan 3	6	9	29	9	5	4
Golfe du Morbihan 5	*nd*	*nd*	*nd*	*nd*	7	7
Pénerf 2	**6**	**12**	**43**	**15**	**7**	**14**
Pénerf 3	19	14	65	9	13	6
Pen Bé 1	12	19	75	29	21	17
Pen Bé 2	**19**	**24**	**51**	**21**	**17**	**21**
Pen Bé 4	10	22	70	12	33	24
Le Croisic 1	*nd*	*nd*	*nd*	*nd*	9	4
North Brittany	13	11	18	8	14	9
South Brittany	9	12	35	13	14	10
Brittany (national stations)	**9**	**10**	**24**	**12**	**14**	**12**
France (whole national network)	**7**	**13**	**19**	**9**	**12**	**11**

Table 4. Statistics and trends observed on different annual parameters in North and South Brittany

	Juveniles	Adults				
	Mortality rate (%)	Mortality rate (%)	Final weight (g)	Biomass production rate	Final Afnor index	Final *Polydora* index
North Brittany (five national stations)						
• 2-factor ANOVA:						
Sites	N.S.	N.S.	***	N.S.	***	***
Years	N.S.	N.S.	*	N.S.	***	*
• Chronological trend slope/year (*t*-test on linear trends)	N.S.	N.S.	N.S.	N.S.	−0.76**	+0.02*
South Brittany (six national stations)						
• 2-factor ANOVA:						
Sites	***	**	***	**	**	***
Years	*	**	***	***	***	*
• Chronological trend slope/year (*t*-test on linear trends)	−5*	+0.7*	−5.3*	N.S.	−0.49***	N.S.

Levels of confidence: N.S., non significant; * significant ($P<5\%$); ** very significant ($P<1\%$); *** highly significant ($P<1$ ‰).

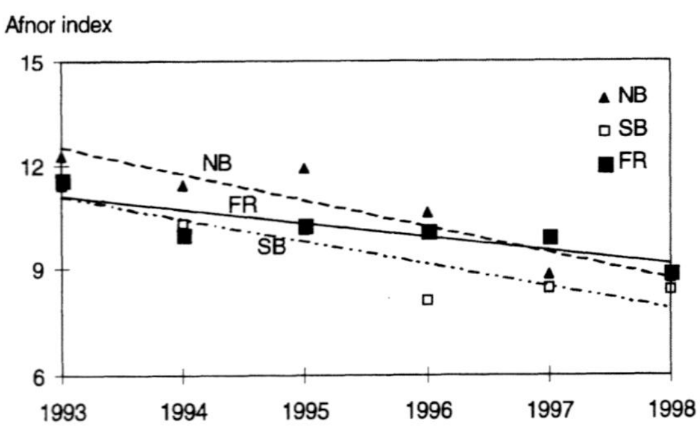

Figure 4. Evolution of the average Afnor index in national stations of North Brittany (NB), of South Brittany (SB) and of France (FR)

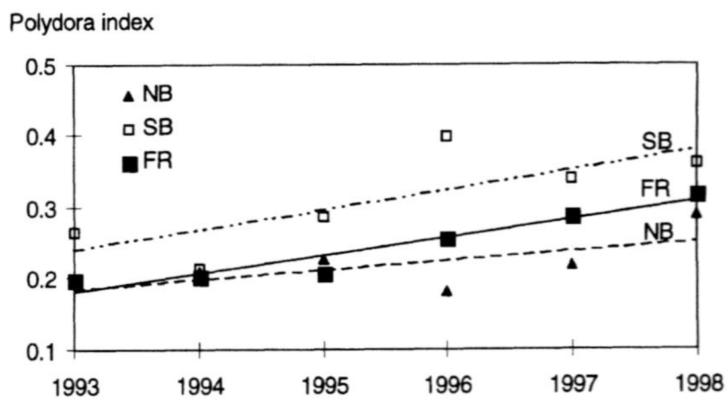

Figure 5. Evolution of the average Polydora index in national stations of north Brittany (NB), of South Brittany (SB) and of France (FR)

and between years in South Brittany but not in North Brittany. No significant trend was observed (Table 4).

Quality

The average of the Afnor index at the final REMORA's sampling in December of each year in Brittany decreased from 11.8 in 1993 to 8.6 in 1998. These declines were significant in both North and South Brittany (−0.75 and −0.49/year, respectively; Table 4). However, when compared (*t*-test on linear trends) to the whole national network (France), no significant difference could be observed between the breton decreasing slopes and the national trend (−0.43/year; Fig. 4).

Table 5. Variations in the final 'adult'osyter whole wet weight (g), from 1993 to 1998, by site, region and for the whole (France) network **bold characters, national stations**

North Brittany	1993	1994	1995	1996	1997	1998
Cancale 1	58.7	68.9	65.6	59.4	56.0	*nd*
Cancale 2	**65.5**	**64.9**	**65.2**	**62.6**	**61.9**	**58.7**
Arguenon 1	*nd*	*nd*	*nd*	*nd*	84.9	74.9
Fresnaie 1	*nd*	*nd*	*nd*	*nd*	53.4	52.1
Paimpol 2	60.8	79.9	65.3	66.7	64.7	*nd*
Paimpol 3	**60.3**	**66.3**	**62.4**	**65.7**	**57.2**	**59.5**
Trieux 1	*nd*	*nd*	*nd*	*nd*	60.4	55.4
Jaudy 1	*nd*	*nd*	*nd*	*nd*	65.8	57.1
Morlaix2	**58.3**	**65.8**	**nd**	**66.8**	**63.9**	**57.5**
Morlaix 3	76.0	79.5	76.9	92.5	61.8	*nd*
Morlaix 4	59.1	66.3	62.3	59.9	53.8	*nd*
Penzé 1	*nd*	*nd*	*nd*	*nd*	73.2	60.9
Aber Wrac'h 1-2	67.8	61.5	*nd*	*nd*	87.9	75.0
Aber Benoit 1	61.2	64.0	64.7	82.0	67.5	*nd*
Aber Benoit 2	**74.8**	**79.0**	**76.5**	**83.0**	**81.6**	**87.6**
Aber Benoit 3	86.7	80.8	77.3	83.6	85.1	*nd*
Brest 1	61.3	77.2	61.8	72.0	*nd*	65.6
Brest 2	56.0	65.7	58.6	68.4	61.4	65.0
Brest 3	**69.4**	**71.8**	**59.3**	**59.2**	**54.9**	**66.0**
Brest 4	*nd*	*nd*	59.4	63.9	51.6	54.4
South Brittany						
Etel 1	76.1	79.4	76.3	68.7	72.4	74.5
Etel 2	**80.7**	**78.5**	**75.8**	**68.9**	**65.9**	**53.3**
Etel 3	68.9	70.5	75.5	63.5	71.5	*nd*
Quiberon 1	**56.8**	**70.9**	**48.1**	**49.6**	**nd**	**49.5**
Quiberon 2	**67.0**	**<u>106.6</u>**	**91.8**	**80.2**	**70.8**	**65.9**
Auray 1	*nd*	*nd*	*nd*	*nd*	73.1	60.9
Auray 2	72.6	90.2	75.4	57.8	67.1	55.0
Golfe du Morbihan 2	**71.2**	**80.0**	**78.1**	**64.3**	**72.9**	**57.3**
Golfe du Morbihan 3	60.0	55.5	60.9	55.2	50.3	52.0
Golfe du Morbihan 5	*nd*	*nd*	*nd*	*nd*	65.0	48.3
Pénerf 2	**70.0**	**72.2**	**68.7**	**54.4**	**55.2**	**48.3**
Pénerf 3	62.7	73.9	71.8	74.6	56.1	70.0
Pen Bé 1	83.0	89.3	79.6	65.8	63.3	58.8
Pen Bé 2	**<u>100.2</u>**	**<u>107.3</u>**	**90.2**	**70.3**	**80.9**	**68.6**
Pen Bé 4	67.2	68.8	75.0	65.5	58.7	67.6
Le Croisic 1	*nd*	*nd*	*nd*	*nd*	79.7	83.0
North Brittany	65.4	70.8	65.8	70.4	65.6	63.5
South Brittany	72.0	80.2	74.4	64.5	66.9	60.9
Brittany (national stations)	**70.4**	**78.5**	**71.6**	**65.9**	**66.5**	**61.1**
France (whole national network)	**64,7**	**73,4**	**69,3**	**62,1**	**65,2**	**60,9**

Table 6. Variations in the final 'adult' osyter biomass production rate, from 1993 to 1998, by site, region and for the whole (France) network **bold characters, national stations**

North Brittany	1993	1994	1995	1996	1997	1998
Cancale 1	1.8	2.0	1.9	1.9	1.6	*nd*
Cancale 2	**2.1**	**2.1**	**1.8**	**2.0**	**1.7**	**1.7**
Arguenon 1	*nd*	*nd*	*nd*	*nd*	2.3	2.2
Fresnaie 1	*nd*	*nd*	*nd*	*nd*	1.6	1.7
Paimpol 2	1.8	2.5	2.0	2.1	1.7	*nd*
Paimpol 3	**1.8**	**2.1**	**1.9**	**2.0**	**1.8**	**1.7**
Trieux 1	*nd*	*nd*	*nd*	*nd*	1.7	1.8
Jaudy 1	*nd*	*nd*	*nd*	*nd*	2.0	1.8
Morlaix2	**1.7**	**2.1**		**2.1**	**2.0**	**1.6**
Morlaix 3	2.3	2.5	2.2	2.9	1.9	*nd*
Morlaix 4	1.8	2.0	1.8	1.9	1.6	*nd*
Penzé 1	*nd*	*nd*	*nd*	*nd*	2.2	1.8
Aber Wrac'h 1-2	2.3	2.0	*nd*	*nd*	2.8	2.1
Aber Benoit 1	1.8	1.8	1.9	2.6	2.0	*nd*
Aber Benoit 2	**1.8**	**2.0**	**2.2**	**2.5**	**2.1**	**2.7**
Aber Benoit 3	1.6	2.4	2.1	2.4	2.1	*nd*
Brest 1	1.9	1.8	1.7	2.3	*nd*	2.0
Brest 2	1.9	2.2	1.0	2.0	1.6	1.9
Brest 3	**2.1**	**2.2**	**1.5**	**1.7**	**1.6**	**2.0**
Brest 4	*nd*	*nd*	1.5	2.0	1.5	1.7
South Brittany						
Etel 1	2.3	2.4	2.0	2.0	2.1	2.4
Etel 2	**2.6**	**2.5**	**2.3**	**2.1**	**2.0**	**1.5**
Etel 3	2.1	2.2	2.1	1.9	2.1	*nd*
Quiberon 1	**1.8**	**2.2**	**1.3**	**1.4**	**nd**	**1.5**
Quiberon 2	**2.1**	**3.4**	**2.6**	**2.1**	**1.5**	**2.0**
Auray 1	*nd*	*nd*	*nd*	*nd*	2.2	1.9
Auray 2	2.2	2.7	2.2	1.9	2.0	1.8
Golfe du Morbihan 2	**2.3**	**2.4**	**1.8**	**2.0**	**2.2**	**1.8**
Golfe du Morbihan 3	1.9	1.7	1.4	1.7	1.6	1.7
Golfe du Morbihan 5	*nd*	*nd*	*nd*	*nd*	2.0	1.5
Pénerf 2	**2.2**	**2.1**	**1.3**	**1.5**	**1.7**	**1.4**
Pénerf 3	1.7	2.1	0.8	2.3	1.6	2.2
Pen Bé 1	2.4	2.4	0.7	1.6	1.7	1.6
Pen Bé 2	**2.7**	**2.7**	**1.5**	**1.9**	**2.2**	**1.8**
Pen Bé 4	2.0	1.8	0.8	1.9	1.3	1.7
Le Croisic 1	*nd*	*nd*	*nd*	*nd*	2.4	2.7
North Brittany	1.9	2.1	1.8	2.2	1.9	1.9
South Brittany	2.2	2.4	1.6	1.9	1.9	1.8
Brittany (national stations)	**2.1**	**2.3**	**1.8**	**1.9**	**1.9**	**1.8**
France (whole national network)	**2.0**	**2.1**	**1.9**	**1.8**	**2.0**	**1.8**

Table 7. Afnor index in 'adult' oysters at final sampling, from 1993 to 1998, by site, region and for the whole (France) network **bold characters, national stations**

North Brittany	1993	1994	1995	1996	1997	1998
Cancale 1	7.1	8.1	8.1	8.1	9.3	*nd*
Cancale 2	**10.8**	**9.3**	**9.8**	**8.5**	**9.9**	**6.7**
Arguenon 1	*nd*	*nd*	*nd*	*nd*	9.7	9.3
Fresnaie 1	*nd*	*nd*	*nd*	*nd*	9.3	10.0
Paimpol 2	11.8	9.1	10.5	9.3	9.2	*nd*
Paimpol 3	**9.4**	**9.1**	**12.2**	**9.0**	**7.9**	**7.7**
Trieux 1	*nd*	*nd*	*nd*	*nd*	6.8	7.9
Jaudy 1	*nd*	*nd*	*nd*	*nd*	8.1	7.6
Morlaix2	**11.4**	**12.1**		**11.4**	**9.1**	**9.7**
Morlaix 3	9.8	12.3	11.3	14.5	6.8	*nd*
Morlaix 4	8.3	11.1	9.4	11.3	7.8	*nd*
Penzé 1	*nd*	*nd*	*nd*	*nd*	8.6	11.4
Aber Wrac'h 2	*nd*	*nd*	*nd*	*nd*	9.8	12.6
Aber Benoit 1	12.3	12.7	14.6	13.5	10.3	*nd*
Aber Benoit 2	**16.5**	**14.4**	**14.9**	**14.4**	**10.7**	**11.7**
Aber Benoit 3	15.7	11.7	15.8	16.8	11.2	*nd*
Brest 1	8.2	9.7	13.5	9.6	*nd*	7.3
Brest 2	9.4	9.3	5.4	8.9	5.8	6.1
Brest 3	**13.1**	**12.1**	**10.7**	**9.7**	**6.5**	**8.1**
Brest 4	*nd*	*nd*	9.3	8.9	6.8	7.0
South Brittany						
Etel 1	8.9	8.9	10.1	8.3	7.7	8.4
Etel 2	**9.5**	**9.1**	**10.0**	**8.2**	**8.5**	**8.1**
Etel 3	9.6	9.5	9.1	7.1	7.9	*nd*
Quiberon 1	**12.4**	**9.4**	**9.6**	**7.6**	*nd*	**8.1**
Quiberon 2	**10.1**	**13.5**	**10.6**	**9.4**	**8.4**	**9.4**
Auray 1	9.7	8.2	9.1	9.1	7.0	6.3
Auray 2	*nd*	*nd*	*nd*	*nd*	9.1	8.8
Golfe du Morbihan 2	**11.9**	**9.7**	**9.5**	**7.1**	**8.2**	**7.2**
Golfe du Morbihan 3	7.6	7.2	9.2	6.9	6.0	6.8
Golfe du Morbihan 5	*nd*	*nd*	*nd*	*nd*	8.0	6.7
Pénerf 2	**11.9**	**8.0**	**8.9**	**8.0**	**7.4**	**7.7**
Pénerf 3	8.3	7.7	10.4	8.7	6.8	5.9
Pen Bé 1	11.1	9.5	12.7	7.7	7.5	7.0
Pen Bé 2	**12.8**	**12.4**	**12.4**	**8.3**	**9.7**	**10.0**
Pen Bé 4	11.6	9.7	12.5	8.6	6.7	8.9
Le Croisic 1	*nd*	*nd*	*nd*	*nd*	10.6	7.2
North Brittany	11.1	10.9	11.2	11.0	8.6	8.8
South Brittany	10.4	9.4	10.3	8.1	8.0	7.8
Brittany (national stations)	**11.8**	**10.8**	**10.8**	**9.2**	**8.6**	**8.6**
France (whole national network)	**11,5**	**9,9**	**10,0**	**10,1**	**9,9**	**8,9**

This decline appeared in 1996 in South Brittany and one year after in North Brittany, which from year to year remained around one point over South Brittany's value (Table 7). Because of this, most oysters from stations of South Brittany did not reach the label 'Spéciales' (Afnor index over 9) while several areas of North Brittany continued to qualify: e.g., Aber Benoit, L'Arguenon, La Fresnaye, Morlaix.

Breton areas showed a high heterogeneity with regards to the *Polydora* infestation, and no sector is free of this parasite (Table 8). The variability was more due to sites than to years (Table 4). However light increases were observed in North Brittany from 1993 (index=0.19) to 1998 (index=0.29) with a significant increase of +0.02/year) and in South Brittany (from 0.27 to 0.36; but not significant). Such a steady increase from year to year (Fig. 5) was observed also in the whole REMORA network (Fleury et al., 1999).

Discussion

The aim of the network REMORA is to provide a simple annual assesment of oyster performances in the main French farming areas and, over a period, chronological series which give averages (references), potential chronological trends and typology of sites.

The analysis of the results in Brittany provided some references: 84% of the results were in the range of 60–80 g in final weight and 80% were under 20% of annual mortality. On the other hand, a significant decrease in growth was observed in South Brittany. Decreases in quality were also observed from year to year (decrease in the Afnor index and increase in the Polydora one). These trends and the large range of values between stations did not allow any calculation of reference figures for these indices.

Decreasing trends in growth rate in South Brittany and in quality in both North and South Brittany could suggest the possibility of some long term changes in productivity. Despite the fact that the aim of REMORA is to collect data and not to explain, some studies could be undertaken by facing REMORA data to results of other networks, such as REPHY (the IFREMER phytoplankton monitoring network; Belin & Berthomé, 1991) which monitors temperature, salinity, chlorophyll and other food supply in oyster areas. Various authors have compared oyster growth to environmental factors. Mann (1979) established experimentally that at sustained elevated temperatures, whole weight growth of *Crassostrea gigas* increased faster than the meat weight. Maurer (1989) in Arcachon (France), also found that whole weight growth was correlated to sea water temperature while other parameters of growth and quality were more correlated to clorophyll concentration in the sea water. In the same way, Brown (1988), in British Columbia (Canada), established a significant correlation of oyster growth rate to chlorophyll (mainly) and to temperature (secondary). He added that growth rates were also highly site-specific depending on food supply and salinity conditions.

Actually, local conditions in topography or climatology have also to be taken into account. In France, Bacher et al., (1990) and Raillard (1991) in Marennes-Oléron as much as Lazure (1992) in Bourgneuf tried to explain the spatial variations of oyster growth through ecosystem models.

Without modelling or correlations to environmental factors, the network REMORA, by itself and despite the annual variations, pointed out differences between stations, enough to classify various types of sites. In the same way, comparisons between North and South Brittany showed some contrasts, especially in mortality and growth rates, which were more stable in North Brittany between sites and from year to year.

Such characterization of sites may be explained afterwards by geography or meteorology. For instance the more heterogeneous results in South Brittany may come from oyster farming areas being more located in estuaries and near the influence of the large rivers Loire and Vilaine. Dormoy (1995) and Jourdain (1996) discriminated REMORA stations of North Brittany and of South Brittany from both climatic and topographic factors; they presumed that the limiting factor for oyster growth in North Brittany was sea temperature (an increase of 3°C from one year to another would increase the final total weight of 10 g), while in South Brittany the limiting factor was rain (the same increase of 10 g/oyster would be obtained when annual rain increased of 100 mm).

Such a multivariate analysis of oyster performances in REMORA faced to meteorological and environmental conditions has to be carried out again with longer data series and a larger network (whole of France).

At last, it is also worthwhile to notice that some trends cannot be attributed to annual or geographic variations, and have to be explained by human practices or local management: allocation of new concessions, location of these concessions to facilitate water renewal, oyster density in bags, avoidance of predators

Table 8. Polydora index in 'adult' oysters at final sampling, from 1993 to 1998, by site, region and for the whole (France) network **bold characters, national stations**

North Brittany	1993	1994	1995	1996	1997	1998
Cancale 1	0.07	0.01	0.08	0.04	0.08	*nd*
Cancale 2	**0.12**	**0.00**	**0.14**	**0.23**	**0.15**	**0.13**
Arguenon 1	*nd*	*nd*	*nd*	*nd*	0.32	0.29
Fresnaie 1	*nd*	*nd*	*nd*	*nd*	0.29	0.03
Paimpol 2	0.19	0.42	0.45	0.13	0.31	*nd*
Paimpol 3	**0.19**	**0.28**	**0.23**	**0.09**	**0.18**	**0.12**
Trieux 1	*nd*	*nd*	*nd*	*nd*	0.25	0.13
Jaudy 1	*nd*	*nd*	*nd*	*nd*	0.23	0.06
Morlaix2	**0.20**	**0.16**		**0.14**	**0.29**	**0.43**
Morlaix 3	0.28	0.43	0.18	0.13	0.45	*nd*
Morlaix 4	0.30	0.23	0.25	0.17	0.22	*nd*
Penzé 1	*nd*	*nd*	*nd*	*nd*	0.40	0.19
Aber Wrac'h 2	*nd*	*nd*	*nd*	*nd*	0.19	0.31
Aber Benoit 1	0.38	0.42	0.33	0.31	0.25	*nd*
Aber Benoit 2	**0.40**	**0.49**	**0.48**	**0.44**	**0.43**	**0.59**
Aber Benoit 3	0.82	0.55	0.60	0.73	0.56	*nd*
Brest 1	0.43	0.16	0.25	0.41	*nd*	0.43
Brest 2	0.81	0.52	0.74	0.38	*nd*	0.37
Brest 3	**0.06**	**0.11**	**0.05**	**0.00**	**0.03**	**0.18**
Brest 4	*nd*	0.10	0.10	0.03	0.14	0.12
South Brittany						
Etel 1	0.11	0.23	0.18	0.17	0.30	0.57
Etel 2	**0.28**	**0.27**	**0.26**	**0.24**	**0.45**	**0.34**
Etel 3	0.37	0.31	0.28	0.23	0.39	*nd*
Quiberon 1	**0.20**	**0.11**	**0.33**	**0.22**	*nd*	**0.15**
Quiberon 2	**0.14**	**0.12**	**0.26**	**0.34**	**0.27**	**0.27**
Auray 1	0.29	0.10	0.31	0.35	0.42	0.51
Auray 2	*nd*	*nd*	*nd*	*nd*	0.13	0.41
Golfe du Morbihan 2	**0.23**	**0.13**	**0.11**	**0.44**	**0.130.33**	
Golfe du Morbihan 3	0.28	0.38	0.51	0.43	0.23	0.14
Golfe du Morbihan 5	*nd*	*nd*	*nd*	*nd*	0.22	0.46
Pénerf 2	**0.38**	**0.37**	**0.48**	**0.55**	**0.49**	**0.36**
Pénerf 3	0.29	0.29	0.38	0.22	0.58	0.20
Pen Bé 1	0.05	0.47	0.33	0.78	0.73	0.67
Pen Bé 2	**0.38**	**0.29**	**0.28**	**0.59**	**0.37**	**0.73**
Pen Bé 4	0.29	0.03	0.20	0.40	0.68	0.36
Le Croisic 1	*nd*	*nd*	*nd*	*nd*	0.37	0.40
North Brittany	0.33	0.28	0.30	0.23	0.26	0.24
South Brittany	0.25	0.24	0.30	0.38	0.38	0.39
Brittany (national stations)	**0.23**	**0.21**	**0.26**	**0.30**	**0.280.33**	
FRANCE (whole national network)	**0,19**	**0,20**	**0,21**	**0,26**	**0,29**	**0,32**

and competitors, etc. REMORA, the aim of which is mainly descriptive, cannot explain such trends by itself; but it may pinpoint atypic trends and question the industry and the managers.

References

AFNOR, 1985. Huîtres creuses: Dénominations et classifications. AFNOR/NF V 45-056, 5 pp.

Agius, C., V. Jaccarini and D. A. Ritz, 1978. Growth trials of *Crassostrea gigas* and *Ostrea edulis* in inshore waters of Malta (central Mediterranean). Aquaculture 15: 195–218.

Bacher, C., A. Menesguen, O. Raillard, M. Héral, J. M. Deslous-Paoli & D. Razet, 1990. Modélisation de l'écosystème de bassin de Marennes-Oléron; IFREMER/RIDRV 90–33, 125 pp.

Belin, C. & J. P. Berthomé, 1991. REPHY: The French monitoring network of phytoplankton. Proceedings of Symposium on Marine biotoxins, Paris, 30–31 January 1991: In Fremy, J. M. (ed), 189–194.

Belin, C. & B. Raffin, 1997. Evolution des espèces phytoplanctoniques toxiques sur le littoral Loire Bretagne de 1984 à 1995: résultats du REPHY (réseau de surveillance du phytoplancton et des phycotoxines). Rapport IFREMER.

Berthomé, J. P., J. Prou & A. Bodoy, 1986. Performances de croisssance de l'huître creuse, *Crassostrea gigas*, dans le bassin d'élevage de Marennes-Oléron entre 1979 et 1982. Haliotis 15: 183–192.

Brown, J. R., 1988. Multivariate analyses of the rôle of environmental factors in seasonal and site-related growth variation in the Pacific oyster *Crassostrea gigas*. Mar. Ecol. Prog. Ser. 45: 225–236.

Brown, J. R. & E. B. Hartwick, 1988a. Influences of temperature, salinity and available food upon suspended culture of the Pacific oyster, *Crassostrea gigas*. 1. Absolute and allometric growth. Aquacultrure 70: 231–251.

Brown, J. R. & E. B. Hartwick, 1988b. Influences of temperature, salinity and available food upon suspended culture of the Pacific oyster, *Crassostrea gigas*. II. Condition index and survival. Aquaculture 70: 253–267.

Catherine M., F. Dumont, C. Ménanteau & A. Pézeron, 1996. Abnormalities of the *Crassostrea gigas* oyster shells due to *Polydora* sp., observed in Pen-Bé and Mesquer harvesting areas (Loire-Atlantique) from 1988 to 1989. Rapport IFREMER, 42 pp.

Dormoy, J. M., 1995. Déterminisme de la croissance de l'huître creuse *Crassostrea gigas* en Bretagne. méthodologie d'élaboration d'un modèle global. IFREMER /RIDRV 95-26, 40 pp.

Fleury, P. G., F. Ruelle, S. Claude, H. Palvadeau, S. Robert, F. d'Amico, C. Vercelli & J. M. Chabirand, 1999. REMORA, résultats des stations nationales, année 1998. Rapport IFREMER DRV/RA/RST 99-03, 43 pp.

Haure, J. & J. P. Baud, 1990. Croissance, engraissement et mortalité de *Crassostrea gigas* en baie de Bourgneuf. Comparaison de quatre secteurs ostréicoles. IFREMER/RIDRV 90.11, 25 pp.

Imaï, T. & S. Sakaï, 1961. Study of breeding of Japanese oyster, Tohoku J. Agric. Res. 1(2): 23–27.

Jourdain, C., 1996. Facteurs de différenciation de la croissance de l'huître creuse *Crassostrea gigas* dans les principaux bassins ostréicoles français entre 1990 et 1995. Mémoire de stage ENSA Rennes/IFREMER La Trinité-sur-mer, 40 pp.

Lawrence, D. R. & G. I. Scott, 1982. The determination and use of condition index of oysters. Estuaries 5(1): 23–27.

Lazure, P., 1992. Etude de l'hydrodynamique de la baie de Bourgneuf. IFREMER/DEL 92.24-RA, 20 pp.

Le Bec, C. & J. Mazurié, 1992. The growth of *Crassostrea gigas* in Brittany (France), setting of monitoring study. Haliotis 22: 61–70.

Littaye-Mariette, A. & J. Mazurié, 1993, Réseau de suivi de la croissance de l'huître creuse en Bretagne; synthèse des quatre années 1989 à 1992. IFREMER /RIDRV 93-39, 57 pp.

Mann, R., 1979. Some biochemical and physiological aspects of growth and gametogenesis on *Crassostrea gigas* and *Ostrea edulis* grown at sustained elevated temperatures. J. Mar. Biol. Ass. U.K. 59: 95–110.

Maurer, D., 1989. Approche des relations entre la croissance de l'huître *Crassostrea gigas* et le milieu dans le bassin d'Arcachon. IFREMER/RIDRV 89.34, 33 pp.

Raillard, O., 1991. Etude des interactions entre les processus physiques et biologiques intervenant dans la production de l'huître *Crassostrea gigas* du bassin de Marennes-Oléron: essais de modélisation. PhD, University of Paris 6.

Walne, P. R. & R. Mann, 1975. Growth and biochemical composition in *Ostrea edulis* and *Crassostrea gigas*. Proc. 9th Eur. Mar. Biol. Symp. 1975: 587–607.

Hydrobiologia **465:** 209–217, 2001.
G. Burnell (ed.), Coastal Shellfish – A Sustainable Resource.

The role of bivalve molluscs as tools in estuarine sediment toxicity testing: a review

Pamela A. Byrne & John O'Halloran*
Department of Zoology & Animal Ecology, University College Cork, Lee Maltings, Prospect Row, Cork, Ireland
**Author for correspondence: Tel. +353-21-4904193. Fax+353-21-4270562. E-mail: j.ohalloran@UCC.ie*

Key words: Scrobicularia plana, Tapes semidecussatus, cultured clams, whole sediment toxicity testing, estuaries

Abstract

Estuarine sediments frequently are repositories and therefore potential sources of anthropogenic contaminants. Many organic and metallic chemical compounds released into aquatic systems bind to particulates and so accumulate in the sediments, thus, sediments become repositories of contaminants in estuaries. These may also cause contamination through diffusion of porewater, resuspension of particulates and dispersal of benthic fauna. There is a need to assess the biological affects of these anthropogenic contaminants because they may be toxic to infauna and bottomfish. Sediment toxicity bioassays are a means for carrying out such an assessment and primarily provide data on toxicity by measuring the effects on the test organism. Existing sediment toxicity bioassays rely on a battery of aquatic toxicity tests, which are based on the extraction of pore water, and elutriate from sediments and then subjecting these sediment phases to toxicity testing regimes. Two estuarine bivalve molluscs, *Scrobicularia plana* and *Tapes semidecussatus* were used to assess the ecotoxicity of field-collected sediments from estuarine and coastal areas around the Irish and English Coast over a 3-year study period. A variety of endpoints were measured during the study including survival in air, behaviour, animal condition, biochemistry, soft tissue metal concentrations, lysosomal membrane integrity and histopathology. Of these endpoints, the most sensitive were survival, survival in air, lysosomal membrane integrity, behaviour and histopathology.

Introduction

While the open ocean waters are a great resource to mankind, estuaries and coastal marine waters rank among the most ecologically sensitive and economically important of all ecosystems. The coastal zone is clearly at risk to anthropogenic activities, especially those associated with waste disposal and the release of toxic pollutants. The highest concentrations of contaminants occur in estuarine and shallow coastal marine systems, especially those in close proximity to heavily industrialized metropolitan centers (Kennish, 1997).

Estuarine ecosystems are a physiologically rigorous environment for most invertebrates because of the wide range of salinity (14–30‰), tidal inundation, changes in temperature and oxygen availability found along the reach of the estuary (DeWitt et al., 1989). In many published estuarine studies, marine species have been employed after artificially increasing the interstitial salinity of estuarine sediments (DeWitt et al., 1989). Burton et al. (1989) exposed the freshwater cladoceran, *Daphnia magna* to whole sediment and elutriates collected from two harbour areas. Artificially increasing or decreasing the salinity of sediments may cause dilution of the contaminant load and chemically alter the bioavailable forms of the contaminants (DeWitt et al., 1989). This may also influence the homeostatic responses of the bioassay organisms. Many species that might be appropriate for sediment testing do not meet the selection criteria, recommended by the American Society for Testing Materials because historically little emphasis has been placed on developing testing procedures for benthic organisms (Ingersoll, 1997). At present, there is no standardised method for testing the toxicity of whole sediments using bivalve molluscs.

The impact of environmental contamination is frequently apparent only after major effects on the populations of aquatic organisms are observed, such as mass mortality and changes in the diversity of ecological communities. If these types of events are to be avoided or ameliorated by effective environmental management, then the level of risk of damage to aquatic animals or plants needs to be assessed and used to predict the possible consequences before the event occurs. The use of whole sediment bioassays are a means to assess this impact and can provide regulatory agencies with the information required for the management of aquatic ecosystems when combined with chemical and biological surveys. This review discusses the use of two bivalve species, *Scrobicularia plana* and *Tapes semidecussatus* as potential bioassay organisms for use in whole-sediment toxicity testing.

Sediment toxicity tests

Sediment toxicity tests can provide rapid information on the potential toxicity of contaminants to benthic organisms. The objective of sediment toxicity tests is to determine whether the sediment is harmful to benthic organisms. These tests are becoming increasingly important in regulation and assessment programmes throughout the world (Chapman et al., 1992). Sediment toxicity tests measure interactive toxic effects of complex contaminant mixtures. Furthermore, the tests do not require knowledge of specific pathways of interactions among sediment and test organisms (Kemp & Swartz, 1988). Various methods have been developed to evaluate sediment toxicity. These procedures range in complexity from short-term lethality tests that measure the effects of individual contaminants on single species to long-term tests that determine the effects of chemical mixtures on the structure and function of communities (Ingersoll, 1997). The evaluated sediment phase may include whole sediment, suspended sediment, elutriates or sediment extracts. The amount of sediment tested ranges from a few millilitres to over 800 l (Lamberson & Swartz, 1988). The test organisms include algae, macrophytes, fish, epibenthic and benthic invertebrates and pelagic invertebrates (Burton, 1992).

Most methods used for measuring toxicity of contaminants in sediments rely on relatively short term (<10-day) exposures (Nebeker et al., 1984, Dillon and Gibson, 1986). These test procedures have a number of limitations:

- only lethal responses are measured;
- the experimental design is usually limited to static tests;
- the commonly tested species may not be the most sensitive and may not include the most sensitive life stages;
- elutriate and suspended sediment phases are often tested instead of whole sediment resulting in the loss of data regarding the whole sediment toxicity;
- when the whole sediment is tested, the species are often not in direct contact with the sediment (Anderson et al., 1984).

Ideally a sediment toxicity test should be rapid, simple and inexpensive if the objective of the study is to screen a large number of samples (Giesy & Hoke, 1989). Acutely lethal sediment can be identified by existing methods. However, concentrations of contaminants in sediments may not be lethal but may interfere with the ability of an animal to develop, grow or reproduce (Chapman et al., 1985). A better understanding of the sublethal effects of chemicals in sediment is needed to identify areas with moderate contamination and evaluate chemicals that do not elicit acutely lethal responses.

Sediment bioassay organisms

Bivalve molluscs possess many physiological attributes that render them suitable for consideration as bioassay organisms for whole estuarine sediment toxicity testing. They are widespread and sedentary. They have a wide geographical distribution, are sensitive to a number of contaminants, tolerant of a wide range of abiotic factors and easily cultured and maintained in the laboratory (Depledge & Hopkin, 1995). In most cases, bivalve molluscs are good accumulators of organometallic and heavy metal contaminants from their surrounding environment due both to their behaviour and mode of feeding (Bryan et al., 1985).

Scrobicularia plana

The potential of *Scrobicularia plana* and *Tapes semidecussatus* as indicators of whole sediment toxicity was assessed during this 3-year study. Both of these bivalve species are predominantly deposit-feeders and therefore in direct contact with their surrounding environment. They inhabit the soft sediments of estuarine ecosystems and have the ability to withstand wide ranges in salinity (4–30‰). *S. plana* is widespread and can be found intertidally buried in mud and in large numbers from Scotland and Norway in the North to Senegal, West Africa, in the South (Essink et al., 1991). However, in recent years, the numbers of this species have been declining at a number of coastal areas and it is suspected that tri-butyl tin (TBT) may be responsible (Langston, 1996). Bryan et al. (1980) have shown *S. plana* to be a good accumulator of heavy metals due to the slow rate of exchange of metals with its environment. Also, Ruiz et al. (1994) have shown this species to be suitable for assaying the toxicity of TBT. However, as this species is not a commercially cultured and field-collection difficult, its use it routine toxicity testing by research institutions and regulatory agencies has been avoided. The results of initial experiments to assess the ability of *S. plana* to determine the toxicity of field-collected sediments from Irish estuarine ecosystems revealed that burrowing ability was adversely affected on exposure to sediments contaminated with high levels of copper and zinc (Byrne & O'Halloran, 1999). However, field-collection of this species is labour intensive and the collection of a large number of individuals is difficult. Also, there is no information available on pre-test history and age class is difficult to determine. Therefore, the use of this species as a routine bioassay organism is not recommended at this time. However, if this species was cultured its use as a bioassay organism for assessing the toxicity of field-collected sediments is recommended.

Tapes semidecussatus

The Manila Clam, *T. semidecussatus*, is also a soft sediments dwelling bivalve mollusc, indigenous to the Indo-Pacific region (Partridge, 1977) and is available commercially in Ireland as well as in Europe and attains lengths of up to 4 cm in 24 months when reared under optimum conditions. It fulfills many of the criteria recommended for species selection in toxicity testing. It is available all year round, easy to maintain in the laboratory, economically relevant and is tolerant to abiotic factors and has a wide geographical distribution (Rand et al., 1997). The use of this species as a whole sediment bioassay organism was assessed through its exposure to a number of different field-collected sediments from both Irish and English estuaries (Byrne & O'Halloran, 1999, 2000, in prep.).

Sediment toxicity test endpoints

The effects of the exposure of an organism to a toxic substance, either in the aqueous or sediment phase, are evaluated through the use of endpoints. Endpoints are biological alterations in an organism, which are selected prior to testing, and are based on processes important to the survival, growth, behaviour and reproduction of a species.

Biomarkers of pollution induced stress are also used in toxicity testing. Of the current tools available for this purpose, the biomarker approach appears to offer considerable promise (Moore, 1990; Moore & Simpson, 1992). An ecotoxicological biomarker is a biochemical, cellular, physiological or behavioural variation that can be measured in tissue or body fluid samples or at the level of whole organisms that provide evidence of exposure to and/or effects of, one or more chemical pollutants (and/or radiations) (Depledge, 1993). Molecular or cellular changes ('early warning' distress signals) can be indicative of exposure to contaminants, induced injury or perhaps both (Moore, 1993). The underlying basis of contaminant-induced pathological change is damage or perturbation of living processes at the molecular and subcellular levels of biological organisation (Slater, 1978). Consequently, detection of changes at these most basic levels should, in theory, provide markers of early onset damage, which if sustained will lead to cell injury and pathology with an eventual deterioration in the health of the population. The use of biomarkers in biological assessment of contamination can provide a very powerful set of tools for environmental surveillance with the added potential to provide evidence of exposure to bioavailable contaminants as well as mechanistic linking of the causative processes (Moore, 1993).

Biomarkers

Recently the biomarker concept has been extended from purely biochemical measurements to include those of cellular pathology, physiological processes and behaviour of organisms exposed to varying concentrations of pollutants (Depledge et al., 1992). This enhancement creates the possibility of using a hierarchy of biomarker measurements (Depledge, 1993). Initially, effects of pollutants might be detected by relatively non-specific biomarkers, usually high in the hierarchy (e.g., behavioural and physiological biomarkers). Recent technological developments now permit characteristic changes in behaviour associated with pollutant exposure to be determined (e.g., distribution of time spent by animal moving at different speeds, distanced travelled, frequency of turning, avoidance and/or burrowing speed of the animal) (Depledge et al., 1992). Detection of abnormalities with these non-specific biomarkers at a site at risk from pollution might then justify the measurements of more costly, lower hierarchy, specific biochemical and cellular biomarkers (e.g. intracellular granules, tissue lesions) to seek to identify the class of pollutants responsible for the exposure (Depledge & Hopkin, 1995). The magnitude of biomarker responses together with determination of tissue residue concentrations of pollutants would contribute to the overall assessment of pollution impact.

Endpoints used during this study

Throughout this study a number of different endpoints, including biomarkers, were used to investigate the effects of laboratory exposure to field-collected estuarine sediments (Fig. 1). Of these endpoints, the most sensitive were survival in air, behaviour, histopathology (lipofuscin accumulation) and assessment of lysosomal membrane integrity (neutral red retention assay). These endpoints are relatively easy to measure and provide information at the whole organism and cellular level of biological organisation.

Survival

The effect of field-collected estuarine sediments from around the Irish coast on the survival of *T. semidecussatus* is shown in Table 1. The relative pollution of each test sediment is also indicated in Table 1. These data are derived from three different studies undertaken over a three-year study period and indicate the acute toxicity of sediments contaminated with high levels of metals and oil refinery effluent over a 21-day bioassay exposure period.

Throughout the study, reference and control sediments were used to provide a measure of test acceptability, evidence of organism health and a basis for interpreting data obtained from test sediments. Prior to initiation of sediment bioassays, chemical and physical profiles of reference and control sediments were undertaken. In Ireland, reference (Youghal & Ballymacoda) and control (Sterile Sand, BDH) sediments were collected from coastal areas known to contain background levels of contaminants. These sediments also had similar physical and chemical characteristics to test sediments collected from estuarine and coastal areas known to be contaminated with industrial and domestic inputs (Anon, 1999). In England, Bantham (reference) estuarine site was used. This site has been previously used in laboratory studies as a reference and a control estuarine site (S. Bamber, pers. commun.).

Sediments containing high levels of metals, especially zinc were acutely toxic to *T. semidecussatus*. However, as test sediments were not analysed for all potentially toxic contaminants, it is difficult to determine the true cause of toxicity. Test sediments assayed from the southwest coast of England were not acutely toxic to this species (Byrne, 2000).

Behaviour

Behavioural modification is one of the most sensitive indicators of environmental stress (Eisler, 1979; Olla et al., 1980) and may affect direct survival. When an organism comes into contact with a contaminant, it either remains unaware of the change in environmental quality, or chemosensory perception initiates a suite of behavioural responses (Hebel et al., 1997). Animals may show a number of behavioural responses, including avoidance of the contaminant, moving away from the contaminant source, reduced feeding and reproduction. The swimming behaviour of neonates of the freshwater rotifer, *Brachionus calyciflorus*, was adversely affected when they were exposed to copper at a concentration of 2 μg Cu/l for 2 h (Charoy & Janssen, 1999).

In clams, burrowing behaviour is adaptive and allows the animal to escape predation. Both *S. plana* and *T. semidecussatus* initiate the burrowing response on introduction to sediments under laboratory conditions. However, the burrowing response can be affected by sediment quality (McGreer, 1979; Pearson et al., 1981; Nipper et al., 1998). The burrowing behaviour of both *S. plana* and *T. semidecussatus* was significantly affected when these species were exposed to field-collected estuarine sediments containing high levels of metals (Figs 2 and 3). In all sediment bioassays undertaken, the proportion of animals buried in test sediments after 24 h exposure was lower in test sediments with high levels of contaminants, especially copper and zinc. For example, there was a significantly lower number of animals buried in test sediments collected from Dublin Bay when compared to the proportion of animals buried reference and control sediments (Byrne and O'Halloran, 1999, 2000). The use of burrowing behaviour as an endpoint is sensitive and can be used as a rapid screening tool in determining the toxicity of sediments. In most cases behavioural modification is an easily measured endpoint and one that only requires little technical expertise.

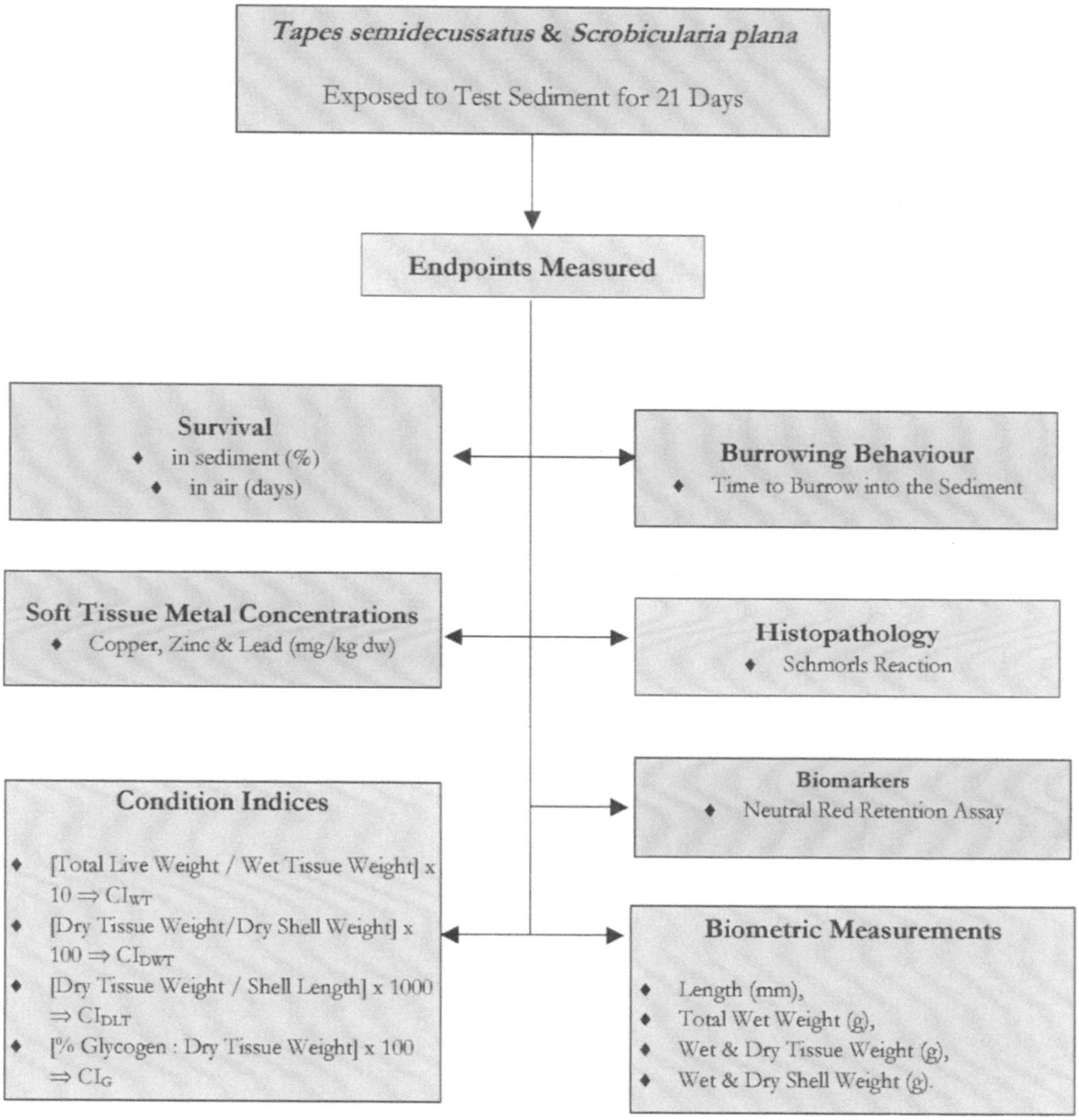

Figure 1. A flow diagram indicating the endpoints used during the bioassay study.

Lysosomes

In many organisms cellular responses to environmental stress may be detected before stress becomes apparent in the physiology of the whole organism (Lin & Steichen, 1994). Cellular responses can provide a rapid and sensitive indicator of environmental alterations. Pathological alterations in lysosomes have been especially useful in the indication of adverse environmental impact as many of the tissues of molluscs are extremely rich in lysosomes (Moore, 1990). This organelle, found in the digestive diverticula, served as a major site of intracellular digestion and is the main interface between the organism and its environment (Owen, 1972). When marine mussels are exposed to contaminants, there is an excessive build up of lipofuscin in the lysosomal compartment. Lipofuscin is a product of peroxidative reactions on autophagocytosed lipoprotein membranes and can be detected using a Schmorls reaction as described by Pearse (1985). During this study, this reaction was used to invest-

Table 1. The percent survival of *Tapes semidecussatus* exposed to field-collected estuarine sediments from around the Irish Coast over a bioassay exposure period of 21 days

Test sediment	Pollution level (1–5)	% Survival of *Tapes semidecussatus*
Reference Sediment, Ballymacoda	5	100.0
Reference Sediment, Youghal	5	100.0
Control Sediment, Sterile Sand (BDH Co.)	5	100.0
Cork Harbour, Blackrock	4	100.0
Cork Harbour, Douglas	3	85.0
Cork Harbour, Whitegate (Location of Oil Refinery)		
Whitegate Bay (Reference Site)	4	100.0
At Discharge Pipe	2	42.5
Opposite Discharge Pipe	1	82.5
Left of Discharge Pipe	2	80.0
Right of Discharge Pipe	3	92.5
Bantry Bay	2	13.3–100.0
Dungarvan Bay	3	87.5–100.0
Dublin Bay, Alexandra Basin (Site 1)	2	55.6–100.0
Dublin Bay, Alexandra Basin (Site 2)	1	2.2–90.0
Dublin Bay, Alexandra Basin (Site 3)	1	0.0–100.0
Dublin Bay, Alexandra Basin (Site 4)	1	6.7–100.0

(Note: The pollution score (based on chemical analysis of sediments) indicates the level of pollution compared to the control and reference sediments and ranges from 1–5 with 5 being the least contaminated of test sediments assayed. Range of percent survival is reported when sediments were assayed on two separate occasions)

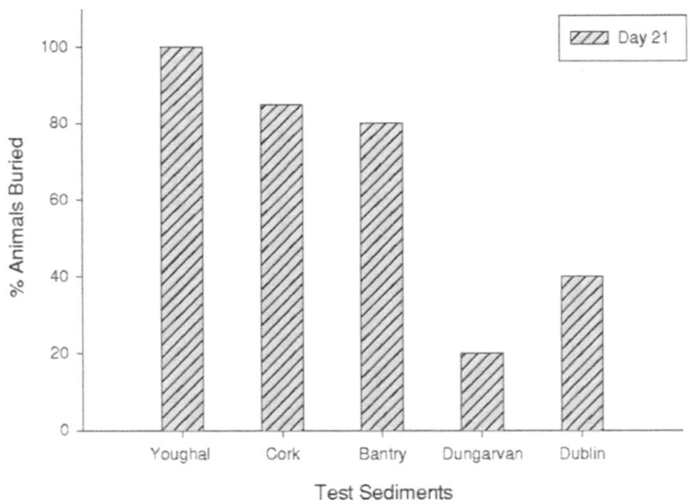

Figure 2. The percent individuals of *Scrobicularia plana* buried in test sediments (1 replicate/treatment) on day 21 of the whole-sediment bioassay after 24 h exposure to test sediments.

igate the effects of field-collected sediments on the lysosomes of *T. semidecussatus*. Lipofuscin accumulation was detected in the lysosomes of individuals of *T. semidecussatus* exposed to test sediments containing high levels of metals (Plates 1–4) (Byrne & O'Halloran, 1999, in press).

Individuals exposed to uncontaminated control and reference sediments did not accumulate lipofuscin granules in the digestive diverticula. This endpoint enabled the detection of the cellular effects of contaminants in cases where the test sediments assayed were not acutely toxic to this species. Histological

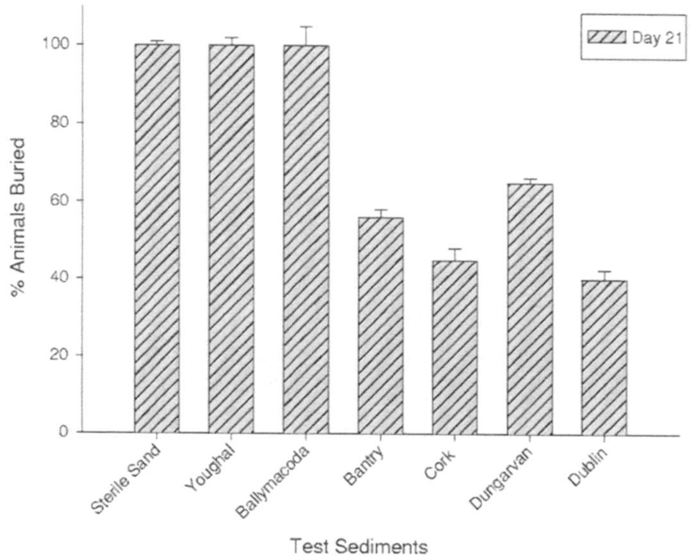

Figure 3. The mean (±se) % individuals of *Tapes semidecussatus* buried in test sediments on day 21 of the bioassay after 24 h exposure to test sediments.

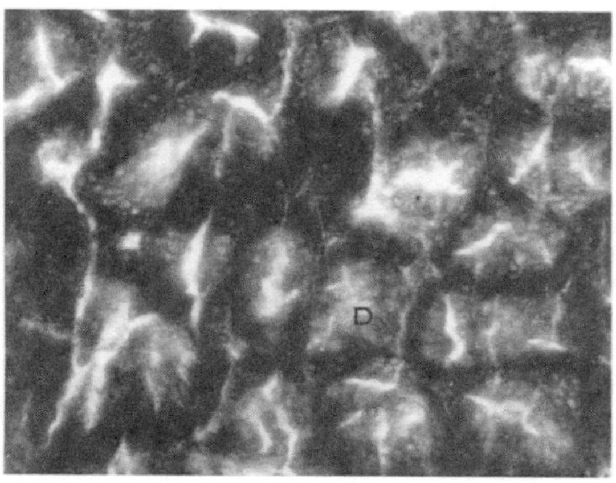

Plate 1. A 7-μm histological section of the digestive diverticula of a clam exposed to control sediments for 21 days (D, digestive diverticula).

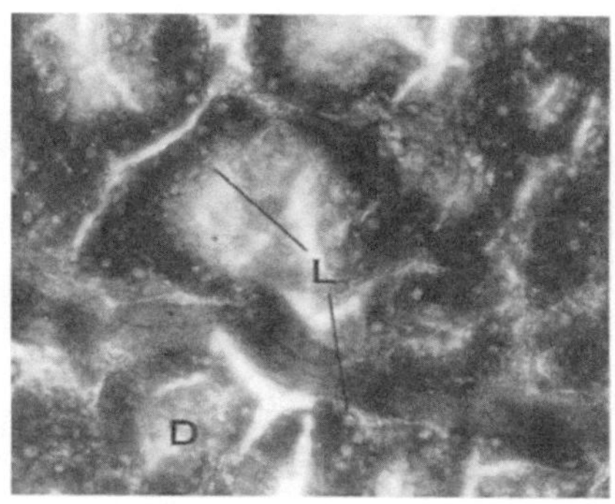

Plate 2. A 7-μm histological section of the digestive diverticula of a clam exposed to Dublin Bay test sediments for 21 days (D, digestive diverticula; L, lipofuscin granules).

sections (7 μm) taken from individual clams exposed to reference sediments (Ballymacoda) and contaminated sediments (Cork, Bantry and Dublin) contained varying degrees of lipofuscin accumulation (Plates 1–4).

The assessment of lysosomal membrane integrity was also used in this study to determine the effect of contaminated field collected sediments at the cellular level of biological organisation. The Neutral Red Retention Assay (Lowe & Pipe, 1994) assesses the retention time of the neutral red dye by the lysosomal membrane. A reduction in the retention time of the dye by animals exposed to estuarine sediments for a period of 7 days was observed during this study (Byrne, 2000). Both of these cellular biomarkers are considered to be important in the assessment of the sublethal toxicity that can occur when animals are exposed to sediments with high levels of organic and inorganic contaminants.

Conclusions

The Manila clam, *T. semidecussatus* is a suitable organism for consideration in whole-sediment bioassays of estuarine ecosystems and should be incorporated

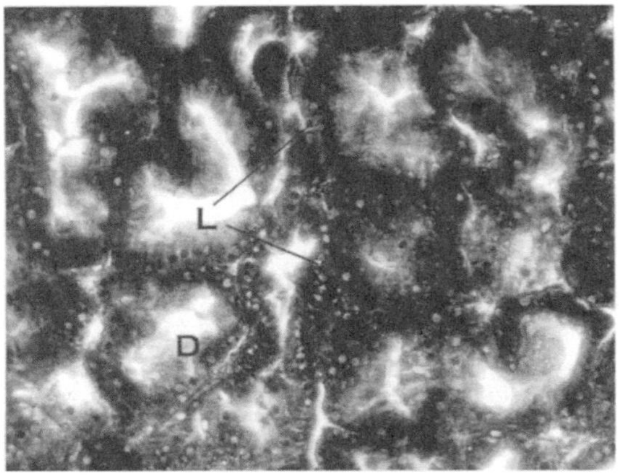

Plate 3. A 7-μm histological section of the digestive diverticula of a clam exposed to Bantry Bay test sediments for 21 days (D, digestive diverticula; L, lipofuscin granules).

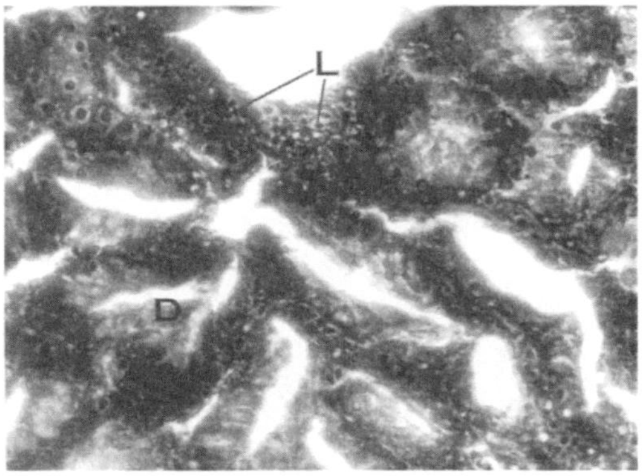

Plate 4. A 7-μm histological section of the digestive diverticula of a clam exposed to Dungarvan Bay test sediments for 21 days (D, digestive diverticula; L, lipofuscin granules).

into a suite of bioassays used for the routine assessment of aquatic ecosystems by the regulatory agencies. This species, cultured throughout Europe, and available commercially in Ireland can be used year round and is easy to maintain in the laboratory. It is an ecologically important species, positioned at the lower end of the food chain and it is a food source for many species of estuarine birds. It is economically important with a large export market to many European countries. While not indigenous to European estuaries, this species has been cultured very successfully for many years in Ireland and Europe.

A bioassay exposure of 21 days is recommended as this species has the ability to respire anaerobically for long periods of time, and shorter exposure periods may lead to false-positive results (Byrne & O'Halloran, 2000). A number of endpoints should be used to assess the toxicity of field-collected sediments at the whole organism (survival/mortality, survival in air and burrowing behaviour) and cellular (lipofuscin accumulation and lysosomal membrane integrity) levels in order to fully determine both acute and chronic toxicity. Other endpoints may be employed depending on the information required. These include growth or changes in biometric measurements and condition indices (Byrne & O'Halloran, in prep.).

The use of commercially cultured species in routine ecotoxicity testing, by regulatory agencies, merits further investigation as these resources are readily available in most countries and are easily accessible. This study has shown that *T. semidecussatus* can be used to determine the toxicity of field-collected estuarine sediments containing high levels of metals (Byrne, 2000) and also sediments exposed to oil refinery effluent (Byrne & O'Halloran, in prep.). Whole-sediment toxicity tests are also of benefit to regulatory agencies in defining sites for remediation or as an early warning system for sites, which are sensitive to contamination.

References

Anderson, J., W. Birge, J. Gentile, J. Lake, J. Rodgers, Jr. & R. Swartz, 1984. Biological effects, bioaccumulation, and ecotoxicology of sediment-associated sediments. In: Dickson K. L., A. W. Maki & W.A. Brungs (eds), Fate and Effects of Sediment-Bound Chemicals in Aquatic Systems. Pergamon, New York: 267–297.

Anon., 1999. Irelands marine and coastal areas and adjacent seas: an environmental assessment. Prepared by the Quality Status Response (QSR) Team of the Marine Institute on behalf of the Depts. of Environment & Local Governments and Marine & Natural Resources.

Bryan, G. W., W. J. Langston & L. G. Hummerstone, 1980. The use of biological indicators of heavy-metal contamination in estuaries – with special reference to an assessment of the biological availability of metals in estuarine sediments from south-west Britain. Marine Biological Association United Kingdom Occ. Publ. 1: 1–73.

Bryan, G. W., W. J. Langston, L. G. Hummerstone & G. R. Burt, 1985. A guide to the assessment of heavy-metal contamination in estuaries using biological indicators. Marine Biological Association of the United Kingdom Occ. Publ. 4: 1–92.

Byrne, P. A., 2000. Assaying Sediment Toxicity in Irish Estuarine Ecosystems using Bivalve Molluscs. PhD Thesis, National University of Ireland, Cork.

Byrne, P. A. & J. O'Halloran, 1999. Aspects of assaying sediment toxicity in Irish Estuarine ecosystems Mar. Poll. Bull. 39 (1-12): 97–105.

Byrne, P. A. & J. O'Halloran, 2000. Acute and chronic toxicity of the field-collected estuarine sediments to the Manila clam, *Tapes semidecussatus*. Environ. Toxicol. 15: 456–468.

Burton, G. A. Jr., 1992. Sediment Toxicity Assessment. Lewis, Chelsea, MI.

Burton, G. A., B. L. Stemmer, K. L. Winks, P. E. Ross & L. C. Burnett, 1989. A multitrophic level evaluation of sediment toxicity in Waukegan and Indiana harbours. Environ. Toxicol. Chem. 8: 1057–1066.

Chapman, P. M., R. N. Dexter, R. M. Kocan & E. R. Long, 1985. An overview of biological effects testing in Puget Sound, Washington: methods, results, and implications. In: Cadrwell, R. D., R. Purdy & R. C. Bahner (eds), Aquatic Toxicology and Hazard Assessment: Seventh Symposium, ASTM STP 854. Philadelphia, ASTM: 344–363.

Chapman, P. M., R. C. Swartz, B. Roddie, H. L. Phelps, P. van den Hurk & R. Butler, 1992. An international comparison of sediment toxicity tests in the North Sea. Mar. Ecol. Prog. Ser. 91: 253–264.

Charoy, C. & C. R. Janssen, 1999. The swimming behaviour of *Brachionus calyciforus* (Rotifer) under toxic stress. II. Comparative sensitivity of various behavioural criteria. Chemosphere 38 (14): 3247–3260.

Depledge, M. H., 1993. The rational basis for the use of biomarkers as ecotoxicological tools. In: Fossi, M. C. & C. Leonzio (eds), Nondestructive Biomarkers in Invertebrates, Lewis Publishers, Boca Raton, FL: 261–285.

Depledge, M. H. & S. P. Hopkin 1995. Methods to assess effects on brackish, estuarine, and near-coastal water organisms. In: Lindhurst, R. A., P. Bourdeau & R. G. Tardiff (eds), Methods to Assess the Effects of Chemicals on Ecosystems. Chap. 7: 125–149.

Depledge, M. H., J. J. Amaral-Medes, B. Daniel, R. Halbrook, P. Kloepper-Sams, M. M. Moore & D. B. Peakall, 1992. The conceptual basis of the biomarker approach. In: Peakall D. B. & L R. Shugart (eds), Biomaarkers: Research and Application in the Assessment of environmental Health NATO ASI Series H: Cell Biology, Vol. 68, Springer, Berlin: 15–29.

Depledge, M. H., A. Aagaard & P. Gyorkos, 1995. Assessment of trace metal toxicity using molecular, physiological and behavioural biomarkers. Mar. Poll. Bull. 31 (1–3): 19–27.

DeWitt. T. H., R. C. Swartz & J. O. Lamberson, 1989. Measuring acute toxicity of estuarine sediments. Environ. Toxicol. Chem. 8: 1035–1048.

Dillon, T. M. & A. B. Gibson, 1986. Bioassessment methodologies for the regulatory testing of freshwater dredged material. Miscellaneous Paper EL-86-6, U.S. Army Engineer Waterways Experiment Station. Vicksburg, MS: U.S. COE.

Eisler, R., 1979. Behavioural responses of marine poikilotherms to pollutants. Phil. Trans r. Soc. London. B, 286: 507–521.

Essink, K., J. J. Beukema, J. Coosen, J. A. Craeymeersch, J. P. Ducrotoy, H. Michaelis & B. Robineau, 1991. Population dynamics of the bivalve mollusc *Scrobicularia plana* da Costa: comparisons in time and space. Proc. ECSA 19 Symposium 162–172.

Giesy, J. P. & R. A. Hoke, 1989. Freshwater sediment toxicity bioassessment: rationale for species selection. J. Great Lakes Res. 15: 539–369.

Hebel, D. K., M. B. Jones & M. H. Depledge, 1997. Responses of crustaceans to contaminant exposure: a holistic approach. Estuar. Cstl. Shelf Sci. 44: 177–184.

Ingersoll, C. G., 1997. Chapter 8: Sediment Test. In: Rank, G. M. (ed.), Fundamentals of Aquatic Toxicology – Effects, Environmental Fate and Risk Assessment, 2nd edn. Taylor & Francis, London, UK.

Kemp, P.F. & R. C. Swartz, 1988. Acute toxicity of interstitial and particle-bound cadmium to marine infaunal amphipods. Mar. Environ. Res. 26: 135–153.

Kennish, M. J., 1997. Heavy metals. In: Kennish, M. J. & P. L. Lutz, (eds), Practical Handbook of Estuarine and Marine Pollution Chapter 6. CRC Press, Bacon Raton, FL: 253–329.

Lamberson, J. O. & R. C. Swartz, 1988. Use of bioassays in determining the toxicity of sediment to benthic organisms. In: Evans, M. S. (eds), Toxic Contaminants and Ecosystems Health: A Great Lakes Focus. Wiley, New York: 257–279.

Langston, W. J., 1996. Recent developments in TBT Ecotoxicology 3: 6: 179–187.

Lin, S. & D. J. Steichen, Jnr., 1994. A method for determining the stability of lysosomal membranes in the digestive cells of *Mytilus edulis*. Mar. Ecol. Prog. Ser. 115: 237–241.

Lowe, D. M. & R. K. Pipe, 1994. Contaminant induced lysosomal membrane damage in marine mussel digestive cells: an in vitro study. Aquat. Toxicol. 30, 357–365.

McGreer, E. R., 1979. Sublethal effects of heavy metal contaminated sediments on the bivalve *Macoma balthica* (L.). 10: 259–262.

Moore, M. N., 1990. Lysosomal cytochemistry in marine environmental monitoring. Histochem. J. 22: 187–191.

Moore, M. N., 1993. Biomarkers of contaminant exposure and effect: a way forward in marine environmental toxicology. Sci. Total Environ. Suppl. 1335–1343.

Moore, M. N. & M. G. Simpson, 1992. Molecular and cellular pathology in environmental impact assessment. Aquat. Toxicol. 22: 313–322.

Nebeker, A. V., M. A. Cairns, J. H. Gakstatter, K. W. Mauleg, G. S. Schuytema & D. F. Krawczyk, 1984. Biological methods for determining toxicity of contaminated freshwater sediments to invertebrates. Environ. Toxicol. Chem. 3: 617–630.

Nipper, M. G., D. S. Roper, E. K. Williams, M. L. Martin, L. F. Van Dam & G. N. Mills, 1998. Sediment toxicity and benthic communities in mildly contaminated mudflats. Environ. Toxicol. Chem. 17: 3: 502–510.

Olla, B. L., W. H. Pearson & A. L. Studholme, 1980. Applicability of behavioural measures in environmental stress assessment. Rapp. P.-v. Reun. Cons. Int. Explor. Mer. 179: 162–173.

Owen, G. 1972. Lysosomes, peroxisomes and bivalves. Sci. Prog., Oxford 60: 299–318.

Partridge, J. K. 1977. Studies on *Tapes decussatus* in Ireland. PhD Thesis. National University of Ireland.

Pearse E. A. G. 1985. Pigments and pigment precursors. In: Histochemistry, Theoretical and Applied. Churchill Livingstone, Edinburgh, London, Melbourne and New York: 916 pp.

Pearson, W. H., D. L. Woodruff, P. C. Sugarman & B. L. Olla, 1981. Effects of oiled sediment on predation on the littleneck clam, *Protothaca staminea*, by the Dungeness crab, Cancer magister. Estuar. Cstl. Shelf Sci. 13: 445–454.

Rand, G.M., P. G. Wells & L. S. McCarty, 1997. Introduction to Aquatic Toxicology. In: Rand, G. M. (ed.), Fundamentals of Aquatic Toxicology – Effects, Environmental Fate and Risk Assessment, 2nd edn., Chapter 1. Taylor & Francis, London, UK.

Ruiz, J. M., G. W. Bryan & P. E. Gibbs, 1994. Bioasssaying the toxicity of tributyltin-(TBT)-polluted sediment to spat of the bivalve *Scrobicularia plana*. Mar. Ecol. Prog. Ser. 113: 119–130.

Slater, T. F. 1978. Biochemical studies on liver injury. In: Slater, T. F. (ed.), Biochemical Mechanisms of Liver Injury. Academic Press, London, New York: 1–44.

Hydrobiologia **465:** 219–224, 2001.
G. Burnell (ed.), Coastal Shellfish – A Sustainable Resource.

Subject index

FSC
www.fsc.org
MIX
Papier | Fördert gute Waldnutzung
FSC® C083411

Kolibri 360 GmbH
Ferdinand-Jühlke-Straße 7
99095 Erfurt, Deutschland
produktsicherheit@kolibri360.de